# Azaheterocycles Based on $\alpha,\beta$-Unsaturated Carbonyls

Valentin A. Chebanov • Sergey M. Desenko •
Thomas W. Gurley

# Azaheterocycles Based on $\alpha,\beta$-Unsaturated Carbonyls

Springer

Valentin A. Chebanov
State Scientific Institution "Institute for Single Crystals" of National Academy of Sciences of Ukraine
Lenin Ave. 60, 61001 Kharkiv
Ukraine
chebanov@isc.kharkov.com

Sergey M. Desenko
State Scientific Institution "Institute for Single Crystals" of National Academy of Sciences of Ukraine
Lenin Ave. 60, 61001 Kharkiv
Ukraine
desenko@isc.kharkov.com

Thomas W. Gurley
Manning Wood LLC
1481 Limonis St.
Uniontown, OH 44685
USA
tgurley@manningwood.com

ISBN: 978-3-540-68361-2 e-ISBN: 978-3-540-68367-4

Library of Congress Control Number: 2008929333



*Cover design:* WMXDesign GmbH

Printed on acid-free paper

9 8 7 6 5 4 3 2 1

springer.com

# Preface

This book is devoted to heterocyclizations of aliphatic and aromatic $\alpha,\beta$-unsaturated carbonyls with various binucleophiles leading to three-, five-, six and seven-membered partially hydrogenated nitrogen-containing heterocycles. During the last decade interest in these classes of organic compounds has been experiencing a scientific renaissance owing to their significant role in biological processes in living cells and diverse effects on physiological activities. In addition, such compounds are also more prevalent from the viewpoint of "classical" problems of organic chemistry, among them reactivity, chemo- and regioselectivity, tautomerism, conformational analysis and features of their electronic structure. The character of these problems in the case of partially hydrogenated heterocycles differs sufficiently from that for heteroaromatized and perhydrogenated heterocyclic compounds and investigations in this field very often lead to interesting and unusual results.

Extensively characterized cyclocondensations of $\alpha,\beta$-unsaturated carbonyls, their synthetic equivalents and their precursors are the most widespread, facile and generally valid pathway to dihydroazaheterocycles. The popularity and significance of this synthetic approach is based on the high reactivity and availability of unsaturated carbonyl compounds and the precise selectivity of the heterocyclization reactions in comparison with that involving $\beta$-dicarbonyls. The recent development of combinatorial high-throughput methods and the use of new energy sources such as microwaves and ultrasound to enhance reactions have also increased interest in $\alpha,\beta$-unsaturated carbonyls and their reactions.

The main aim of this monograph is a comprehensive review and organization of the known literature data devoted to the reactions of $\alpha,\beta$-unsaturated ketones, their synthetic equivalents and their precursors utilized in the synthesis of nitrogen-containing heterocycles. The book is separated into four chapters and an Addendum, and contains nearly 900 literature references. Each chapter describes the synthesis and chemical and other interesting properties and features of certain classes of heterocyclic compounds.

The first chapter is devoted to the formation and subsequent modification of three-membered heterocycles—aziridines. Synthesis and properties of aziridinyl ketones, bi- and tricyclic aziridine derivatives, cycloaddition and photochemical reactions are described. The second chapter deals with five-membered heterocycles. Heterocyclizations of unsaturated ketones leading to six-membered heterocycles are the topic of the third and largest chapter of the book. Besides the reactions of unsaturated ketones with diverse 1,3-binucleophiles and the chemical properties of the partially hydrogenated azines that are formed, the problem of tautomerism is addressed as well. The last chapter includes data on the reactions of 1,2-diamines with unsaturated carbonyls which occur in a "classical" manner with the formation of diazepines and exhibit some unusual directions. The book also contains an Addendum describing general as well as special synthetic procedures relating to the chemistry of $\alpha,\beta$-unsaturated carbonyl compounds.

We are very thankful to many scientists who have taken an active role in the writing of this book. The chapters were prepared in very close collaboration with Valery Orlov, Nadezhda Kolos, Fedor Yaremenko and Alexander Zbruyev. Their efforts, remarks, corrections and additions were very useful in enhancing the quality of the book. A large part of the Addendum was written by Sergey Komykhov. A lot of help was obtained from our colleagues working in the Department of Heterocyclic Compounds Chemistry of the State Scientific Institution "Institute for Single Crystals" NAS of Ukraine as well as from our friends around the world—Nikolay Gorobets, Eugeny Gladkov, Vyacheslav Saraev, Yana Sakhno, Kirill Kobzar, Bogdan Khanetsky, Katerina Gura, Dmitry Sysoev and many others. We also would like to thank the Karl Franzens University of Graz (Austria) and the University of Constance (Germany) and personally Gert Kollenz, Oliver Kappe and Ulrich Groth for allowing us to access their extensive scientific databases and electronic and university libraries.

The book was written with partial support from the National Academy of Sciences of Ukraine, INTAS and the US Fulbright Scholar Program, and the German and Austrian Academic Exchange Services (DAAD and OeAD).

The book is oriented to chemists working in the field of synthesis and both in experimental and in theoretical investigations of nitrogen-containing heterocycles, university lecturers and both graduate and undergraduate students.

June 2008

Valentin A. Chebanov
Sergey M. Desenko
Thomas W. Gurley

# Contents

**Introduction** . . . . . . . . . . . . . . . . . . . . . . . . . . . . . . . . . . . . . . . . 1

**1 Three-Membered Azaheterocycles** . . . . . . . . . . . . . . . . . . . . . . . . . . . 5
1.1 Synthesis of Aziridinyl Ketones and Their Chemical Properties . . . . . . . . . . . . . . . . . . . . . . . . . . . . . . . . . . 5
1.2 Synthesis of Bi- and Tricyclic Aziridine Derivatives . . . . . . . . . 15
1.3 Chemical Properties of Bi- and Tricyclic Aziridines . . . . . . . . . 20
References . . . . . . . . . . . . . . . . . . . . . . . . . . . . . . . . . . . . . . . . . . 33

**2 Five-Membered Azaheterocycles** . . . . . . . . . . . . . . . . . . . . . . . . . . . 37
2.1 Reactions with Diazoalkanes . . . . . . . . . . . . . . . . . . . . . . . . . . 37
2.2 Reactions of Hydrazines and Hydroxylamines . . . . . . . . . . . . . 41
2.3 Structure, Chemical and Physicochemical Properties of Pyrazolines 50
References . . . . . . . . . . . . . . . . . . . . . . . . . . . . . . . . . . . . . . . . . . 55

**3 Six-Membered Azaheterocycles Based on 1,3-Binucleophiles** . . . . . . . . 61
3.1 Reactions of Enamines . . . . . . . . . . . . . . . . . . . . . . . . . . . . . . 62
3.2 Reactions of Urea and Its Analogues . . . . . . . . . . . . . . . . . . . . 65
3.3 Reactions of Malonic Acid Derivatives . . . . . . . . . . . . . . . . . . 76
3.4 Reactions of Aminoazoles . . . . . . . . . . . . . . . . . . . . . . . . . . . . 83
3.5 Reactions of Anilines and Aminoazines. . . . . . . . . . . . . . . . . . . 100
3.6 Chemical Properties of Dihydroazines. . . . . . . . . . . . . . . . . . . . 108
3.7 Tautomerism of Pyridine and Pyrimidine Dihydro Derivatives 130
References . . . . . . . . . . . . . . . . . . . . . . . . . . . . . . . . . . . . . . . . . . 136

**4 Cyclocondensations of 1,2-Diamines** . . . . . . . . . . . . . . . . . . . . . . . . 149
4.1 Synthesis of 1,5-Dihydroazepines. . . . . . . . . . . . . . . . . . . . . . . 149
4.2 Three-Component Condensation of 1,2-Diamines with Ketones 157
4.3 Reactions of Dihydrobenzodiazepines. . . . . . . . . . . . . . . . . . . . 163
4.4 Five- and Six-Membered Heterocycles Based on *ortho*-Diamines 170
References . . . . . . . . . . . . . . . . . . . . . . . . . . . . . . . . . . . . . . . . . . 176

**Addendum: Selected Synthetic Procedures**. . . . . . . . . . . . . . . . . . . . . . . . 181

**Index** . . . . . . . . . . . . . . . . . . . . . . . . . . . . . . . . . . . . . . . . . . . . . . 205

# Introduction

Specific questions of the chemistry of α,β-unsaturated carbonyl compounds are described in several publications, including two books. Properties of chalcones are discussed in the monograph of Dhar [1]. However, the syntheses of azaheterocycles in this book are described only occasionally since it was published before systematic investigations in the appropriate fields began. Another book [2], written in Russian, devoted to azaheterocycles based on aromatic unsaturated ketones is inaccessible for the most part to many scientists as well as containing out-of-date and incomplete data. Some aspects of the application of α,β-unsaturated ketones in synthetic organic chemistry, partially in the synthesis of certain classes of nitrogen-containing heterocycles, are described in several summary reviews [3, 4, 5, 6, 7, 8]. There are also reviews dealing with properties and features of target azaheterocycles [9, 10, 11, 12, 13].

The presence of two electrophilic reaction centers in the molecule of α,β-unsaturated carbonyls is responsible for their ability to participate in the synthesis of heterocycles. Such compounds can react as ambident electrophiles owing to delocalization of electron density in a C=C–C=O system. The addition of nucleophiles to these molecules can proceed in one of two main directions—via attack of the carbonyl group (1,2-addition) or involving the β-carbon (1,4-addition).

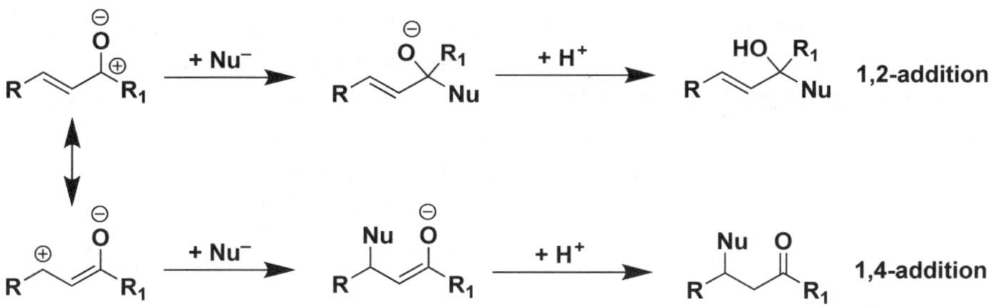


Features of ambident behavior of enones and enales and their treatment with CH-acids have been discussed in a review [14]. The relative reactivity of carbanions in 1,2- and 1,4-addition reactions is considered from the viewpoint of the perturbation of the molecular orbitals theory [15]. On the basis of this theory, considering the electronic structure of the enone fragment (the maximum effective positive charge is on the carbonyl group, while the maximum localization of LUMO (lowest unoccupied molecular orbital) is on the $\beta$-carbon [16]) for the case of 1,2-addition charge control of the reaction takes place. Correspondingly, 1,4-addition occurs under orbital control. Consequently, with all things being equal, localization of the charge at the nucleophilic group and decreasing of the HOMO (highest occupied molecular orbital) energy favor addition of the nucleophile to the carbonyl group. On the other hand, increasing of the charge delocalization and increasing of the HOMO energy causes orbital control of the 1,4-addition [14, 17].

Sufficient differences in the nature of these two electrophilic centers of enones are reflected in the high regioselectivity of their reactions with mono- and binucleophiles. This fact clearly discriminates $\alpha$,$\beta$-unsaturated carbonyls from other bielectrophilic compounds, for instance, $\beta$-diketones. However, the application of this advantage requires the determination and subsequent systematization of the factors influencing the directions of the reactions.

It should be noted that in the case of the reaction of $\alpha$,$\beta$-unsaturated ketones, 1,4-addition is more preferable. It is known [1, 18, 19, 20, 21, 22, 23] that their treatment with aromatic, secondary and most primary amines leads to the formation of the appropriate $\beta$-adducts. Unsaturated aldehydes, in contrast, usually react with primary and aromatic amines, forming the appropriate azomethines [24, 25, 26, 27, 28, 29, 30].

On the other hand, reactions of $\alpha$,$\beta$-unsaturated ketones with hydrazines and hydroxylamines involve, in the first stage, a carbonyl group [13, 31, 32, 33, 34, 35, 36, 37, 38, 39, 40] (see Sect. 2.3 for more details). This fact is usually explained by enhancing the nucleophilicity of the hydrazine and hydroxylamine derivatives owing to the presence of an unshared pair being adjacent to the attacking atom ($\alpha$-effect) [41, 42, 43]. However, *N*-amino-substituted heterocycles also have enhanced nulcleophilicity relative to carbonyls, although in this case the $\alpha$-electron pair is incorporated in the heteroaromatic system [44].

The conditions influence the direction of the reactions between the nucleophile and unsaturated carbonyl compounds as well. For example, the reaction of hydrazides of some organic acids with chalcone in the presence of acetic acid involves the carbonyl group of the unsaturated ketone, while basic catalysis (piperidine) promotes $\beta$-addition [45]. An analogous influence of the acidity on the direction is observed in reactions with binucleophiles [46]. Numerous similar examples are given in the appropriate chapters of this book.

The problems of $\beta$-amination of aromatic unsaturated ketones and their mechanism are described in [31, 47, 48]. The multistage character of this process and high sensitivity both to the nature of the reaction components and to the reaction conditions should be noted. For the same reactions in the literature, different data concerning kinetics are described owing to the influence of trace contaminants in the reaction mixture [47, 48].

$\alpha,\beta$-Unsaturated ketones may have a *trans* and *cis* structure with *S-trans* or *S-cis* conformation of the enone fragment.

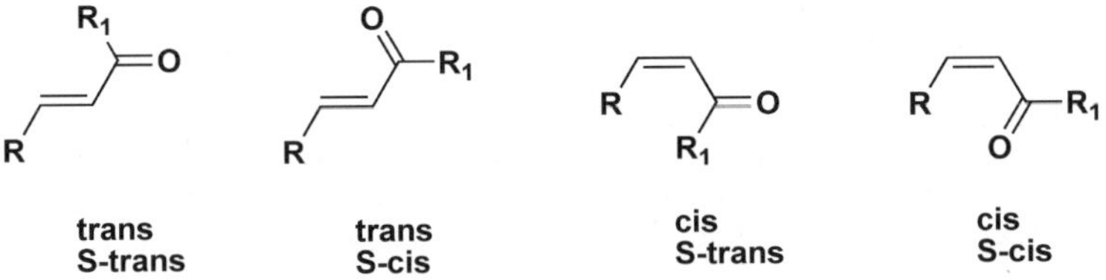


When the reactivity of the $\alpha,\beta$-unsaturated carbonyls is considered, their stereochemistry is very important. However, literature data concerning this problem are very rare because of the high rate of isomerization and conformational processes in enones. Some examples of such an influence are discussed in this monograph as well.

## References

1. Dhar, D. N. *The Chemistry of Chalcones and Related Compounds*, Wiley-Interscience: New York, 1981.
2. Desenko, S. M.; Orlov, V. D. *Azaheterocycles Based on Aromatic Unsaturated Ketones*, Folio: Kharkov, 1998.
3. Sausins, A.; Duburs, G. *Khim. Geterotsikl. Soedin.*, 1992, *4*, 435.
4. Sausins, A.; Duburs, G. *Heterocycles*, 1988, *27*, 269.
5. Insuasty, B.; Quiroga, H.; Meier H. *Trends Heterocycl. Chem.*, 1997, *5*, 83.
6. Desenko, S. M. *Khim. Geterotsikl. Soedin.* 1995, 147.
7. Chebanov, V. A.; Desenko, S. M. *Curr. Org. Chem.*, 2006, *10*, 297.
8. Druzhinin, S. V.; Balenkova, E. S.; Nenajdenko, V. G. Tetrahedron, 2007, *63*, 7753
9. Sausins, A.; Duburs, G. *Heterocycles*, 1988, *27*, 291.
10. Sausins, A. E.; Dubur, G. Y. *Khim. Geterotsikl. Soedin.,* 1993, 579.
11. Sausins, A. E.; Dubur, G. Y. *Khim. Geterotsikl. Soedin.*, 1992, 781.
12. Weis, A. L.; Van der Plas, H. C. *Heterocycles*, 1986, *24*, 1433.
13. Levai, A. *Khim. Geterotskl. Soedin.*, 1997, 579.
14. Yanovskaya, L. A.; Kryshtal', G. V.; Kul'ganek, V. V. *Usp. Khim.*, 1984, 53, 1280.
15. Carey, F. A.; Sundberg, R. J. *Advanced Organic Chemistry. Part A: Structure and Mechanisms*, Springer: New York, 2007.
16. Yanovskaya, L. A.; Umirzakov, B.; Kucherov, V. F.; Yakovlev, I. P.; Zolotarev, B. M.; Chizhov, O. S.; Vorontsova, L. G.; Fundyler, I. N.; Grunfest, M. G.; Kolodyazhny, Y. V.; Osipov, O.A. Terahedron, 1973, 29, 4321.
17. Kryshtal', G. V.; Burshtein, K. Y.; Kul'ganek, V. V.; Yanovskaya, L. A. *Izv. Akad. Nauk Ser. Khim.*, 1984, 2541.
18. Yang, L.; Xu, L.-W.; Zhou, W.; Li, L.; Xia, C.-G. *Tetrahedron Lett.*, 2006, *47*, 7723.
19. Shankar, R.; Jha, A. K.; Singh, U. S.; Hajela, K. *Tetrahedron Lett.*, 2006, *47*, 3077.
20. Lesma, G.; Danieli, B.; Sacchetti, A.; Silvani, A. *J. Org. Chem.*, 2006, *71*, 3317.
21. Gandelman, M.; Jacobsen, E. N. *Angew. Chem. Int. Ed.*, 2005, *44*, 2393.

22. Zahouily, M.; Bahlaouan, B.; Rayadh, A.; Sebti, S. *Tetrahedron Lett.*, 2004, *45*, 4135.
23. Cimarelli, C.; Palmieri, G.; Volpini, E. *Tetrahedron*, 2001, *57*, 6808.
24. Monleon, L. M.; Grande, M.; Anaya, J. *Tetrahedron*, 2007, *63*, 3017.
25. Dalpozzo, R.; Nino, A. D.; Nardi, M.; Russo, B.; Procopio, A. *Synthesis*, 2006, 1127.
26. Man, C. L.; Tin, W. C.; Wong, K.-Y.; Tak, H. C. *J. Org. Chem.*, 2007, *72,* 923.
27. Moonen, K.; Stevens, C. V. *Synthesis*, 2005, *20*, 3603.
28. Joffe, A. L.; Thomas, T. M.; Adrian, J. C. *Tetrahedron Lett.*, 2004, *45*, 5087.
29. Chatani, N.; Kamitani, A.; Murai, S. *J. Org. Chem.*, 2002, *67*, 7014.
30. Eisch, J. J.; Dluzniewski, T. *J. Org. Chem.*, 1989, *54*, 1269.
31. Weissberger, A. *The Chemistry of Heterocyclic Compounds*, Wiley, New York, 1967.
32. Barton D., Ollis W. D. (Eds.). *Comprehensive Organic Chemistry: The Synthesis and Reactions of Organic Compounds*, vol. 4. *Heterocyclic Compounds*, Pergamon Press: Oxford, 1979.
33. Lavrushin, V. F.; Bezugly, V. D.; Belous, G. G.; Tishchenko, V. G. *Zh. Obshch. Khim.*, 1964, *34*, 7.
34. von Auwers, K.; Muller, K. *Chem. Ber.*, 1921, *54*, 1000.
35. Suchchar, S. P.; Singh, A. K. *J. Indian Chem. Soc.*, 1985, *62*, 142.
36. Ferres, H.; Hamdam, M. S.; Jackson, W. R. *J. Chem. Soc. B*, 1971, 1892.
37. Bezuglyi, V. D.; Kotok, L. A.; Shimanskaya, N. P.; Bondarenko, V. E. *Zh. Obsch. Khim.*, 1969, *39*, 2167.
38. Raiford, L. C.; Manley, R. H. *J. Org. Chem.,* 1940, *5*, 345.
39. Bayer, H.; Dieterle, H. *Chem. Ber.*, 1911, *44*, 2697.
40. Tsukerman, S. V.; Danil'chenko, E. V.; Lavrushin, V. F. *Ukr. Khim. Zh.*, 1976, *42*, 383.
41. Grekov, A. P.; Veselov, V. Y. *Russ. Chem. Rev.,* 1978, *47*, 631.
42. Fina, N. J.; Edwards, J. O.; *Int. J. Chem. Kinet.*, 1973, *5*, 1.
43. Smith, M. B.; March, J. *March's Advanced Organic Chemistry: Reactions, Mechanisms and Structure*, Wiley-Interscience: Hoboken, 2007, p. 495.
44. Pozharskii, A. F.; Kuz'menko, V. V.; Bumber, A. A.; Petrov, É. S. Terekhova, M. I.; Chikina, N. L.; Nanavyan, I. M. *Khim. Geterotsikl. Soedin.*, 1989, 221.
45. Al-Farkh, Y. A.; Al-Hajjar, F. H.; Al-Shamali, F. S.; Hamoud, H. S. *Chem. Pharm. Bull.* 1979, *27*, 257.
46. Orlov, V. D.; Papiashvili, I. Z. *Khim. Geterotsikl. Soedin.*, 1985, 241.
47. Menger, F. M.; Smith, J. H. *J. Am. Chem. Soc.*, 1969, *91*, 4211.
48. Kinatowski, S.; Grabarkiewitz-Szczesna, J. *Bull. Acad. Pol. Sci. Ser. Sci. Chim.*, 1979, *27*, 1.

# Chapter 1
# Three-Membered Azaheterocycles

One of the features of $\alpha,\beta$-unsaturated ketones is the presence of two electrophilic centers. Because of this feature, reactions with binucleophiles can proceed as a 1,2-addition or as a 1,4-addition. Regarding three-membered nitrogen-containing heterocycles formed from $\alpha,\beta$-unsaturated ketones and their derivatives, the unsaturated ketone acts either as a 1,2-bielectrophile (substituted ethylene), which leads to the formation of ethyleneimines, or as a 1,4-bielectrophile, giving rise to either bi- or tricyclic aziridines. Hence, the present chapter is divided into two parts, one which is entirely dedicated to aziridinyl ketones and the other to bi- and tricyclic aziridines.

Fused aziridines are interesting compounds owing to the fact that the strained three-membered ring can easily open and cause dipolar cycloaddition reactions as well as their photochromic properties. Therefore, most of this chapter covers the chemical and photochemical properties of bi- and tricyclic aziridines. Some properties of aziridinyl ketones are also reviewed, in particular, reactions leading to aziridinyl anils.

## 1.1 Synthesis of Aziridinyl Ketones and Their Chemical Properties

Methods of obtaining aziridine derivatives from $\alpha,\beta$-unsaturated ketones can be divided into two basic groups: one-pot synthesis directly from an unsaturated ketone or stepwise synthesis involving the initial modification of a double bond.

In practice the most common method is the one based upon obtaining $\alpha,\beta$-dihalogen derivatives of unsaturated ketones with their subsequent interaction with ammonia or primary amines, known as the Gabriel reaction (Scheme 1.1).

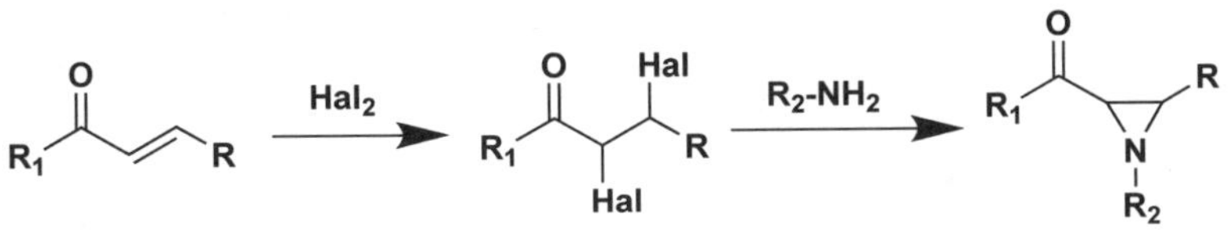


**Scheme 1.1**

The synthesis of 1,3-diaryl-2,3-dibromopropan-1-one and its interaction with ethanolic [1, 2, 3, 4, 5, 6, 7, 8, 9] or methanolic [10] ammonia solutions is often described in the literature. Examples of two aziridinyl ketones **2** and **4** synthesized from the corresponding dibromides **1** and **3** are shown in Scheme 1.2.

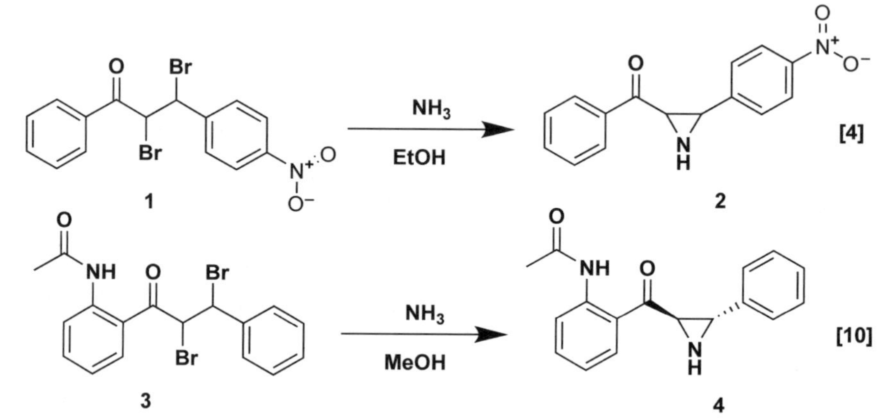


**Scheme 1.2**

For synthesis of N-substituted aziridinyl ketones, primary amines such as methylamine [11, 12, 13], cyclohexylamine [8, 11, 14, 15, 16, 17] and benzylamines [17, 18, 19, 20] are introduced in the reaction instead of ammonia. These reactions can be carried out in different solvents, such as alcohols, benzene, toluene, dimethylformamide, etc. On the basis of this chemistry, aziridinyl ketones containing either one or more three-membered cycles can be synthesized (e.g., compounds **7** and **10**; Scheme 1.3).

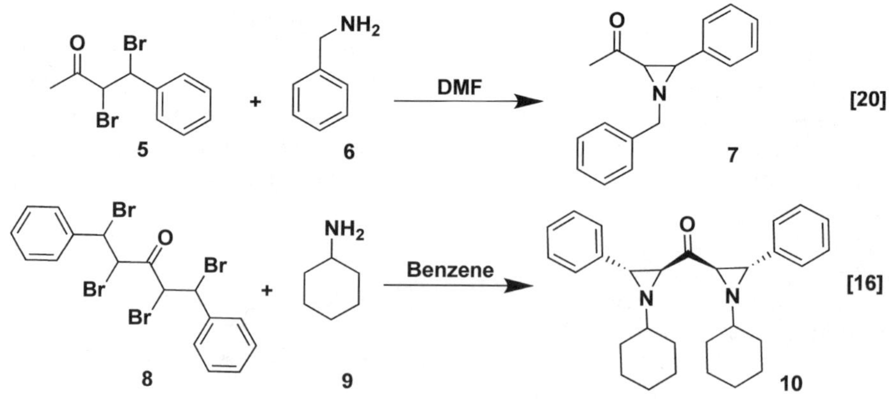


**Scheme 1.3**

It is known that the first stage of the reaction of $\alpha,\beta$-dibromoketones with amines is their dehydrobromination leading to $\alpha$-bromo derivatives [9, 21]. $\alpha$-Bromochalcones are mentioned in the literature and also can be used for the synthesis of aziridinyl ketones [8, 11, 22, 23, 24, 25]. For example, Khomutov et al. [25] carried out the synthesis of the dodecafluoro derivative **13** by the interaction of the corresponding $\alpha$-bromoketone **11** with hexylamine **12** (Scheme 1.4).

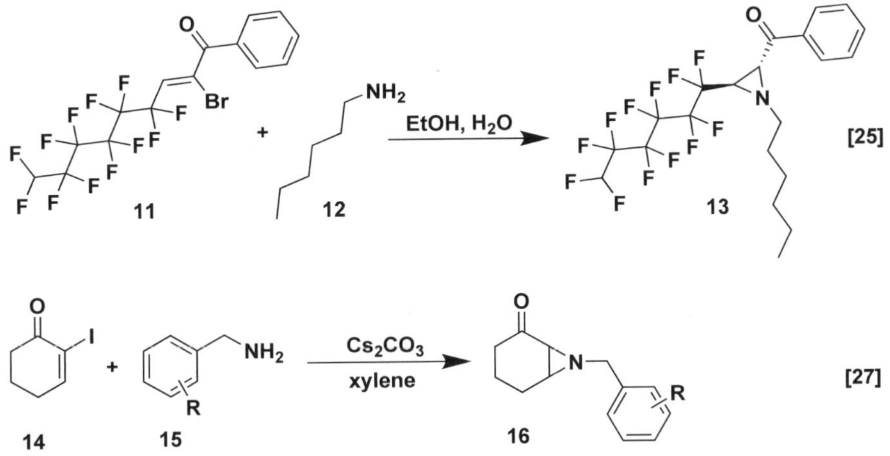


**Scheme 1.4**

Besides $\alpha,\beta$-dibromopropan-1-ones, there are references related to using $\alpha,\beta$-dichloro [26], $\alpha$-chloro [19] and $\alpha$-iodo derivatives [27] in the Gabriel reaction. Bicyclic aziridines **16** were obtained by the treatment of $\alpha$-iodocyclohexene **14** with benzylamines **15** in the presence of cesium carbonate in xylene [27] (Scheme 1.4).

Another synthetic method for the preparation of aziridinyl ketones involves the initial modification of unsaturated ketones, with formation of $\beta$-methoxyamino derivatives, followed by treatment with either metal alcoholates [11, 28, 29, 30, 31], or hydroxylamine hydrochloride and then potassium hydroxide [32]. An obvious drawback of this approach is the possibility of obtaining an exclusively unsubstituted nitrogen atom for the aziridinyl ketones. Among the advantages are high yields for these reactions. For example, Jin et al. [31] recorded yields of aziridine **18** of 99%. In other publications the yields of target compounds were reported to be around 90%.

Additionally, Cromwell et al. [33] described obtaining the *N*-cyclohexyl derivative of aziridinyl ketone **21** by reacting $\beta$-amino adduct **19** with *N*-bromocyclohexylamine **20** (Scheme 1.5).

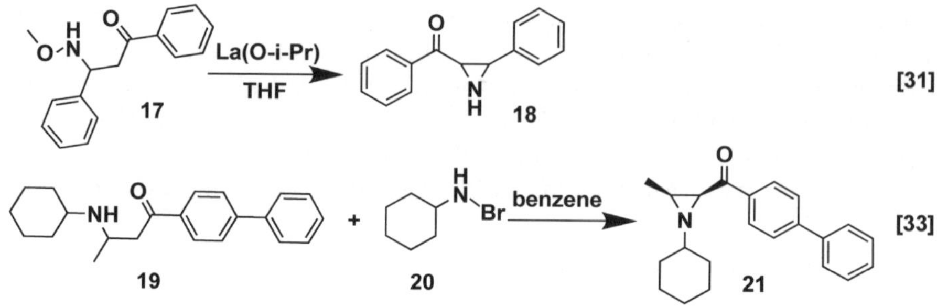


Scheme 1.5

As already mentioned, besides multistage reactions there are also synthetic methods for obtaining aziridinyl ketones from *α*,*β*-unsaturated carbonyl compounds without the need for initial modification. Among such methods, the Southwick reaction consisting of the interaction of unsaturated ketones

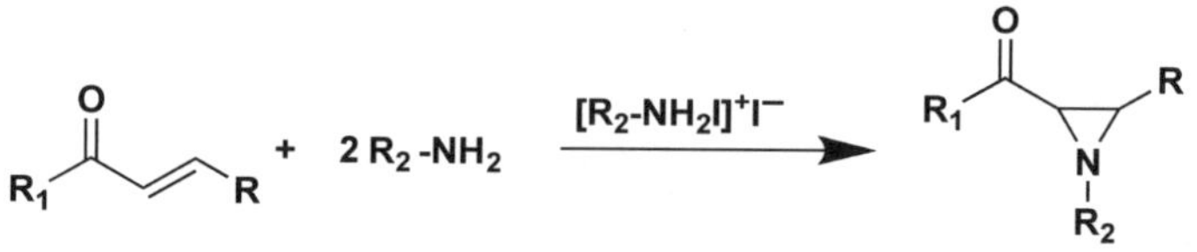


Scheme 1.6

with amines and an iodine–amine complex should be mentioned (Scheme 1.6).

For example, Southwick and Christman [34] described obtaining *trans*-aziridine **18** from chalcone **22** (Scheme 1.7). Reactions of **22** with cyclohexylamines and benzylamines forming the corresponding N-substituted derivatives are mentioned in the same publication. Zvonok et al. [35] studied the interaction of cyclohexylamine **9** with 6-phenyl-2,3-epoxyhex-5-en-4-ones **23**, leading to compounds **24** (Scheme 1.7).

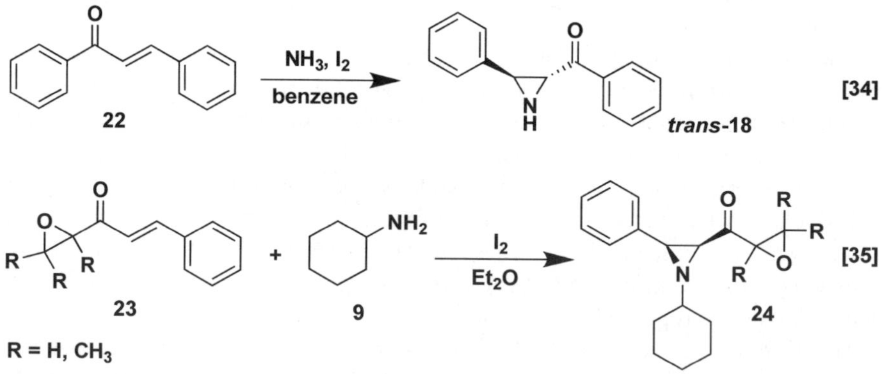


Scheme 1.7

It should be noted that the stereochemistry of the Gabriel and Southwick reactions depends essentially on both the structure of the starting compounds and the reaction conditions. It was established [36, 37, 38, 39, 40, 41] that increasing the *trans* to *cis* isomer ratio is promoted by the volume extension of the substituent in the $\beta$-position of the starting enone, whereas volume extension of the N-substituent leads to the opposite effect. The preferred formation of *cis*-aziridinyl ketones in the Gabriel reaction is also promoted by the use of methanol as a solvent instead of benzene at the stage where the 2,3-dibromopropan-1-one interacts with amines. In the Southwick reaction the opposite effect of the solvent is observed. A more detailed analysis of the mechanism and stereochemistry of the reactions mentioned is given in [36, 37, 38, 39, 40, 41]; however, as Tarburton et al. [41] themselves note, the models proposed do not explain all the features of the interaction.

One more method used in the last decade is the interaction of unsaturated ketones with *N*-tosyliminoaryliodinanes in the presence of monovalent copper complexes [42, 43, 44, 45]. A serious disadvantage of this method is low yields. For example, in the reaction of cyclohexenone **25** with compound **26** in the presence of $Cu(CH_3CN)_4ClO_4$, the target bicyclo[4.1.0]heptanone **27** was obtained with 21% yield [42] (Scheme 1.8).

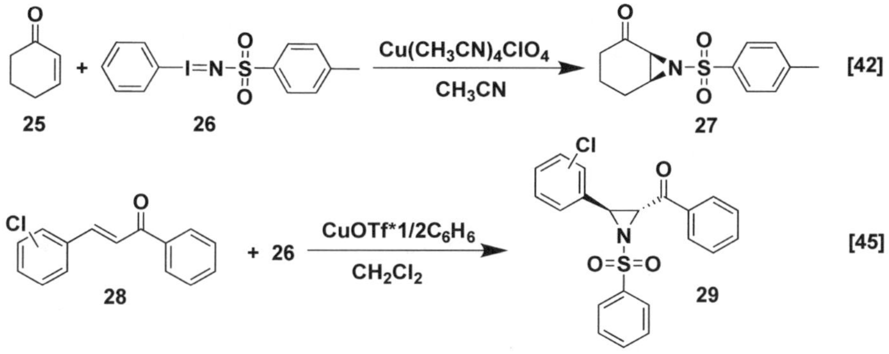


**Scheme 1.8**

In the case of 2-chlorochalcone, 3-chlorochalcone and 4-chlorochalcone **28**, compounds **29** could not be separated completely from the by-products [45] (Scheme 1.8). The same problem was observed in [44].

Bis(acetoxyphenyliodane) can also take part in such reactions [46]. In this case, the yields and purity of aziridinyl ketones are much higher than in the case of the *N*-tosyliminoaryliodinanes.

Among other important practical methods, the reaction of unsaturated ketones, in particular, dibenzoyl ethylene **30**, with *S*,*S*-diphenylsulfimide **31** [47, 48] should be mentioned. The yields of the target aziridines **32** in this case are high and sometimes close to quantitative (Scheme 1.9).

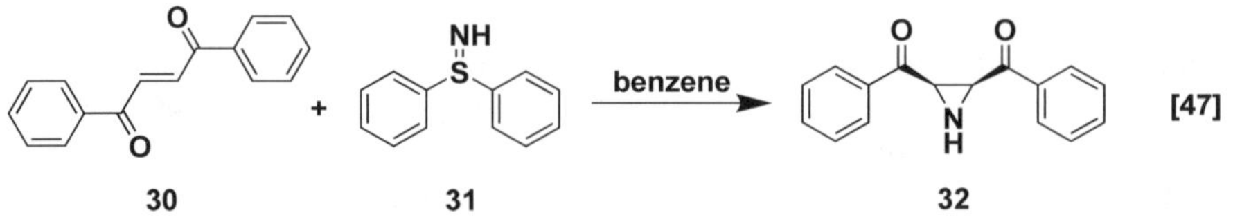


Scheme 1.9

Aziridinyl ketones can be synthesized from unsaturated carbonyls using a series of other methods. For example, azabicyclo[4.1.0]heptanone **27** was obtained from cyclohexenone **25** in its reaction with *N*-bromotoluenesulfonamide sodium salt **33** [49] (Scheme 1.10). The reaction of chalcone with *N*-chlorotoluenesulfonamide in the presence of silver nitrite is described in [50]. *Trans*-Aziridinyl ketone **18** was synthesized by reacting chalcone **22** with *N*,*N*-diamino-1,4-diazoniabicyclo[2.2.2.]octane dinitrate **34** and sodium hydride in 2-propanol [30, 51]. Aziridinyl ketones can be obtained in the reaction of $\alpha,\beta$-unsaturated ketones with *N*,*N*-dichlorosulfonamines [52] and with amines in the presence of lead tetraacetate and trifluoroacetic acid [53] or in the presence of triethylammonium acetate under electrochemical reaction conditions [54].

One-pot synthesis of aziridinyl ketones including the initial dipolar addition of azides to the ethylene bond with subsequent elimination of the nitrogen by photolysis is also possible [55, 56]. For example, in the case of azidocarboxylic acid ethyl ester **35,** 2-oxo-7-azabicyclo[4.1.0]heptane-7-carboxylic acid ethyl ester **37** was synthesized via the formation of the cycloadduct **36** [55] (Scheme 1.10).

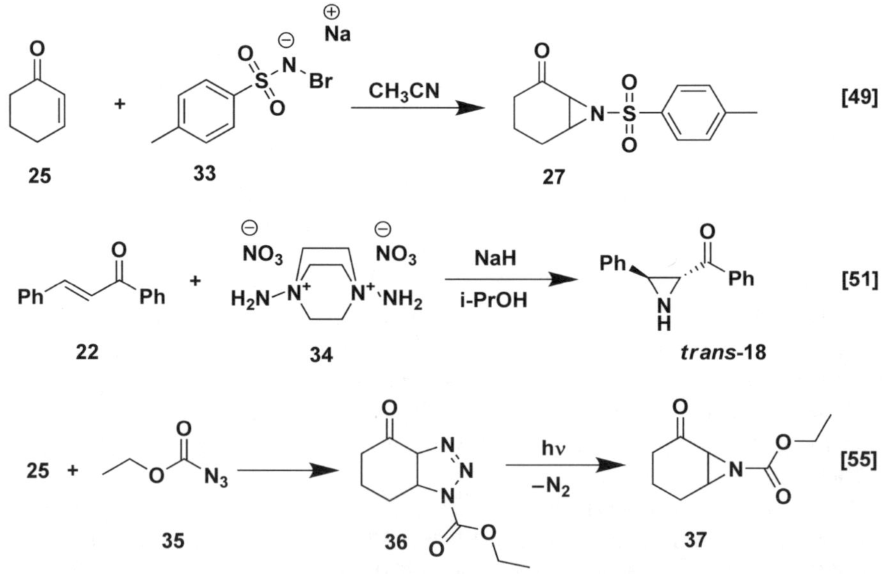


Scheme 1.10

One more method which we would like to mention is the direct aziridination of unsaturated ketones using aminoimides **40** obtained in the reaction of *N*,*N*-dimethylhydrazine **38** with methyl oxirane **39** [57] (Scheme 1.11):

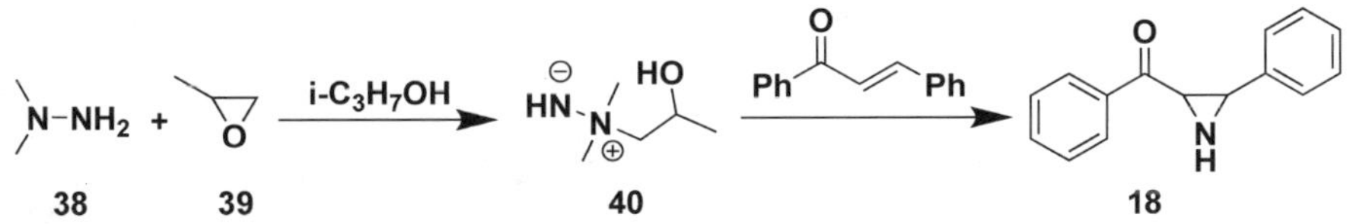


**Scheme 1.11**

Aziridination of (*E*)-chalcones can be carried out with other aminoimines which are formed by deprotonation of *N*-amino-*N*-methylmorpholinium salts [58].

Owing to the presence of two highly reactive fragments (the aziridine cycle and the carbonyl group), aziridinyl ketones are characterized by the diversity of their chemical transformations. They can react as secondary amines, undergo nucleophilic or electrophilic cycle opening, or react as carbonyl derivatives. In particular, reactions involving only the carbonyl group include reduction to alcohols [59, 60, 61] or formation of styrylaziridines **42** by the action of diverse ylidenetriphenylphosphanes **41** [30, 62] (Scheme 1.12).

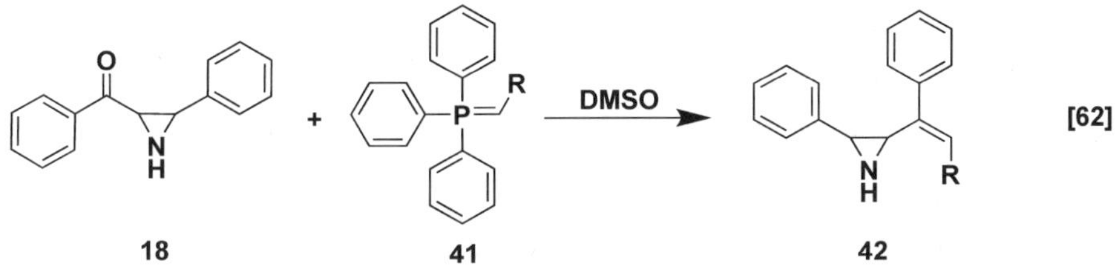


**Scheme 1.12**

The processes involving the aziridine cycle are very diverse. For instance, reactions of alkylation by alkyl halogenides [63], bromoacetic acid derivatives [29, 30] and acetoxypropene [64], are known. The use of arylboronic acids for synthesis of *N*-alkyl derivatives, e.g., compound **45**, is described in [63] (Scheme 1.13). The one-step reaction at room temperature of aziridinyl ketones **46** with chloroacetamides **47** and sulfur in the presence of $Et_3N$ yields mono-thio-oxamidines **48** [65].

Acylation reactions can be carried out by the action of acid anhydrides [66, 67], chloroanhydrides [47, 61], isocyanates [68], isothiocyanates [68, 69] and thioamides [70]. It was shown that the reaction of aziridinyl ketone **49** with chloroacetyl chloride **50** in the presence of $Et_3N$ [61] leads to the formation of *N*-acyl derivative **51**, whereas the alkylation product was not isolated.

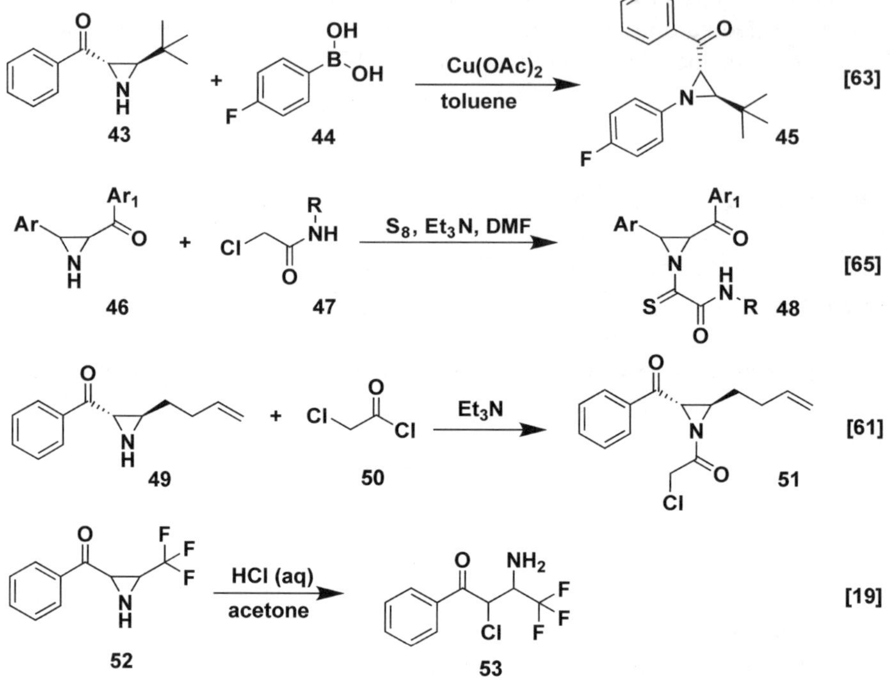


Scheme 1.13

Khomutov and Pashkevich [19] have established that the aziridine ring in compound **52** can open, giving rise to aminochlorobutanole **53**. *N*-Chloro derivatives of aziridinyl ketones can be obtained by the reaction of *tert*-butyl hypochlorite [47] or *N*-chlorosuccinimide [61]. The action of $N_2O_4$ on azirinochalcones in the presence of triethylamine leads to chalcones [71].

Reactions characterizing aziridinyl ketones as polyfunctional compounds are also described in the literature. An example is a widely studied [11, 15, 72, 73, 74] reaction of ketones **18** with phenylhydrazine in which the main products are pyrazole derivatives **55**.

The *cis* and *trans* isomers of **18** behave somewhat differently in this reaction: the *trans* isomer reacts faster at the stage of the aziridine cycle nucleophilic attack, whereas for another isomer this stage is inhibited by an adjacent *cis* substituent (Scheme 1.14). Such inhibition of the cyclization stage results in the formation of acyl hydrazone **56** as a by-product as well as the formation of pyrazole **55** from the intermediate aminopyrazoline by a rapid *trans* elimination. In the case of *trans*-aziridinyl ketone **18**, the slowest step is the *cis* elimination of $RNH_2$, isolating 4-alkylaminopyrazines **54** in most cases.

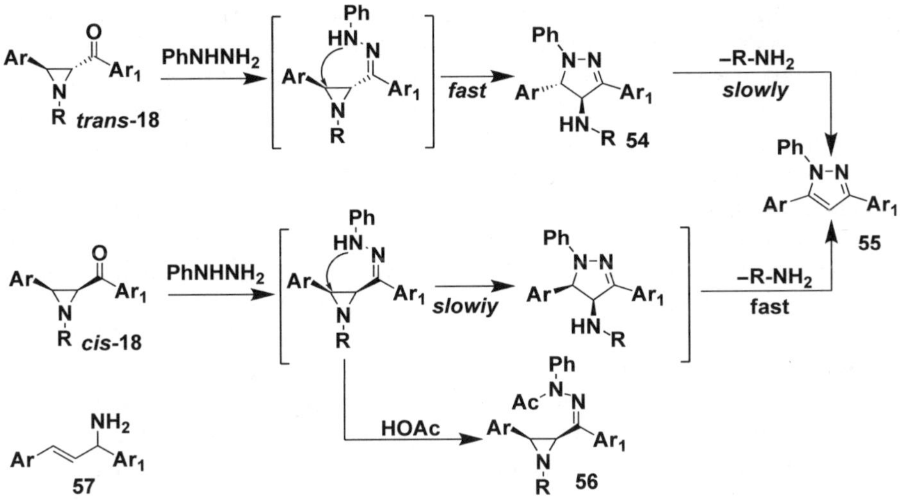


Scheme 1.14

N-Unsubstituted pyrazoles are synthesized by action of hydrazine on compounds **18**. However, the main reaction products in this case are amino chalcones **57** (up to 40%) [75].

An interesting example of heterocyclization involving both the aziridine fragment and the carbonyl group is a reaction of aziridinyl ketones with ammonia and carbonyl compounds, giving rise to 3,5a-dihydro-1*H*-azireno[1,2-*c*]imidazoles. This reaction will be considered in more detail in Sects. 1.2 and 1.3.

Treatment of aziridinyl ketone **18** with propiolic acid ethyl ester **58** leads to two regioisomeric pyrroles **59** and **60** [76] (Scheme 1.15). The reaction of ketone

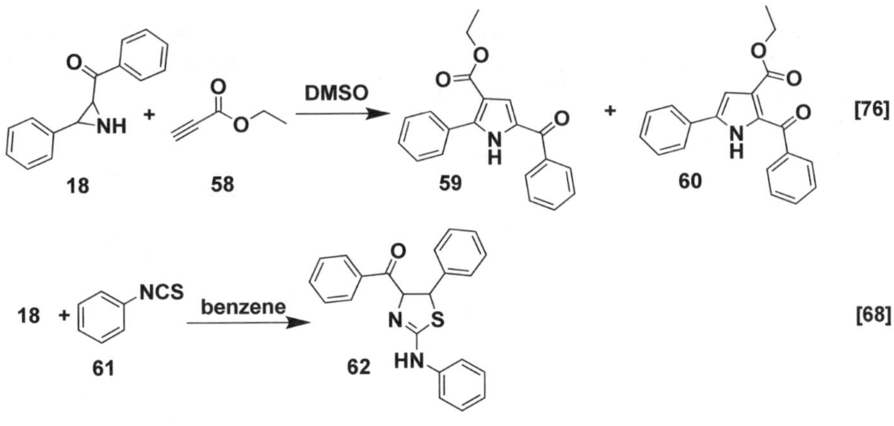


Scheme 1.15

**18** with isothiocyanates, e.g., compound **61**, can lead not only to the formation of thiourea derivatives, but also to thiazolyl ketones like **62** [68].

Methylthiocyanate reacts with $BF_3$ complexes of 3-aroylaziridines, yielding 4-aryl-5-aroyl-2-methylthio-2-imidazoazolines. During this reaction the three-membered ring opens in a regio- and stereospecific manner at the C(2) atom with inversion of the configuration [77].

The most studied type of reaction of aziridines is the transformations via a thermo- or photoinduced bipolar ylide. Among the numerous works related to this question are the reviews by Padwa [78] and Lown [79]; the analysis of reactions of this type which do not involve the aziridine cycle are left out of this present work.

The presence of an aroyl fragment in azomethine ylides obtained from opening of three-membered rings in the case of dipolarophiles with high LUMO (lowest unoccupied molecular orbital) energy or in the absence of an "external" dipolarophile can lead to the possibility of such unusual reactions as intramolecular 1,3-dipolar cycloaddition [80]. Examples of such reactions are the thermal isomerization of aroyl aziridines **63** into a pyrrole derivative **64** [81, 82] or into 2,5-diphenyloxazole **65** (in the presence of diphenyliodonium iodide) [83] (Scheme 1.16).

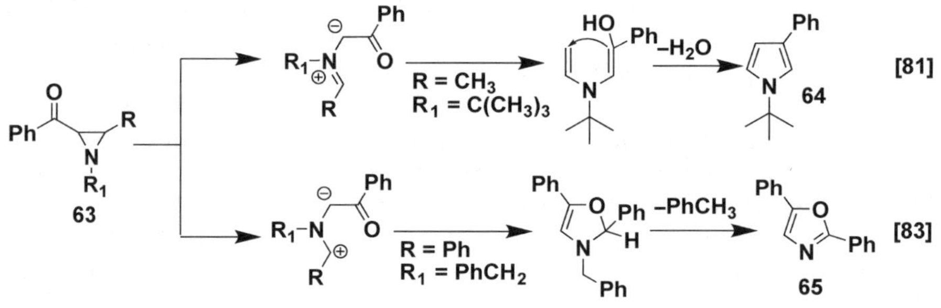


**Scheme 1.16**

The photoactivated intermolecular dimerization reaction of aziridinyl ketone **18** leading to heterocycle **67** after the initial oxidation of piperazine **66** has also been described [84] (Scheme 1.17).

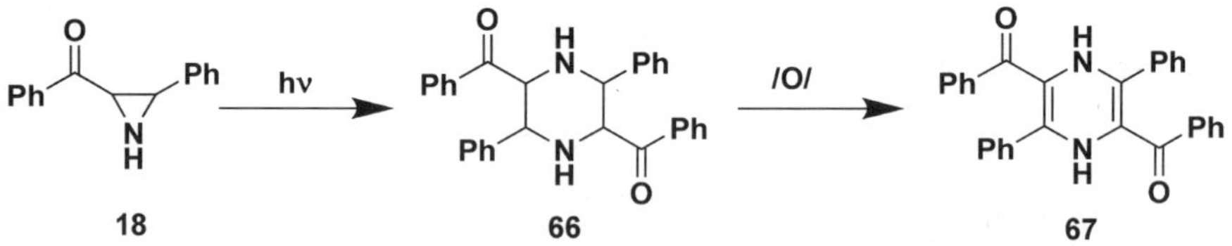


**Scheme 1.17**

An example of azaheterocycle synthesis based upon aziridinyl ketones is also a reductive cyclization of 1-cyclohexyl-2-benzoyl-3-(2-nitrophenyl)aziridine **68** into quinoline **69** [14] (Scheme 1.18).

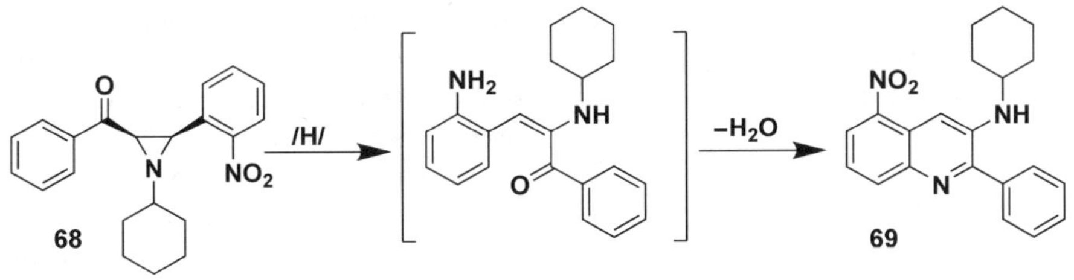


Scheme 1.18

Another process worth mentioning is the *cis–trans* isomerization of aziridinyl ketones. The transformation of *trans*-aziridines into the *cis* form was carried out in a methanol–acetone solution in the presence of triethylbenzylammonium hydrochloride [85]. However, the best yields of the *cis* isomer (up to 90%) were detected by the same authors when water was added to the reaction mixture [86]. The reverse transition from the *trans* configuration into the *cis* form is possible using methanolic solution of sodium methylate [87] or potassium hydroxide [35].

## 1.2 Synthesis of Bi- and Tricyclic Aziridine Derivatives

One of the initial findings about bicyclic aziridines is reported in [88]; the authors carried out the synthesis of 3,5a-dihydro-1*H*-azireno[1,2-*c*]imidazoles by reacting *trans*-2-aryl-3-aroylaziridines with carbonyl compounds in methanol saturated with ammonia in the presence of ammonium bromide. In particular, 2,2-dimethyl-6-(4-nitrophenyl)-4-phenyl-1,3-diazabicyclo[3.1.0]hex-3-ene **72** was obtained in the reaction of ketone **70** with ammonia and acetone **71** (Scheme 1.19).

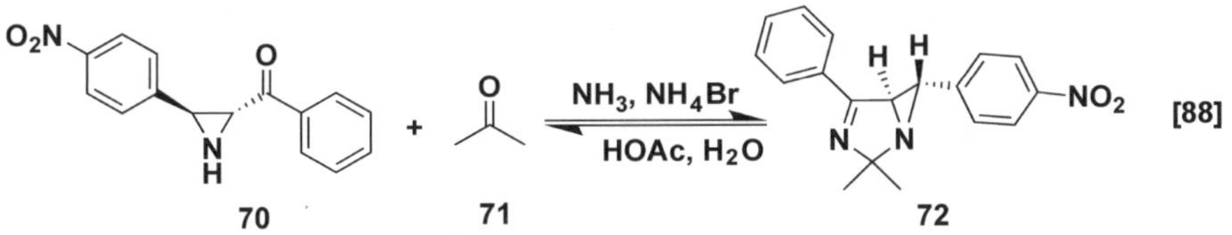


Scheme 1.19

Aldehydes, symmetric and asymmetric ketones, such as formaldehyde, acetaldehyde, substituted benzaldehydes and cyclic ketones, were introduced into the reaction along with acetone. The reaction is reversible: azirenoimidazoles undergo reverse transformation forming *trans*-aziridinyl ketones in acetic acid.

A similar method of synthesis was also used in other works [89, 90, 91, 92]. As mentioned earlier, aziridinyl ketones are easily formed from ammonia and $\alpha,\beta$-dihalogen ketones, enabling azirenoimidazoles to be synthesized in one step from 2,3-dibromopropan-1-ones without separating the intermediate aziridinyl ketones [93, 94, 95]. The method consists of passing ammonia into an alcoholic suspension of $\alpha,\beta$-dihalogen ketones **73** with subsequent addition of an aldehyde or a ketone **75** (Scheme 1.20). But Orlov et al. [95] noted that when using benzaldehyde as a carbonyl and the intermediate aziridinyl ketone is less soluble in alcohol than the starting dibromide (in particular, for 1-aryl-2,3-dibromo-3-(4-nitrophenyl)propan-1-ones), it is advisable to carry out the synthesis in two steps with separating the intermediate aziridinyl ketone.

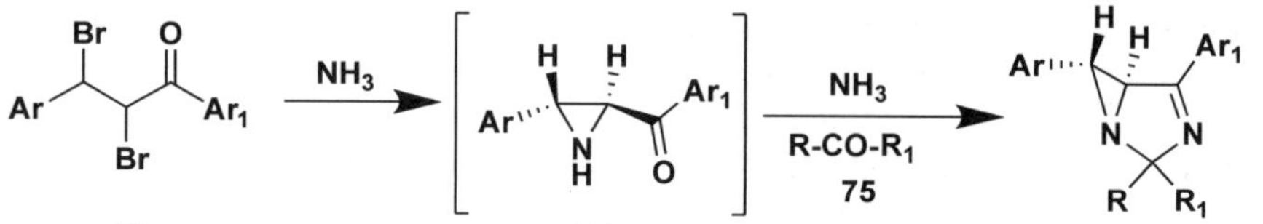


73 74 75

**Scheme 1.20**

The mechanism of azirenoimidazole formation has not been definitively established and various possibilities are discussed in the literature. For example, it is proposed [88] that the first stage of the reaction is the addition of a carbonyl group to the nitrogen atom of aziridinyl ketone **77** forming aminohydrine **78**, which then forms the diamine **79** upon reaction with ammonia, and then the intermediate **79** undergoes cyclization into the final product **80** (Scheme 1.21). An alternative version considered [88] is the reaction of aziridinyl ketone with the previously formed aldimine or ketimine, leading to an analogous intermediate **79**.

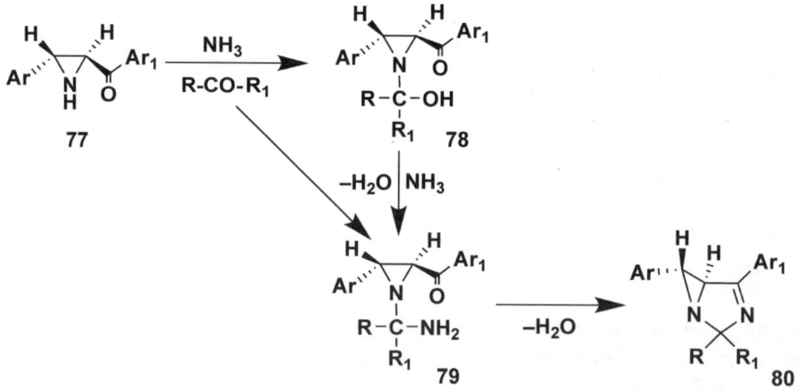


77 78 79 80 [88]

**Scheme 1.21**

It was found later [94] that aziridinyl ketones do not react with aldehydes in the absence of ammonia apparently owing to the strong intramolecular hydrogen bond [96]. Kaluski et al. [94] considered a more probable mechanism where the carbonyl group of the aziridinyl ketone is first attacked by an ammonia molecule, releasing the aziridine imino group, which reacts with the carbonyl compound and subsequently cyclizes into azirenoimidazoles (Scheme 1.22).

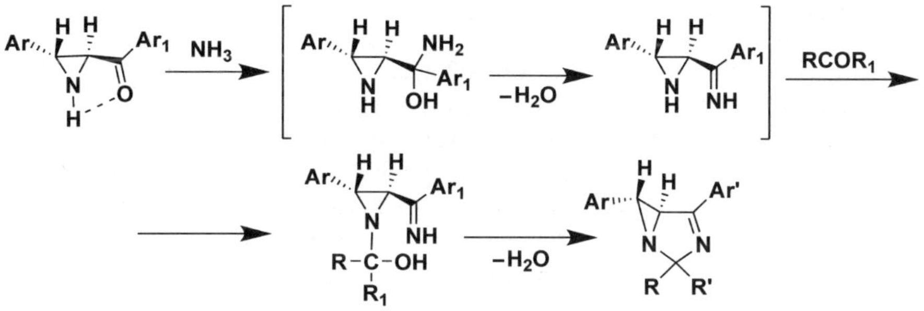


**Scheme 1.22**

Azirenoimidazoles **83** can be also synthesized by the multicomponent reaction of $\alpha$-chloroketones **81** with aldehydes **82** and ammonium acetate under microwave irradiation [97] (Scheme 1.23).

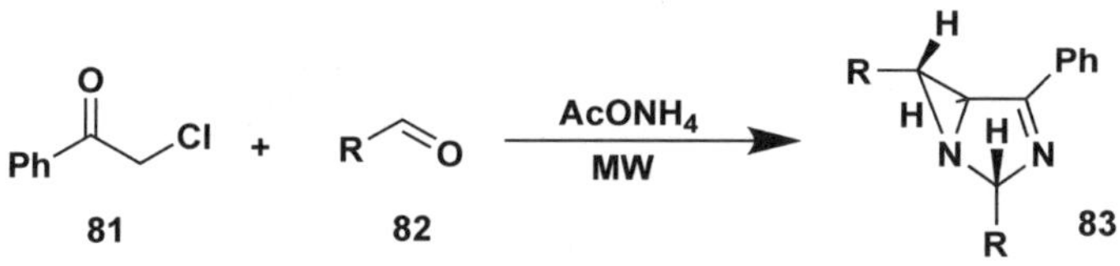


81 82 83

R = Ar, $C(CH_3)_3$

**Scheme 1.23**

The reaction is diastereoselective and in all cases only *exo* isomer of **83** is separated. However, target azirenoimidazoles **83** containing different substituents in positions 2 and 8 cannot be obtained by this method. Other multicomponent and one-pot synthesis procedures for synthesis of similar polycyclic structures are also described [98, 99, 100, 101, 102, 103, 104].

The initial data concerning the formation of *endo* and *exo* isomers in the synthesis of azirenoimidazoles were given in [89]: *endo* and *exo* isomers (**85A** and **85B**, respectively) were obtained in the reaction of *trans*-2-phenyl-3-benzoylaziridine **18** with ammonia and benzaldehyde **84** and separated by fractional crystallization (Scheme 1.24). Data concerning separating these isomers were also given in a series of subsequent works [90, 92, 95, 105].

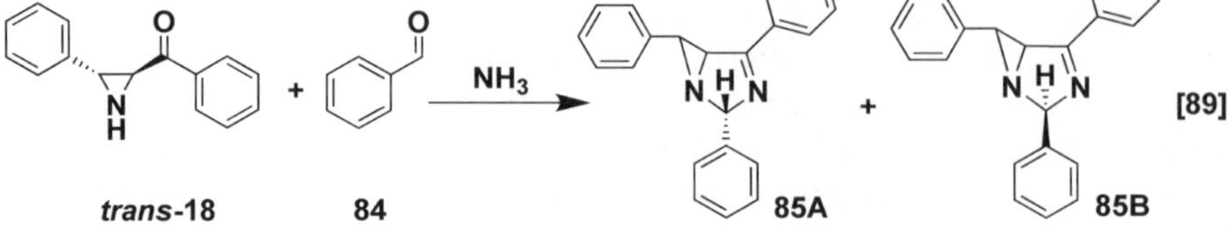


Scheme 1.24

Synthesis of the other classes of polycyclic aziridines is reported in [106]. Derivatives of 1,3,4,6a-tetrahydroazireno[1,2-*a*]pyrazines **88** and 1,1a-dihydro-1,2-diarylazireno[1,2-*a*]quinoxaline **90** were obtained in the reaction of dibromides **86** with ethylenediamine **87** or 1,2-phenylenediamine **89** (Scheme 1.25). The authors showed the possibility of synthesizing dihydroazirenoquinoxalines **90** from $\alpha$-bromochalcones. This can be confirmed by the following reaction of $\alpha$,$\beta$-dihalogen ketones via dehydrobromination.

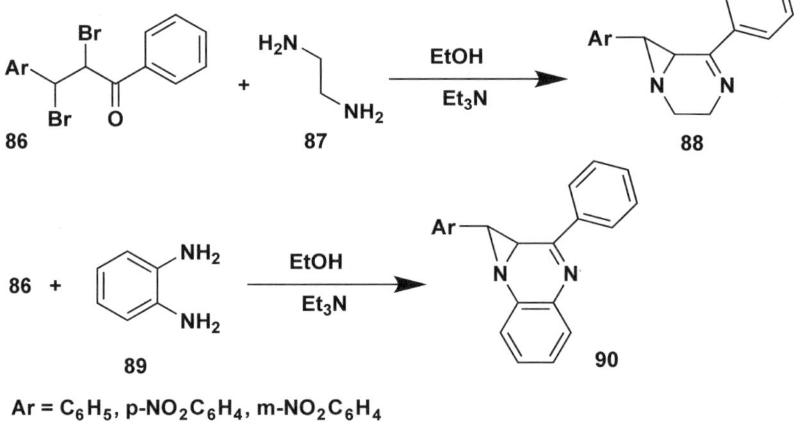


Scheme 1.25

Investigations of the mechanism of the reactions of diamines with 2,3-dibromopropan-1-ones were not reported in early works. An assumption about the mechanism was made only in [107] on the basis of the results of the investigation of a reaction utilizing 4-nitro-1,2-phenylenediamine. The authors propose the formation of $\alpha$-bromochalcones **92** initially, which then form intermediates **93** in the reaction with the amino group of a diamine. Then compounds **93** undergo further transformation into $\beta$-aminoketones **94**, dihydroazirenoquinoxalines **95** or acetylenic ketones **96** (Scheme 1.26). The yields of aziridines **95** did not exceed 10% and the reaction product ratios depended on the conditions. It is noted, in particular, that carrying out the reaction in methanol for 1 h with the use of *N*,*N*-dimethylbenzylamine as a catalyst increased the yield of **95**. It should be noted

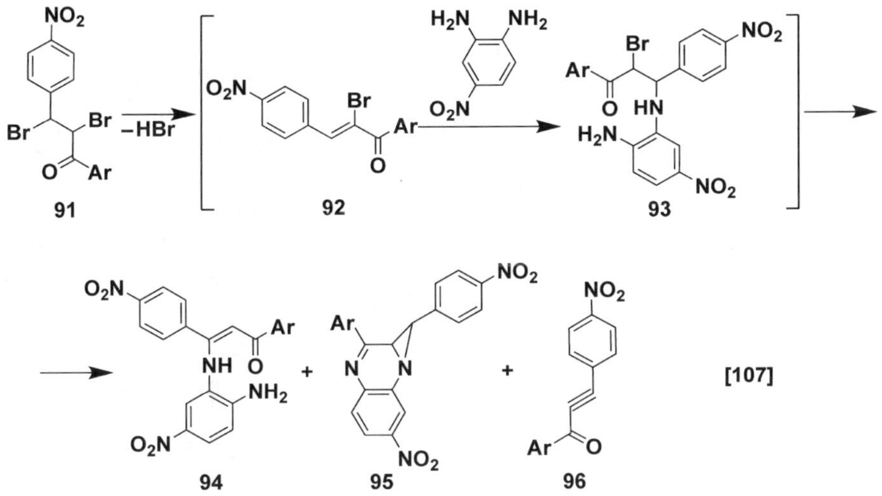


Scheme 1.26

that nucleophilic addition of the second amino group in compounds **94** to the carbonyl group leads to 1,5-benzodiazepines, which are often separated from the reaction mixture as by-products. The question about the formation of fused diazepines in the reactions of 2,3-dibromopropan-1-ones with substituted 1,2-phenylenediamines as well as with some heterocyclic 1,2-diamines is considered in more detail in Chap. 4.

In the case of the interaction of $\alpha,\beta$-dibromoketones with monosubstituted 1,2-phenylenediamines, a question arises about the regioselectivity of the process. There are statements that such reactions proceed with the formation of only one of the possible regioisomers [107, 108]. But Zbruyev et al. [109, 110] showed that the reaction was not regioselective and leding to a mixture of isomers **99A** and **99B** in a different ratio, with isomer **99A** predominating (Scheme 1.27).

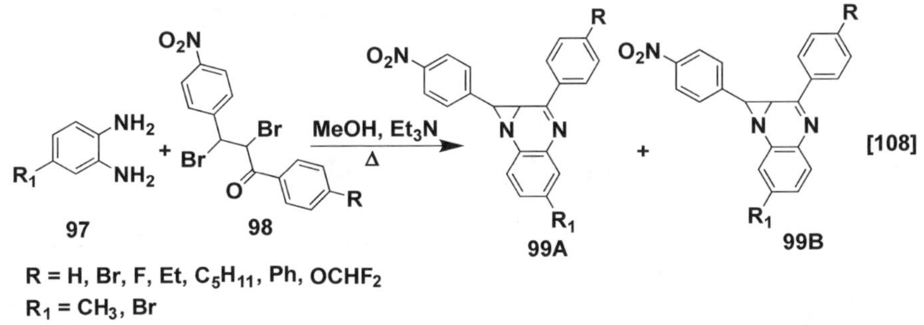


Scheme 1.27

Pashkevich and Khomutov [111] reported the synthesis of aziridines containing polyfluoroalkyl substituents on the aziridine cycle. In the case where $R_F$ is $CF_3$, the only reaction product is bis(aziridinyl ketone) **103** (Scheme 1.28). The results also provide information about the reaction mechanism. As well as in [107], it is assumed that the reaction proceeds via the formation of $\alpha$-bromoketones and $\beta$-aminoketones **101** and **102** where the next step is the intramolecular bromine substitution and heterocyclization into quinoxaline **104**. But in the case where $R_F$ is $CF_3$, the formation of the aziridine cycle is followed by the interaction of the second amino group of the diamine with another molecule of the intermediate **102**, leading to compound **103**.

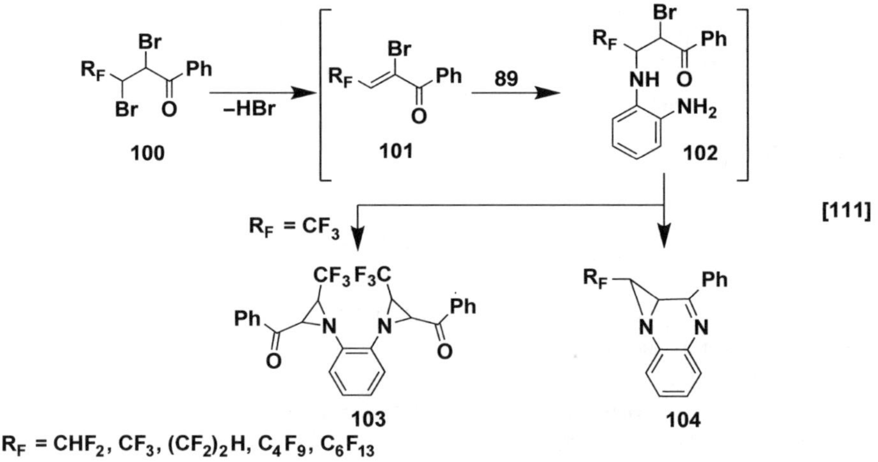


Scheme 1.28

The authors explained these results by a specific electron-withdrawing effect of the $CF_3$ group leading to the activation of the carbon–carbon double bond. Dihydroazirenoquinoxaline **104** containing a $CF_3$ group could be obtained by replacing the bromine in the $\alpha,\beta$-dihalogen ketone with chlorine, which is a worse leaving group.

## 1.3 Chemical Properties of Bi- and Tricyclic Aziridines

The chemical properties of bi- and tricyclic aziridines are determined, first of all, by the presence of the sterically strained three-membered cycle as well as the two nitrogen atoms, which are basic centers. The most interesting reactions are the processes involving the carbon–carbon single bond cleavage of the aziridine cycle which will be considered later.

The possibility of bicyclic aziridine hydrochloride salt formation was first mentioned in [88]. The hydrochlorides of some azirenoimidazoles were obtained by passing dry hydrogen chloride through ether solutions. However, the protonation direction was not established in that work.

The behavior of dihydroazirenoimidazoles and tetrahydroazirenopyrazines in the presence of acids was studied in [112] in detail. The authors note that protonation not accompanied by further transformations takes place under mild conditions (dilute HCl in ethanol at room temperature). 1-(4-Nitrophenyl)-6-phenyl-3,5a-dihydro-1*H*-azireno[1,2-*c*]imidazole and 1-(4-nitrophenyl-6-phenyl-1,3,4,6a-tetrahydroazireno[1,2-*a*]pyrazine hydrochlorides were obtained according to the procedure described in [88].

Orlov et al. [112] concluded that the protonation took place on the azomethine group nitrogen atom on the basis of IR spectral data. In concentrated sulfuric acid, the formation of the only acidolysis product, 1,3-diarylpropan-1,2-one **107**, was observed (Scheme 1.29). Orlov et al. [112] assumed that first protonation took place on the ketimine nitrogen atom (intermediate **106**), but in strong acidic media the aziridine nitrogen atom is protonated as well, making the system unstable and promoting additional transformations.

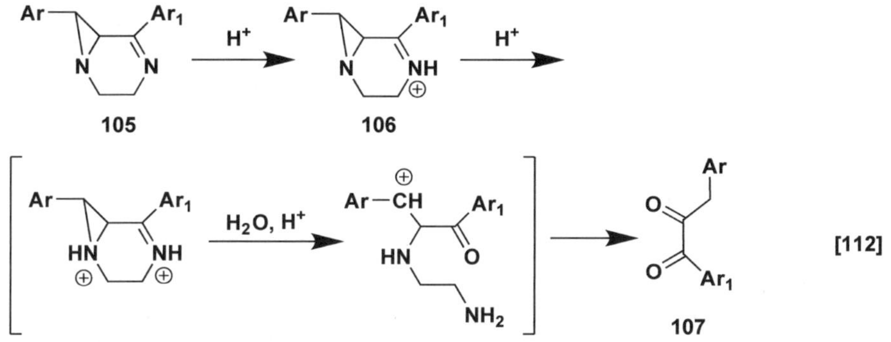


Scheme 1.29

The basic character of the ketimine nitrogen atom in azirenoimidazoles is also shown by its propensity to react with trimethyloxonium tetrafluoroborate, leading to the appropriate salt [113].

The behavior of dihydroazirenoquinoxalines in solvents possessing different polarity in the presence of acids such as HCl, HBr, HI, $HClO_4$, $HBF_4$ and $CF_3COOH$ was studied in [114]. It was shown that even under mild conditions, i.e., HCl, aprotonic solvent, isomerization into the 2-benzylquinoxaline derivatives **109** took place (Scheme 1.30). The authors could not isolate the salts as they had made for azirenoimidazoles and azirenopyrazines [88, 112]. On the basis of the experiments carried out, as well as previous data [112], a conclusion

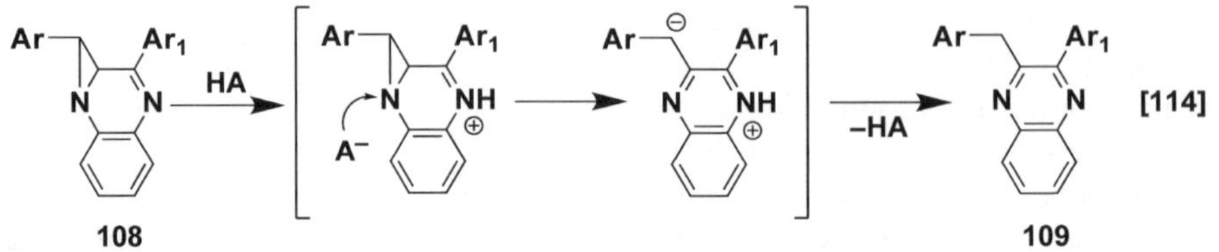


108 109

Scheme 1.30

was drawn about the initial protonation on the azomethine nitrogen atom. The electron-withdrawing influence of the protonated azomethine group on the aziridine cycle increases its sensitivity towards nucleophilic attack by an acid anion, leading to the opening of the three-membered ring. After the attack, the heteroaromatic system is formed, which is the most energetically favorable process.

Thus, acid-catalyzed transformations of fused aziridines include cleavage of the C–N bond of the aziridine cycle. A reaction has also been described where the C–N bond of the imidazole cycle is broken [88]. It was shown that a rearrangement leading to pyrimidine **111** takes place under the action of sodium methylate with azirenoimidazoles **110** (Scheme 1.31).

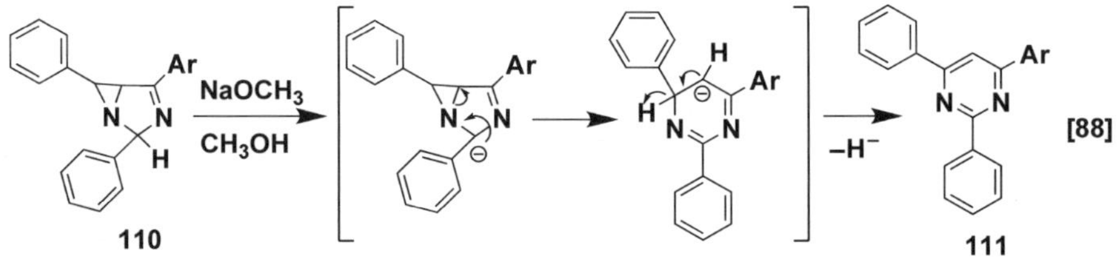


110 111

Ar = $C_6H_5$; 4-$NO_2C_6H_4$

Scheme 1.31

A nitrogen oxide **112** [113] is formed in the reaction of the corresponding azirenoimidazole with *meta*-chloroperoxybenzoic acid. Compound **112** can be easily reduced by sodium borohydride to the corresponding *N*-hydroxy derivative **113** which can be oxidized by *tert*-butylhypochlorite to regenerate the *N*-oxide group (Scheme 1.32).

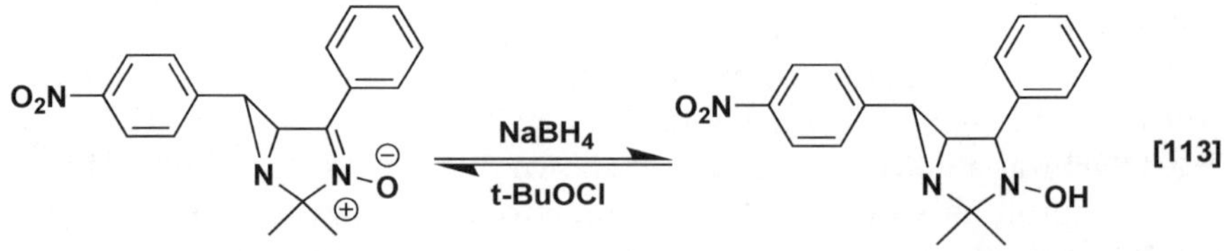


112 113

Scheme 1.32

As already mentioned, most interesting are the properties of the aziridines connected with cleavage of the C–C bond of the sterically strained three-membered cycle, namely, thermo- and photoinduced cycloaddition reactions. Photo- and thermochromism inherent to these compounds are also linked with the breaking of this bond in the literature.

For an aziridine containing four different substituents A, B, C and D at the carbon atom, four different azomethine ylides can be formed: two by conrotatory and two by disrotatory ring opening (Scheme 1.33).

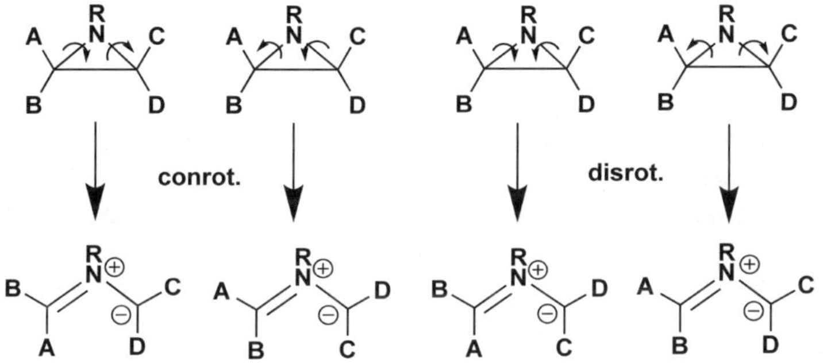


Scheme 1.33

In the event that two substituents at different carbon atoms in the aziridine cycle are hydrogen atoms (for example, A and C in Scheme 1.33), the ylides being formed are called S-shaped, W-shaped, or U-shaped [115] (Scheme 1.34). The reaction of dipolar cycloaddition of S-shaped ylides forms adducts with *trans* orientation of the protons, while W- and U-shaped ylides lead to *cis* adducts (for suprafacial reactions).

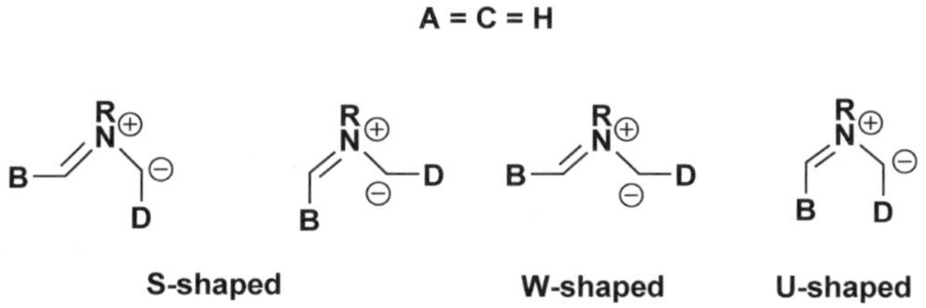


Scheme 1.34

If the aziridine cycle is fused, then rotation of one of the C–N bonds during C–C bond scission becomes possible only in one direction; therefore only two of the four possibilities of ring opening can be realized depending on the rotation direction of the second free C–N bond, and the *cis*-ylides or *trans*-ylides can be formed. The *trans*-substituted aziridines are as shown in Scheme 1.35.

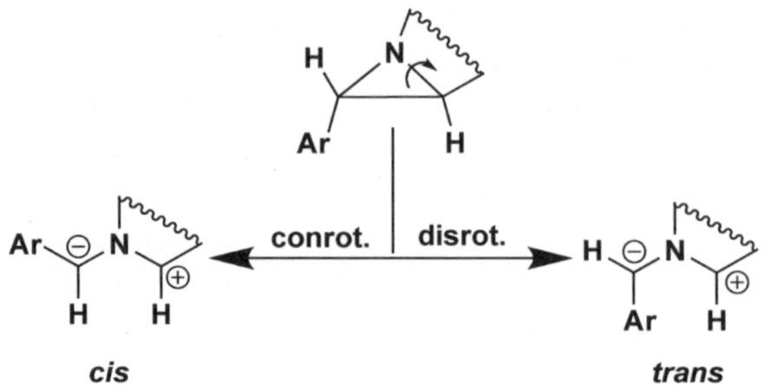


Scheme 1.35

According to orbital symmetry rules, thermal opening of the *trans*-aziridine must take place in a conrotatory manner, leading to *cis*-ylide, and photochemical opening via a disrotatory one gives a *trans*-azomethine ylide.

Most of the chemical properties of fused aziridines described in the literature are concerned with cycloaddition reactions. For example, it was shown in one of the first works [93] that refluxing azirenoimidazole **72** with different compounds containing multiple C–C bonds in *para*-xylene led to the corresponding adducts via cleavage of the C–C bond of the aziridine cycle. In particular, the reaction of **72** with *N*-phenylmaleimide **114** leads to compound **115** and interaction with diethyl acetylene dicarboxylate **116**, diethyl fumarate **118** or 1,4-diarylbut-2-ene-1,4-diones **120** gives rise to the cycloadducts **117**, **119** and **121**, respectively

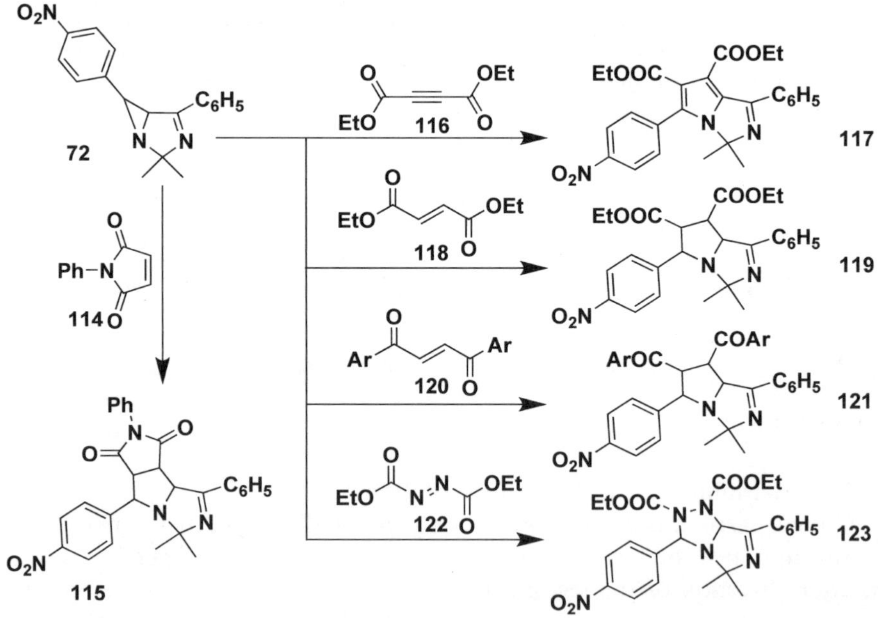


Scheme 1.36

(Scheme 1.36). The heteroatomic diethyl fumarate analogue **122** can also act as a dipolarophile, with its reaction leading to imidazotriazole **123**.

Chemical properties of dihydroazirenoquinoxalines were studied in [106] and it was shown that these compounds react in numerous cycloadditions via cleavage of the C–C bond of the aziridine cycle as well.

Cycloaddition reactions of dihydroazirenoquinoxalines proceed under milder conditions than those for dihydroazirenoimidazoles. As already mentioned, the action of hydrochloric acid on dihydroazirenoquinoxalines **90** results in a rearrangement leading to benzylquinoxalines **109** (Scheme 1.37).

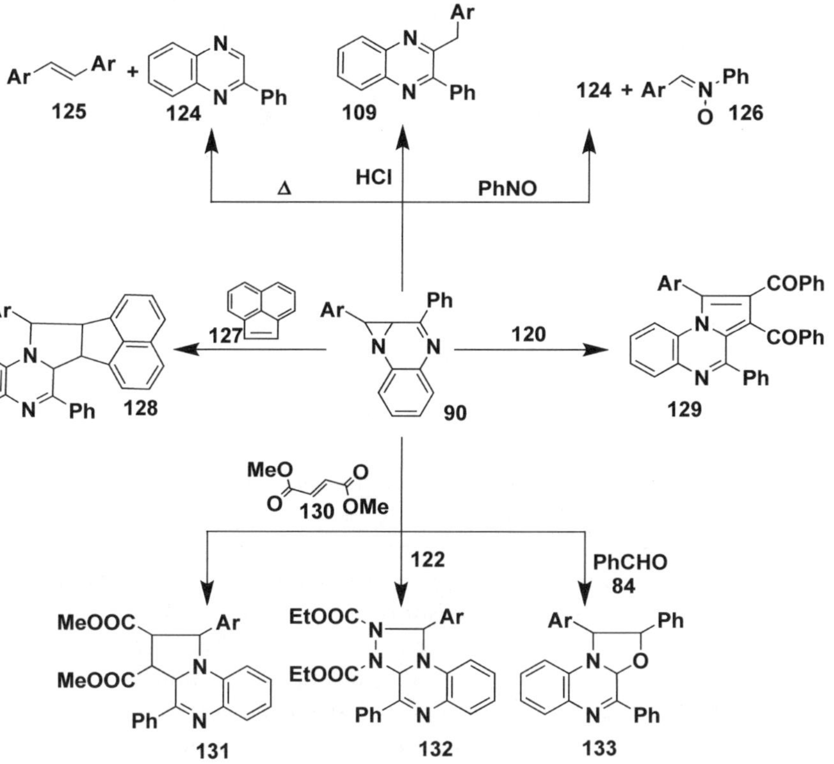


**Scheme 1.37**

Thermal decomposition or action of nitrobenzene gives rise to mixtures of quinoxalines **124** and diarylethylenes **125** or *N*-oxides **126**. As well as for the case of dihydroazirenoimidazoles, reactions with a series of symmetrical dipolarophiles (such as compounds **120**, **122**, **127** and **130**) leading to corresponding cycloadducts **129**, **128**, **131** and **132** have been described. Reactions with asymmetrical dipolarophiles, in particular, with benzaldehyde **84**, proceed chemoselectively and lead in this case to dihydrooxazoloquinoxalines **133**. The same result was obtained by other authors [109] who studied the chemical properties of 5,6-dimethyl-1,1a-dihydroazireno[1,2-*a*]quinoxalines.

The structure of the products enabled Heine and Henzel [106] to propose that the intermediate of the reactions described is an azomethine ylide (Scheme 1.38).

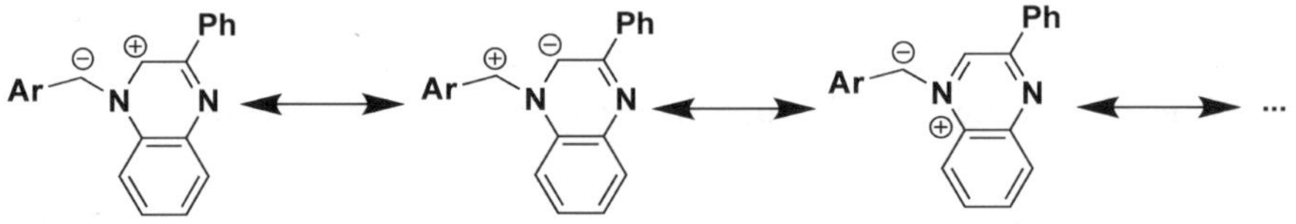

**Scheme 1.38**

If a 4-nitrophenyl substituent is attached to the aziridine cycle of the dyhydroazirenoquinoxaline molecule, charge delocalization involving the nitro group resulting in a quinoid-type structure is possible (Scheme 1.39).

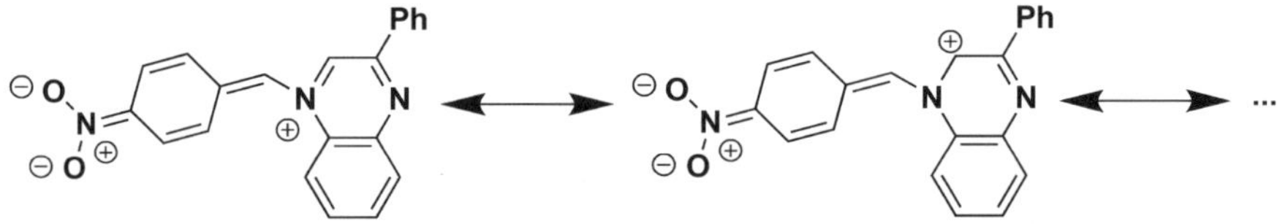

**Scheme 1.39**

The structure of the products reported in [93, 106] was proven by spectral methods and chemical transformations, but the stereochemistry of the products, which provides information about the addition mechanism and intermediate structures, was not considered. These data were obtained in [116].

When a solution of *endo* or *exo* isomers of compound **134** and dipolarophile **135** in ethanol at liquid-nitrogen temperature was irradiated by UV, a red coloration was observed which rapidly disappeared upon heating. A single cycloadduct **137** was found and isolated from the reaction mixture. Padwa and Glazer [116] considered that its formation proceeds via an intermediate **136** with subsequent proton transfer (Scheme 1.40).

It is noted that the stereochemistry of cycloadducts **136** and **137** does not correspond to that predicted on the basis of orbital symmetry rules for photochemical

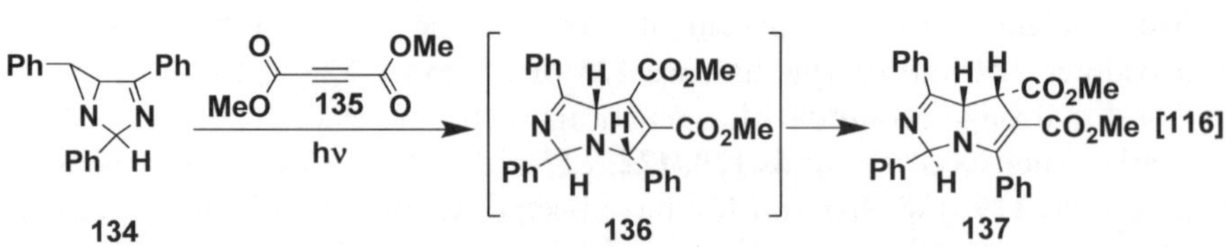

**Scheme 1.40**

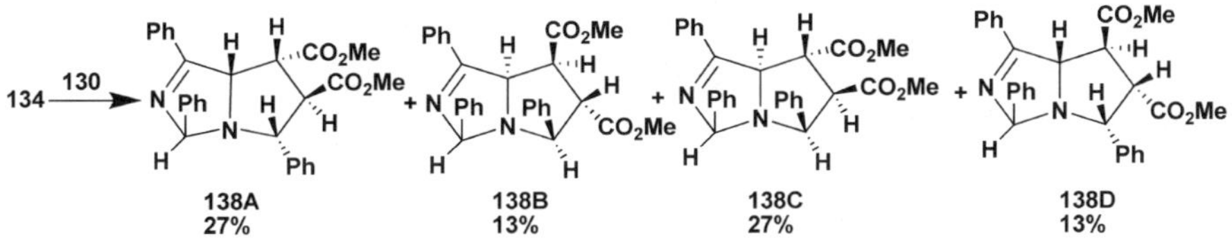

Scheme 1.41

aziridine cycle opening. Cycloadduct **137** was also obtained in a thermal reaction of aziridine **134** with acetylenic diketone **136**, arguing in favor of the formation of *cis*-azomethine ylide in both thermal and photochromic opening. A mixture of four isomeric cycloadducts **138A**–**138D** corresponding to a different stereochemistry of the interaction of the dipolarophile with *cis*-azomethine ylide was obtained in the reaction of aziridine **134** with dimethyl fumarate **130**.

Although the mechanism of the photoinduced aziridine ring opening is not quite clear, the data concerning the stereochemistry of the products are convincing enough evidence that the intermediate of the reaction is a planar azomethine ylide. For instance, the products **138A** and **138C**, differing in the orientation of the phenyl substituent attached to the aziridine cycle, are isolated in equal quantities (Scheme 1.41). Therefore, the rotation of this substituent in the intermediate in either of the directions has equal probability and this is possible only for a planar ylide. Otherwise if the initial attack on the aziridine cycle turned away from imidazole ring plane took place, then a preferred formation of one of the isomers should be expected. The same situation is observed with an other pair of cycloadducts **138B** and **138D**, which are formed in equal quantities as well.

The preservation of the relative configuration of the dipolarophile substituents in all cycloadducts is evidence for the concerted cycloaddition mechanism.

An investigation of the aziridinyl anil transformations in solutions without dipolarophiles addition was carried out in [89], where the behavior of the isomeric azirenoimidazoles **134** in benzene under UV-lamp irradiation was studied. Irradiation of the *endo* isomer at 50°C leads to the formation of *cis*-2,3-dihydro-2,3,5-triphenylpyrazine **140A** (Scheme 1.42).

The process does not occur in a single step: an intermediate enediimine **141A** which cyclizes into **140A** upon subsequent heating to 50°C in the dark is found in the solution.

However, for *exo* isomer of **134** instead of the expected *trans*-2,3-dihydro-2,3,5-triphenylpyrazine **140B**, formation of the *cis* isomer **140A** upon irradiation is observed as well. Padwa and Clough [89] explained this by the existence of an equilibrium mixture of the isomeric enediimines **141A** and **141B** with predominance of isomer **141A**. But the formation of identical reaction products from *endo* and *exo* isomers of **134**, in our opinion, is more easily explained by

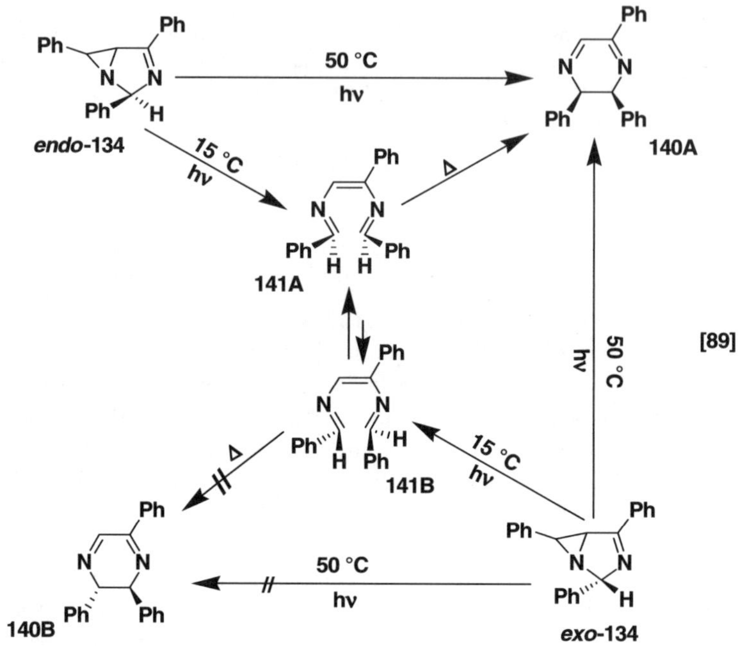


Scheme 1.42

the participation of the planar azomethine ylide which is formed from both isomers as an intermediate.

In continuation of these investigations DoMinh and Trozzolo [90] irradiated a solution of the azirenoimidazole **72** in benzene. Similar to the preceding results, the corresponding enediimine **142** was isolated (Scheme 1.43).

A red coloration forms and slowly disappears in the course of the reaction owing to irradiation. To ascertain the structure of the colored intermediate DoMinh and Trozzolo [90] attempted to catch it chemically. When the dipolarophile **135** was added to the reaction mixture, the coloration immediately disappeared and the corresponding cycloadduct **142** was formed. Irradiation of **72** with tetracyanoethylene **143** gave the adduct **144** in high yield. On the basis of these data, a structure of azomethine ylide formed by aziridine carbon–carbon bond cleavage is assigned to the colored intermediate.

Another reaction pathway is observed in methanol [117]. Irradiation of a methanolic solution of **134** led to imidazoline **147** in 60% yield, with a possible intermediate being azomethine ylide **146** (Scheme 1.44). However, two pathways for the formation of imidazoline **147** are possible. The first one is via initial formation of an enediimine **141** (such reaction is known as a nonphotochemical process [118, 119]). The fact that addition of methanol to compound **141** obtained by irradiation of *cis*-dihydropyrazine or *trans*-dihydropyrazine **140**

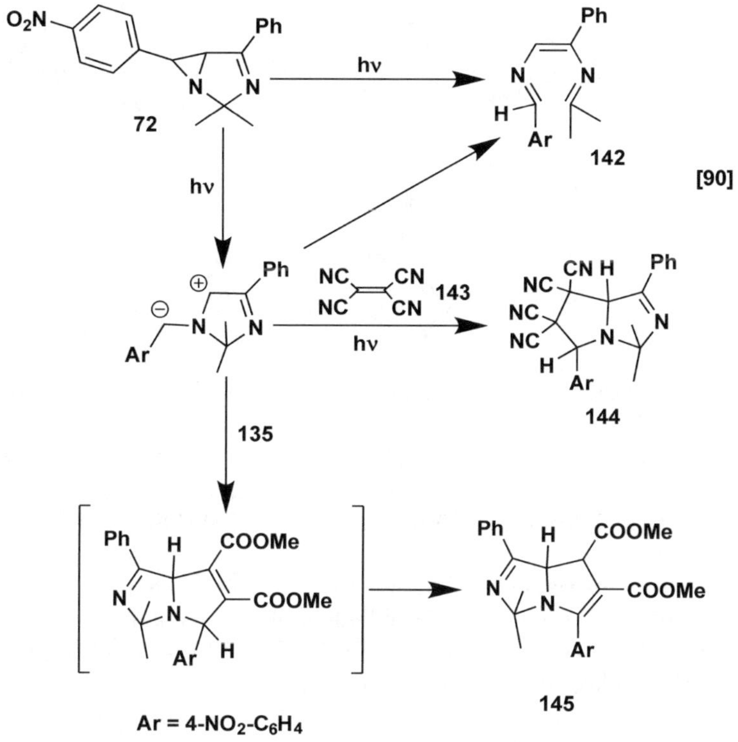


**Scheme 1.43**

leads to imidazoline **147** argues for this pathway. The second presupposes initial formation of the ylide **146** subsequently becoming the imidazoline **147**. On the basis of the results of cycloaddition reaction investigations [116], this path seems more probable. It is possible that ylide and enediimine are in equilibrium and the solvent effect determines the path of further transformations, i.e., in a nonpolar solvent (benzene) cyclization into dihydropyrazine takes place, while a polar solvent, methanol, can stabilize charged intermediates and react itself, promoting the formation of compounds like **147**.

When azirenoimidazoles were first synthesized [88] their photochromic properties were discovered: all the crystalline 6-(4-nitrophenyl)-substituted aziridines, colorless or yellowish, became deep blue upon sunlight irradiation; the coloration disappeared in a few days in the dark. Photochromism was also observed for azirenoimidazole hydrochlorides. Their crystals became red upon irradiation. Photochromic properties were also found for azirenoimidazoles without a nitrophenyl group on the aziridine cycle but they were much less pronounced.

Photochromism is also inherent in tetrahydroazirenopyrazines and dihydroazirenoquinoxalines as well as in azirenoimidazoles. This property is most

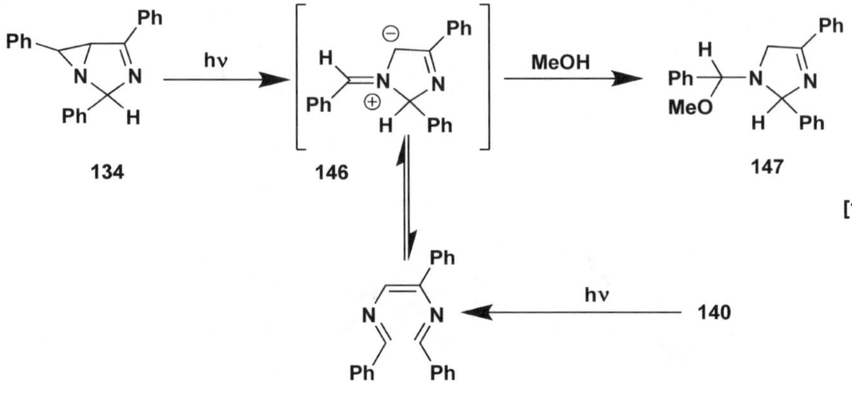


134 146 147 [117] 141 140

Scheme 1.44

pronounced for compounds containing the 4-nitrophenyl substituent on the aziridine cycle.

The fullest investigation of the photochromic properties of various bi- and tricyclic aziridines in the crystalline state and systematization of the data were carried out by Orlov et al. [120] with aziridines containing different substituents R, $R_1$ and the bridge group M (Scheme 1.45).

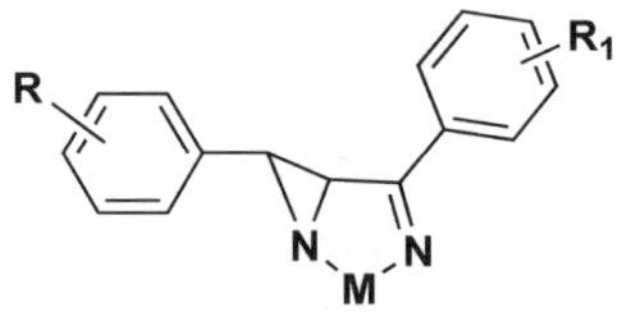


M = $C(CH_3)_2$; $CH_2$-$CH_2$; *o*-$C_6H_4$

Scheme 1.45

It was shown that almost all the compounds studied in [120] possess photochromic properties. The most pronounced influence on photochromic properties is caused by the substituent R found in the ring directly attached to the aziridine cycle, which is explained by its stabilizing effect on the formation of the colored moiety. Electron-withdrawing substituents led to deepening of the coloration. Inductive effects have much less influence than conjugation effects. Introduction of the halogens and even a strong $\sigma$-acceptor, such as the $SO_2CF_3$ group [121], leads to only to small changes in the photochromic properties, whereas compounds containing a nitro group in the *para* position possess the most pronounced photochromic properties. Moving the nitro group to the *meta* position of the aromatic ring, which excludes it from the conjugation, drastically reduces the photochromic effect. The deepest coloration is inherent in the compounds containing the 5-nitro-2-thienyl substituent instead of the 4-nitrophenyl substituent. In the opinion of Orlov et al. [120], it is related to the fact that the thiophene ring is more polarizable than the phenyl ring and, therefore, transmits better the electron-withdrawing effect of the nitro group.

Much less influence on photochromic properties is exerted by the substituent $R_1$, and the relationship between character of the substituent and the position of the absorption band maximum of the colored species is difficult to discover. However, it was shown [120] that strengthening of the electron-withdrawing properties of the substituent $R_1$ led to a reduction of the time for decoloration in the dark.

Photochromic properties also depend on the bridge group M. It was found from an example compound where R is 4-$SO_2CF_3$ [121] that the absorption band maximum of the colored species was redshifted in the series M is $C(CH_3)_2$, $CH_2$–$CH_2$, *o*-$C_6H_4$. Introducing a nitro group as an R substituent results in elimination of the role of the bridge group.

It was noted [120] that the rate of dark decoloration reduces drastically with increasing temperature. In particular, for dihydroazirenoquinoxalines it requires 5–10 min at 100 °C, whereas it takes several days at room temperature.

Photochromic properties of the hydrochlorides of 1-(4-nitrophenyl)-6-phenyl-3,5a-dihydro-1*H*-azireno[1,2-*c*]imidazole **72** and 1-(4-nitrophenyl)-6-phenyl-1,3,4,6a-tetrahydroazireno[1,2-*a*]pyrazine **88** (Ar is 4-$NO_2C_6H_4$) were mentioned in [88, 112]. In contrast to bases becoming blue in the light, the photoinduced form of the salts is red. Orlov et al. [112] explained this by a different electron-density distribution in the photoinduced form. Bases corresponding to the ylide of type **A** are stabilized by the nitro group, enabling the resonance structure **B**, whereas protonation of the azomethine nitrogen atom must stabilize structure **C**, eliminating the stabilizing effect of the nitro group and making the photochromic properties of the salts similar to the properties of bases not containing the nitro group in the *para* position of the phenyl ring (Scheme 1.46).

Starting from the early works devoted to this subject, it is generally accepted that the explanation for the photochromic transformations of the

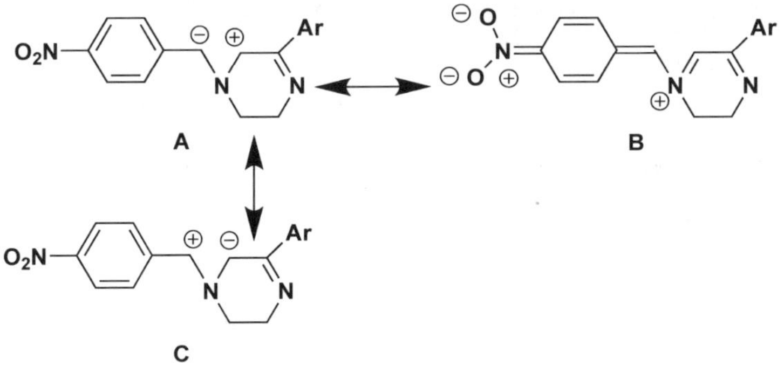


**Scheme 1.46**

fused aziridines in the crystalline state is similar to a cycloaddition reaction where a reversible aziridine ring opening gives rise to the colored azomethine ylides. One of the arguments for this is the analysis of the Raman spectrum of the crystalline irradiated azirenoimidazole **72** [122] containing bands corresponding to a nitrophenyl radical attached to a fragment bearing a negative charge.

It was reported [123] that irradiated single crystals of the azirenoimidazole **72** were highly dichroic. The blue crystals absorbed strongly along one axis, but were transparent in the perpendicular direction. It was reported in the same work that IR spectra of irradiated and subsequently dissolved crystalline azirenoimidazoles closely parallel those taken in rigid glasses at –196°C.

Besides photochromism of aziridinyl anils, possible radiochromic properties are mentioned in the literature [110, 124, 125]. In the opinion of Chebanov et al. [110, 124, 125], the most representative radiochromic compounds are the bi- and tricyclic aziridines containing polyarene fragments. To synthesize polyarene derivatives of aziridinyl anils, a modification of the Suzuki–Miyaura reaction was used (Scheme 1.47).

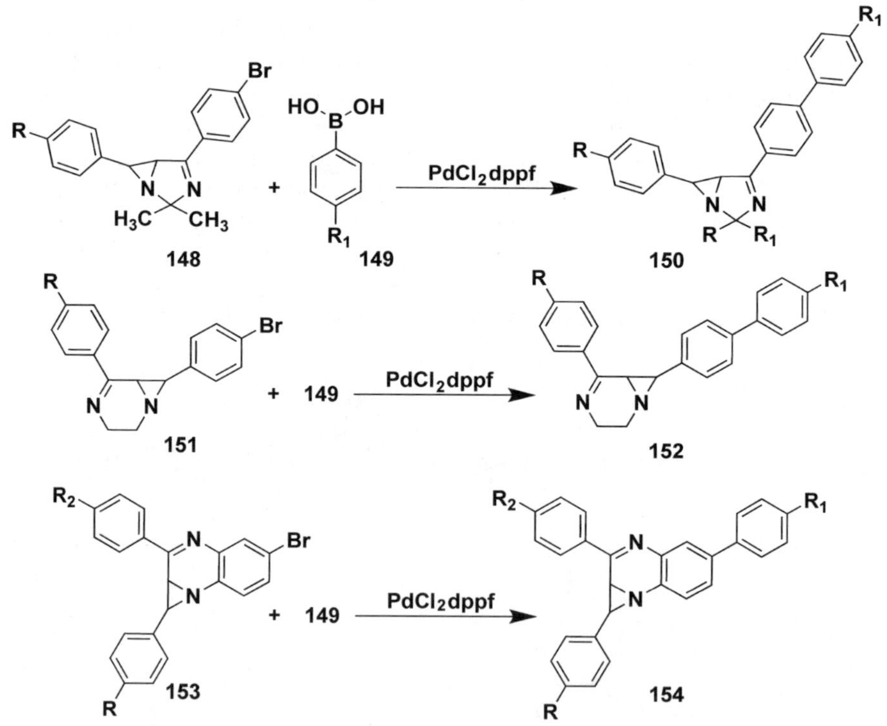


Scheme 1.47

A polyarene fragment can be introduced into any part of the aziridine molecule, e.g., Zbruyev et al. [126] obtained different derivatives of azirenoimidazoles **150**, azirenopyrazines **152** and azirenoquinoxalines **154**.

## References


1. Mahmoodi, N. O.; Zanjanchi, M. A.; Kiyani, H. *J. Chem. Res. Synop.*, 2004, *6*, 438.
2. Deshmukh, G.V.; Wheeler, T. S. *J. Chem. Soc.*, 1939, 96.
3. Weber, F. G.; Liepert, H. *Z. Chem.*, 1985, *25*, 175.
4. Orlov, V. D.; Yaremenko, F. G.; Lavrushin, V. F. *Khim. Geterotsikl. Soedin.*, 1980, 1489.
5. Gabriel, S. *Chem. Ber.*, 1888, *21*, 1049.
6. Wieland, H. *Chem. Ber.*, 1904, 1148.
7. Ruheman, S.; Watson, E. R. *J. Chem. Soc.*, 1904, 1170.
8. Cromwell, N. H.; Babson, R. D.; Harris, C. E. *J. Am. Chem. Soc.*, 1943, *65*, 312.
9. Cromwell, N. H. *Chem. Rev.* 1946, *38*, 83.
10. Tokes, A. L.; Litkei, G.; Janzso, G. *Synth. Commun.*, 1990, *20*, 1905.
11. Cromwell, N. H.; Barker, N.; Vanderhost, H. *J. Am. Chem. Soc.*, 1951, *73*, 1044.
12. Algar, J.; Hickey, A.; Sherry, P. G. *Proc. R. Ir. Acad.*, 1943/1944, 49B, 109.
13. Cromwell, N. H.; Caughlan, J. A. *J. Am. Chem. Soc.*, 1945, *67*, 2235.
14. Cromwell, N. H.; Mercer, G. D.; *J. Am. Chem. Soc.*, 1957, *79*, 3815.
15. Cromwell, N. H.; Mohrbacher, R. J. *J. Am. Chem. Soc.*, 1953, *75*, 6252.
16. Lin, T.; Kingsbury, C. A.; Adelfang, J.; Cromwell, N. H. *J. Heterocycl. Chem.*, 1985, *22*, 1017.
17. Saoudi, A.; Hamelin, J.; Benhaoua, H. *J. Chem. Res. Synop.*, 1996, *11*, 492.
18. Wroblewski, A. E.; Maniukiewicz, W.; Karolczak, W. *J. Chem. Soc. Perkin Trans. 1*, 2000, *9*, 1433.
19. Khomutov, O. G., Pashkevich, K. I. *Izv. Akad. Nauk. Ser. Khim.*, 1996, *3*, 684.
20. Davoli, P.; Forni, A.; Moretti, I.; Prati, F.; Torre, G. *Tetrahedron*, 2001, *57*, 1801.
21. Yaremenko, F.G. *PhD thesis*, Kharkiv State University, Kharkiv, 1981.
22. Cromwell, N. H.; Eby, H. H.; Capps, D. B. *J. Am. Chem. Soc.*, 1951, *73*, 1230.
23. Cromwell, N. H.; Cahoy, R. P. *J. Am. Chem. Soc.*, 1958, *80*, 5524.
24. Derdour, A.; Texier, F. *Can. J. Chem.*, 1985, *63*, 2245.
25. Khomutov, O. G.; Filyakova, V. I.; Pashkevich, K. I. *Izv. Akad. Nauk. Ser. Khim.*, 1994, *2*, 282.
26. Cromwell, N. H.; Wankel, R. A. *J. Am. Chem. Soc.*, 1948, *70*, 1320.
27. Barros, M. T.; Maycock, C. D.; Ventura, M. R. *Tetrahedron Lett.*, 2002, *43*, 4329.
28. Blatt, A. H. *J. Am. Chem. Soc.*, 1939, *61*, 3494.
29. Coldham, I.; Collis, A. J.; Mould, R. J.; Rathmell, R. E. *J. Chem. Soc. Perkin Trans. 1*, 1995, *21*, 2739.
30. Coldham, I.; Collis, A. J.; Mould, R. J.; Rathmell, R. E. *Tetrahedron Lett.*, 1995, *36*, 3557.
31. Jin, X. L.; Sugihara, H.; Daikai, K.; Tateishi, H.; Jin, Y. Z.; Furuno, H.; Inanaga, J. *Tetrahedron*, 2002, *58*, 8321.
32. Kammermeier, T.; Kaiser, A.; Lee, G. S.; Burgemeister, T.; Wiegrebe, W. *Arch. Pharm. (Weinheim)*, 1994, *327*, 207.
33. Cromwell, N. H,; Cahoy, R. P.; Franklin, W. E.; Mercer, C. D. *J. Am. Chem. Soc.*, 1957, *79*, 922.
34. Southwick, P. L.; Christman, D. R. *J. Am. Chem. Soc.*, 1952, *74*, 1886.
35. Zvonok, A. M.; Kuz'nemok, N. M.; Stanishevski, L. S. *Khim. Geterotsikl. Soedin.*, 1986, 456.
36. Leonard, N. J.; Ning, R. Y.; Booth, R. R.; *J. Org. Chem.*, 1965, *30*, 4357.

37. Weissberger, A.; Bach, H. *Chem. Ber.*, 1931, 1095.
38. Cromwell, N. H. *Lect. Heterocycl. Chem.*, 1976, *3*, 1.
39. Cromwell, N. H. *J. Am. Chem. Soc.*, 1959, *81*, 4702.
40. Southwick, P. L.; Shozda, R. J. *J. Am. Chem. Soc*, 1960, *82*, 2888.
41. Tarburton, P.; Woller, P. B.; Banger, R. C.; Doomes, E.; Cromwell, N. H. *J. Heterocycl. Chem.*, 1977, *14*, 459.
42. Molander, G. A.; Stengel, P. J. *Tetrahedron*, 1997, *53*, 8887.
43. Pak, C. S.; Kim, T. H.; Ha, S. J. *J. Org. Chem.*, 1998, *63*, 10006.
44. Suga, H.; Kakehi, A.; Ito, S.; Ibata, T.; Fudo, T.; Watanabe, Y.; Kinoshita, Y.; *Bull. Chem. Soc. Jpn.*, 2003, *76*, 189.
45. Xu, J.; Ma, L.; Jiao, P.; *Chem. Commun.*, 2004, *14*, 1616.
46. Li, J.; Liang, J.-L.; Chan, P. W. H.; Che, C.-M. *Tetrahedron Lett.*, 2004, *45*, 2685.
47. Furukawa, N.; Yoshimura, T.; Ohtsu, M.; Akasaka, T.; Oae, S. *Tetrahedron*, 1980, *36*, 73.
48. Furukawa, N.; Ose, S.; Yoshimura, T. *Synthesis*, 1976, 30.
49. Antunes, A. M. M.; Marto, S. J. L.; Branco, P. S.; Prabhakar, S.; Lobo, A. M. *Chem. Commun.*, 2001, 405.
50. Minakata, S.; Kano, D.; Fukuoka, R.; Oderaotoshi, Y.; Komatsu, M. *Heterocycles*, 2003, 60, *2*, 289.
51. Xu, J.; Jiao, P. *J. Chem. Soc. Perkin Trans. 1*, 2002, *12*, 1491.
52. Chen, D.; Timmons, C.; Guo, L.; Xu, X.; Li, G. *Synthesis*, 2004, 2479.
53. Atkinson, R. S.; Tughan, G. *J. Chem. Soc. Perkin Trans. 1*, 1987, 2803.
54. Siu, T.; Picard, C. J.; Yudin, A. K. *J. Org. Chem.*, 2005, 70, 932.
55. Fioravanti, S.; Pellacani, L.; Tabanella, S.; Tardella, P. A. *Tetrahedron*, 1998, *54*, 14105.
56. Hansen, P. E.; Undheim, K. *Acta Chem. Scand.*, 1973, *27*, 1112.
57. Ikeda, I.; Machii, Y.; Okahara, M. *Synthesis*, 1980, 650.
58. Armstrong, A.; Carbery, D. R.; Lamont, S. G.; Pape, A. R.; Wincewicz, R. *Synlett*, 2006, 2504.
59. Bartnik, R.; Lesniak, S.; Laurent, A. *Tetrahedron Lett.*, 1981, *22*, 4811.
60. Bartnik, R.; Laurent, A.; Lesniak, S. *J. Chem. Res. Miniprint*, 1982, *10*, 2701.
61. Sasaki, M.; Yudin, A. K. *J. Am. Chem. Soc.*, 2003, *125*, 14242.
62. Attia, M. E. M.; Gelas-Mialhe, Y.; Vessiere, R. *Can. J. Chem.*, 1983, *61*, 2126.
63. Sasaki, M.; Dalili, S.; Yudin, A. K. *J. Org. Chem.*, 2003, *68*, 2045.
64. Watson, I. D. G.; Styler, S. A.; Yudin, A. K. *J. Am. Chem. Soc.*, 2004, *126*, 5086.
65. Zavarzin, I. V.; Yarovenko, V. N.; Shirokov, A.V.; Smirnova, N. G.; Es'kov A. A.; Krayushkin, M. N. *ARKIVOK*, 2003, *13*, 205.
66. Wroblewski, A. E.; Karolczak, W. *Pol. J. Chem.*, 1999, *73*, 1191.
67. Egli, M.; Dreiding, A. S. *Helv. Chim. Acta*, 1986, *69*, 1442.
68. Weber, F. G.; Liepert, H.; Koeppel, H. *J. Prakt. Chem.*, 1986, *328*, 612.
69. Weber, F. G.; Marrek, M.; Radeglia, R.; Koeppel, H. *J. Prakt. Chem.*, 1989, *331*, 665.
70. Zavarzin, I. V.; Yarovenko, V. N.; Chernoburova, E. I.; Krayushkin, M. M. *Izv. Akad. Nauk. Ser. Khim.*, 2004, 398.
71. Lee, K.; Kim, Y. H. *Synth. Commun.*, 1999, *29*, 1241.
72. Nagel, D. L.; Cromwell, N. H. *J. Heterocycl. Chem.*, 1974, *11*, 1093.
73. Cromwell, N. H.; Hoeksema, H. *J. Am. Chem. Soc.*, 1949, *71*, 444.
74. Cromwell, N. H.; Graf, M. A. *J. Org. Chem.*, 1952, *17*, 414.
75. Chen, G.; Sasaki, M.; Yudin, A. *Tetrahedron Lett.*, 2006, *47*, 255.
76. Gelas-Mialhe, Y.; Touraud, E.; Vessiere, R. *Can. J. Chem.*, 1982, *60*, 2830.
77. Konovalov, V. A.; Bubel', O. N.; Tishchenko, I. G. *Khim. Geterotsikl. Soedin.*, 1988, 1484.
78. Padwa, A. In: *Organic Photochemistry,* vol. 8, Dekker: New York, 1987, p. 373.
79. Lown, J. W. *Rec. Chim. Progr.*, 1971, *32*, 51.
80. Padwa, A. *Angew. Chem.*, 1976, *88*, 131.
81. Padwa, A.; Dean, D.; Oine, T. *J. Am. Chem. Soc.*, 1975, *97*, 2822.

82. Wartski, L. *Bull. Soc. Chim. France, Part 2*, 1975, 1663.
83. Padwa, A.; Eastman, D.; Hamilton, L. *J. Org. Chem.*, 1968, *33*, 1317.
84. Ramaiah, D.; Muneer, M.; Gopidas, K. R.; Das, P. K.; Rath, N. P.; George, M. V. *J. Org. Chem.*, 1996, *61*, 4240.
85. Tishchenko, I. G.; Bubel', O. N.; Konovalov, V. A. *Khim. Geterotsikl. Soedin.*, 1981, 38.
86. Tishchenko, I. G.; Bubel', O. N.; Konovalov, V. A. *Khim. Geterotsikl. Soedin.*, 1980, 1353.
87. Tanaka, K.; Ohsuga, M.; Sugimoto, Y.; Okafuji, Y.; Mitsuhashi, K. *J. Fluorine Chem.*, 1988, *39*, 39.
88. Heine, H. W.; Weese, R. H.; Cooper, R. A. *J. Org. Chem.*, 1967, *32*, 2708.
89. Padwa, A.; Clough, S.; Glazer, E. *J. Am. Chem. Soc.*, 1970, *92*, 1778.
90. DoMinh, T.; Trozzolo, A. M. *J. Am. Chem. Soc.*, 1970, *92*, 6997.
91. Padwa, A.; Gehrlein, L. *J. Am. Chem. Soc.*, 1972, *94*, 4933.
92. DoMinh, T.; Trozzolo, A. M. *J. Am. Chem.*, 1972, *94*, 4046.
93. Heine, H. W.; Smith, A. B.; Bower, D. *J. Org. Chem.*, 1968, *33*, 1097.
94. Kaluski, Z.; Gresyak-Figas, E.; Vorob'eva, N. P.; Bakumenko, A. I.; Yaremnko, F. G.; Orlov, V. D. *Khim. Geterotsikl. Soedin.*, 1989, 1064.
95. Orlov, V. D.; Kaluski, Z.; Figas, E.; Vorob'eva, N. P.; Bakumenko, A. I.; Yaremenko, F. G. *Khim. Geterotsikl. Soedin.*, 1991, 1060.
96. Orlov, V. D.; Yaremenko, F. G.; Surov, Y. N.; Lavrushin, V. F. *Dokl. Akad. Nauk USSR Ser. B*, 1979, 932.
97. Risitano, F.; Grassi, G.; Foti, F.; Moraci, S. *Synlett*, 2005, *10*, 1633.
98. Weber, L.; Illgen, K.; Almstetter, M. *Synlett* 1999, 366.
99. Dax, S. L.; McNally, J. J.; Youngman, M. A. *Curr. Med. Chem.* 1999, *6*, 255.
100. Risitano, F.; Grassi, G.; Foti, F.; Bilardo, C. *Tetrahedron*, 2000, *56*, 9669.
101. Grassi, G.; Risitano, F.; Foti, F.; Cordaro, M. *Synlett*, 2001, 812.
102. Risitano, F.; Grassi, G.; Foti, F.; Romeo, R. *Synthesis*, 2002, 116.
103. Risitano, F.; Grassi, G.; Foti, F.; Nicolò, F.; Condello, M. *Tetrahedron*, 2002, *58*, 191.
104. Mahmoodi, N. O.; Kiyani, H. *Bull. Korean Chem. Soc.*, 2004, *25*, 1417.
105. Padwa, A.; Glazer, E. *J. Amer. Chem. Soc.*, 1972, *94*, 7788.
106. Heine, H. W.; Henzel, R. P. *J. Org. Chem.*, 1969, *34*, 171.
107. Kolos, N. N.; Orlov, V. D.; Yuzefovskaya, E. Y.; Yaremenko, F. G.; Vorob'eva, N. P.; Shishkin, O. V.; Struchkov, Y. T.; Ivkov, S. M. *Khim. Geterotsikl. Soedin.*, 1995, 950.
108. Orlov, V. D.; Vorob'eva, N. P.; Demenkova, N. N.; Chesnovskii, V. S.; Yaremenko, F. G. *Khim. Geterotsikl. Soedin.*, 1988, 328;
109. Zbruyev A. I., Yaremenko F. G., Chebanov V. A., Desenko S. M., Shishkin O. V., Lukinova E. V., Knyazeva I. V. *Izv. Akad. Nauk. Ser. Khim.*, 2006, 350.
110. Chebanov, V. A.; Zbruyev, A. I.; Vashchenko, V. V.; Desenko, S. M.; Doroshenko, F. O.; Yaremenko, F. G.; Musatov, V. I., Vorob'eva, N. P.; Orlov, V. D. *XX Ukrainian Conference in Organic Chemistry*, Odessa, 2004, 87.
111. Pashkevich, K. I.; Khomutov, O. G. *Zh. Org. Khim.*, 1999, *35*, 106.
112. Orlov, V. D.; Yaremenko, F. G.; Lavrushin, V. F.; *Khim. Geterotsikl. Soedin.*, 1979, 536.
113. Heine, H. W.; Newton, T. A.; Blosick, G. J.; Irving, K. C.; Meyer, C.; Corcoran, G. B. *J. Org. Chem.*, 1973, *38*, 651.
114. Orlov, V. D.; Yaremenko, V. F.; Kolos, N. N.; Pivnenko, N. S.; Lavrushin, V. F. *Khim. Geterotsikl. Soedin.*, 1980, 544.
115. Coldham, I.; Hufton, R. *Chem. Rev.*, 2005, *105*, 2765
116. Padwa, A.; Glazer, E. *J. Org. Chem.*, 1973, *38*, 284.
117. Padwa, A.; Glazer, E. *Chem. Commun.*, 1971, *15*, 838.
118. Beak, P.; Miesel, J. E. *J. Am. Chem. Soc.*, 1967, *89*, 2375.
119. McCoy, G.; Day, A. *J. Am. Chem. Soc.*, 1943, *65*, 2159.
120. Orlov, V. D.; Vorob'eva N. P.; Tischenko, A. A.; Krainov, I. P. *Zh. Prikl. Spektrosk.*, 1991, *54*, 638;

121. Orlov, V. D.; Vorob'eva, N. P.; Tischenko, A. A.; Pikalev, O. M.; Popov, V. I.; Yagupol'skii, L. M. *Khim. Geterotsikl. Soedin.*, 1991, 1179;
122. Sheng, S. J.; Leslie, T. M.; Trozzolo, A. M. *Mol. Cryst. Liq. Cryst.*, 1979, *54*, 175.
123. Trozzolo, A. M.; Sarpotdar, A. S.; Leslie, T. M. *Mol. Cryst. Liq. Cryst.*, 1979, *50*, 201.
124. Chebanov, V. A.; Zbruyev, A. I.; Desenko, S. M.; Doroshenko, A. O.; Vaschenko, V. V. *J. Korean Assoc. Radiat. Prot.*, 2005, *30*, 31.
125. D'yakonenko, V. V.; Zbruyev, A. I.; Chebanov, V. A.; Desenko, S. M.; Shishkin, O. V. *Russ. J. Struct. Chem.*, 2005, *46*, 1152.
126. Zbruyev, A. I.; Vashchenko, V. V.; Andryushchenko, A. Y; Desenko, S. M.; Musatov, V. I.; Knyazeva, I. V.; Chebanov, V. A. *Tetrahedron*, 2007, *63*, 4297.

# Chapter 2
# Five-Membered Azaheterocycles

It is known that pyrazolines and pyrazoles are an important class of pharmacophores [1, 2, 3, 4, 5, 6]. Heterocycles containing these fragments are important targets in synthetic and medicinal chemistry since they are key moieties in numerous biologically active compounds possessing tranquilizing, muscle relaxant, psychoanaleptic, anticonvulsant and antidepressant activities [7, 8, 9, 10, 11, 12]. This includes kinesin spindle protein [12] and monoamine oxidase [13, 14] inhibitors as well as antimycobacterial [15, 16] and anti-inflammatory [17] agents.

On the other hand, pyrazoline and pyrazole derivatives are successfully used for material science tasks. For example, triarylpyrazolines are used in the synthesis of green electroluminescent polymers for light-emitting diodes [18]. Some aromatic substituted 2-pyrazolines are effective organic luminophores [19, 20], fluorophores [21] and scintillating materials [20, 22].

One of the practically important synthetic pathways to aryl-containing pyrazole derivatives is a reaction of α,β-unsaturated ketones with dipolar molecules or 1,2-binucleophiles. There are dozens of stereotypical publications on this topic and in this chapter we will not cite all existing references related to the synthesis and properties of pyrazole derivatives based on unsaturated carbonyl compounds but will cover the most general and some specific examples as well as mechanisms.

## 2.1 Reactions with Diazoalkanes

α,β-Unsaturated ketones usually having a high-polarity double bond are able to react with dipolar molecules to give cyclic compounds. The first references to reactions of diazolalkanes with unsaturated ketones seem to have appeared at the beginning of the twentieth century [23]. In 1937, Smith and Pings [24] investigated chalcone **1** in similar reactions. The reaction products isolated were a mixture of 1-pyrazoline **2** and 2-pyrazoline **3** (Scheme 2.1). Compound **2** when heated is easily isomerized into heterocycle **3**. Upon more extreme heating, nitrogen is eliminated from both 1-pyrazoline and 2-pyrazoline, yielding dypnone. A similar

V.A. Chebanov et al., *Azaheterocycles Based on α,β-Unsaturated Carbonyls*, 

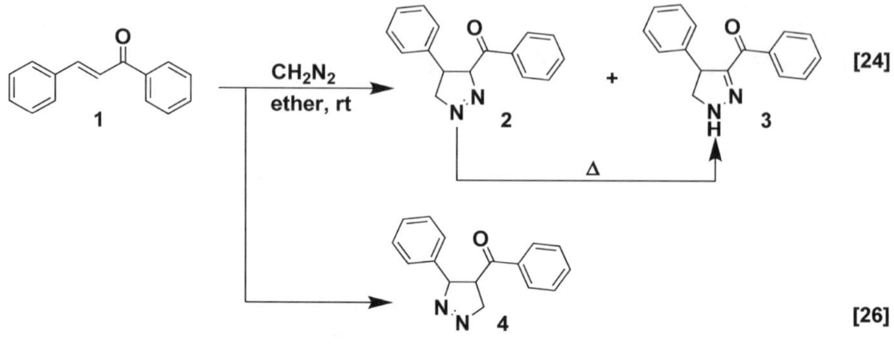


**Scheme 2.1**

direction for the thermal decomposition of pyrazolines and isomerization was described in [25]. In a later investigation [26], the formation of 4-benzoyl-3-phenyl-1-pyrazoline **4** in the reaction of chalcone and diazomethane was established. The important feature is the similarity of the melting points of heterocycles **3** and **4**.

Reactions of diazomethane with various α,β-unsaturated ketones were described in [27, 28]. For instance, Mustafa and Freifel [27] showed that treatment of chalcones **5** with ethereal diazomethane under normal conditions affected addition to the double bond of the ketone, forming pyrazolines **6** (Scheme 2.2). Alternative directions leading to heterocycles **7** as well as to methylation of the hydroxyl group were not observed. Similar results were described in [28]. On the other hand, Aleksandrova et al. [29] reported the formation of 4-aryl-3-(2-furoyl)-2-pyrazolines **9** when furyl analogues of chalcones **8** react with diazomethane.

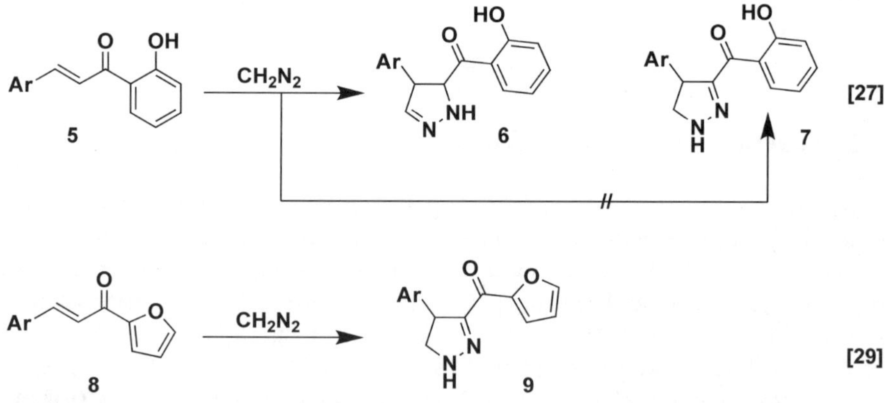


**Scheme 2.2**

Thus, the examples described above illustrate an inconsistency of the literature data which had been published before the 1980s regarding the structure of the reaction products of α,β-unsaturated ketones and diazomethane. This contradiction is associated both with the ambiguity of the reaction direction (pathways a and b in Scheme 2.3) and with the possibility of prototropic rearrangements (**A**–**E** and **F**–**J** in Scheme 2.3).

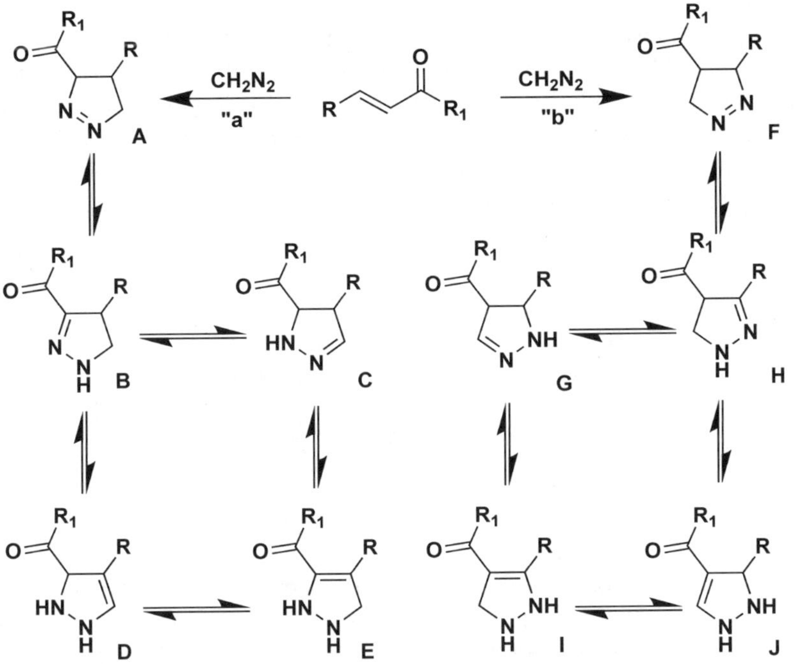


**Scheme 2.3**

Detailed investigations of reactions of diazomethane with unsaturated ketones and the appropriate reaction products were made by Lévai et al. [25, 30, 31, 32]. It is unambiguously established that for the reactions studied, only pathway a was observed.. This leads to the formation of isomers **A**, which spontaneously rearrange into the sole reaction products—2-pyrazolines **B** (Scheme 2.3).

Particularly in [32] it was shown that cycloaddition of the chalcones **10** and diazomethane is a regioselective process providing 3-aroyl-4-aryl-2-pyrazolines **12**, most likely via heterocycles **11**. These are the sole products irrespective of the bulkiness and/or the electronic influence of the two aryl moieties of the starting α,β-unsaturated ketones **10** (Scheme 2.4).

During the same time period, the discovery of another pathway, b, with the formation of pyrazolines **G** (Scheme 2.3) was also described [33, 34]. However, these results were refuted in [30].

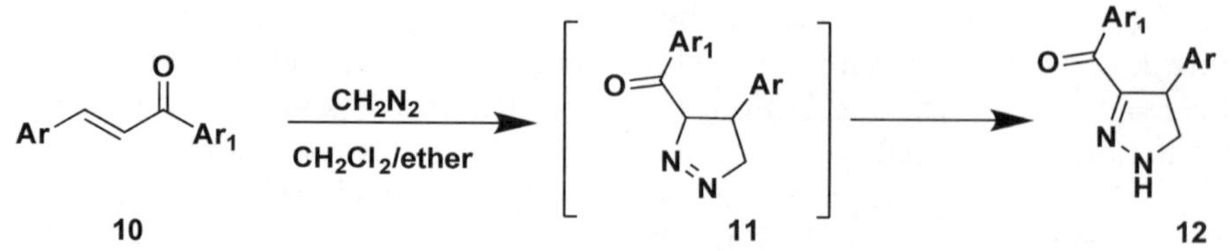


Scheme 2.4

The hypothesis about the formation of compounds like **B** was also proven by other authors [35, 36]. For example, the Padmavathi et al. [36] studied the reactions of heterocyclic analogues of chalcones **14** obtained from the appropriate diaroylacetones **13** and diazomethane (Scheme 2.5). It was shown that this treatment gives a solid which was identified as (4′-aryl-4′,5′-dihydro-1′*H*-pyrazol-3′-yl)-(4-aryl-1*H*-pyrrol-3-yl)methanone **15**.

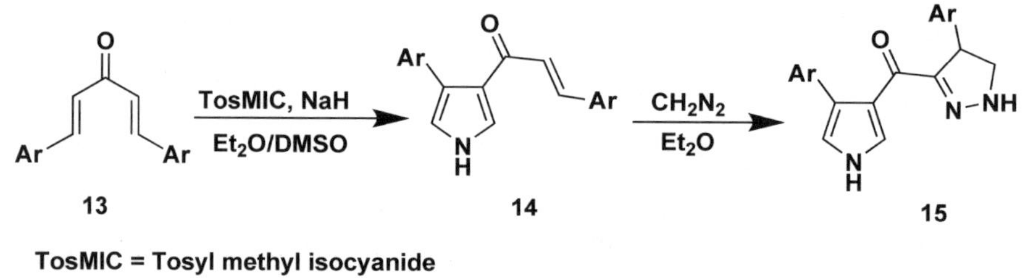


Scheme 2.5

Reactions of diazoalkanes with so-called exocyclic unsaturated ketones, for instance, arylidene derivatives of indanone **16** [39, 40], tetralone **18** [41, 42, 43, 44], chromanone **21**, thiochromanone **23** [41, 42, 43], flavanone **22** and thioflavanone **24** [33, 34, 41, 42, 43] lead to the appropriate spiropyrazolines **17**, **20** and **25–28** (Schemes 2.6, 2.7).

It was shown [43] that reactions of *E* and *Z* isomers of **21–24** proceed as stereospecific processes. Depending on the configuration of the starting cyclic α,β-unsaturated ketone, the reaction yields either *trans* isomers **25–28** (*E* configuration of **21–24**) or their *cis* forms (*Z* configuration of **21–24**).

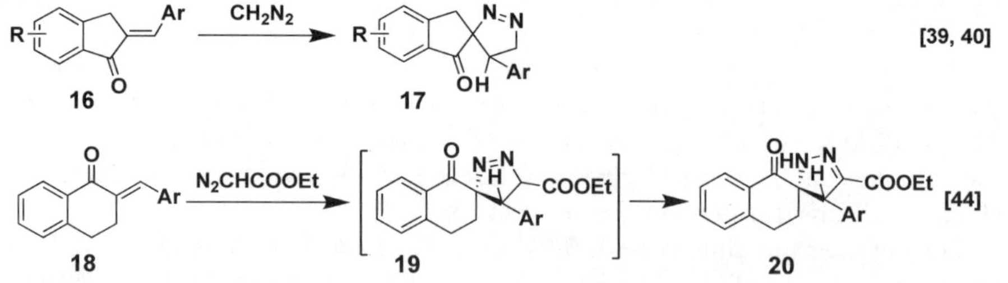


Scheme 2.6

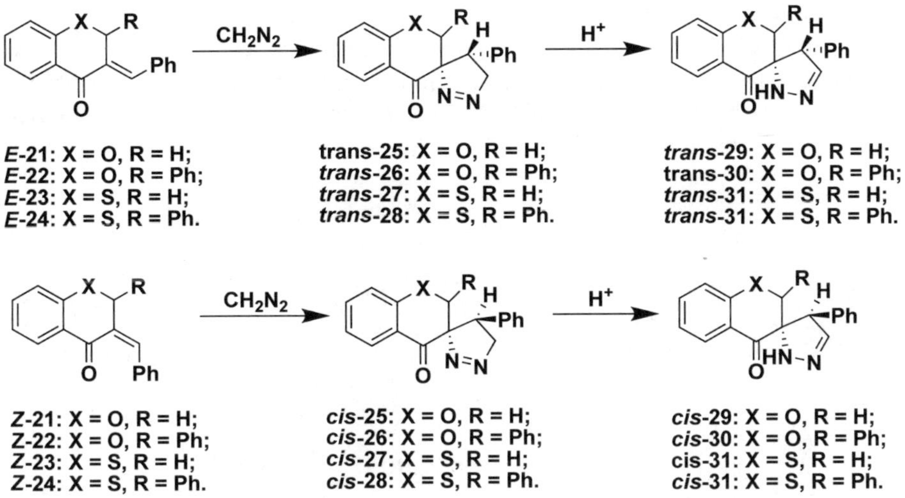


**Scheme 2.7**

In contrast to the reaction products of the 1,3-dipolar cycloaddition to linear enones (chalcones and their derivatives), spiro-1-pyrazolines **25**–**28** are stable heterocycles. Their isomerization into spiro-2-pyrazolines **29**–**31** can be carried out under acidic conditions at room temperature for 24 h in the case of the *cis* isomers, and for more than 1 month in the case of *trans* isomers [30, 41, 42, 43].

## 2.2 Reactions of Hydrazines and Hydroxylamines

The most usable and well-known pathway for the synthesis of five-membered nitrogen-containing heterocycles is condensation involving α,β-usaturated carbonyls and 1,2-binucleophilic compounds, e.g., derivatives of hydrazine and hydroxylamine (Scheme 2.8). The procedure based on these reactions was successfully applied for a long period [45, 46, 47, 48, 49, 50, 51, 52, 53].

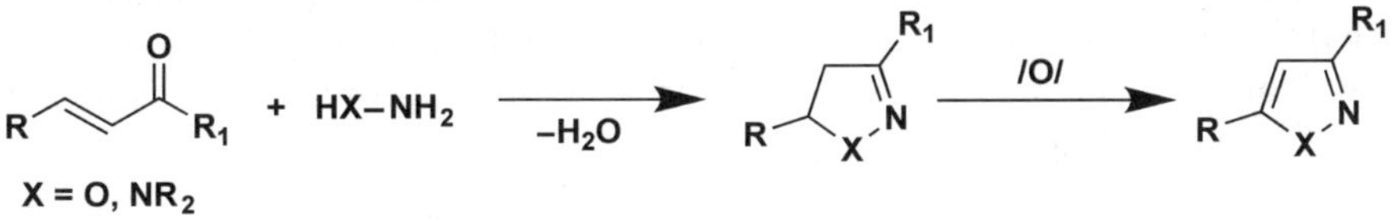


**Scheme 2.8**

These reactions are carried out most efficiently under acidic conditions—in the presence of mineral or organic acids—or without catalysis [35]. For example, Palaska et al. [12] and Fathi et al. [54] reported the synthesis of pyrazolines **34** in high yields by the reaction of unsaturated ketones **32** and phenylhydrazine

**33** in glacial acetic acid. However, *N*′-nicotinoyl-substituted pyrazolines **37** are obtained from the treatment of the appropriate chalcones **35** and isonicotinyl hydrazide **36** in ethanol with catalytic amounts of acetic acid [15] (Scheme 2.9). Reactions with another hydrazide **39** leading to pyrazolines **40** are also carried out in acetic acid [16].

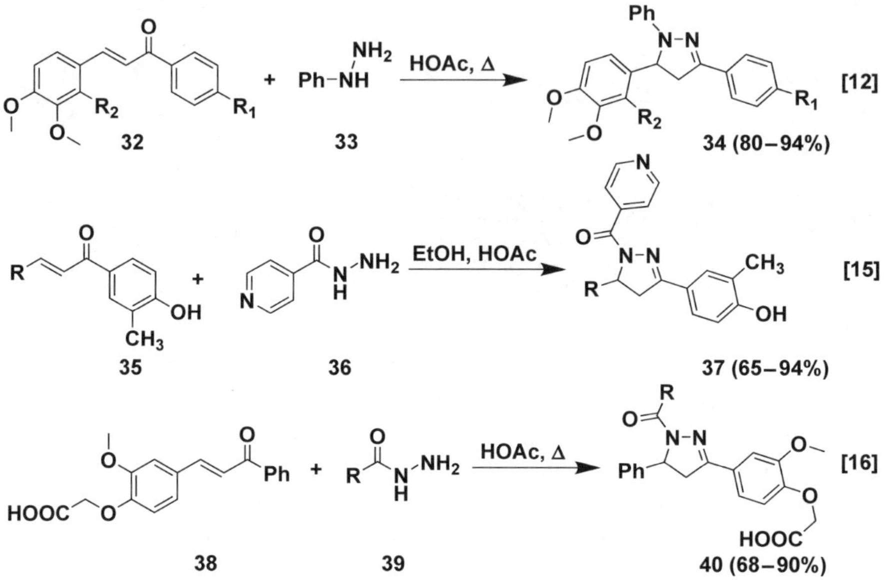


Scheme 2.9

To accelerate the reactions rates and to increase their yields, sometimes microwave-assisted procedures are applied. The first mention of using a solvent-free microwave procedure was in [55]. The authors described the synthesis of 1,3,5-triarylpyrazoline by the cyclization of chalcones with phenylhydrazine on a basic alumina solid support. The target heterocycles were synthesized under microwave irradiation in high yields (up to 85%) in 1–2 min instead of 3 h in the case of thermal activation. Another publication [56] deals with the rapid (2–12 min) solvent-phase cyclization of naphthyl-substituted chalcones **41** and hydrazines **42** in a microwave field yielding the appropriate pyrazolines **43** quantitatively (Scheme 2.10).

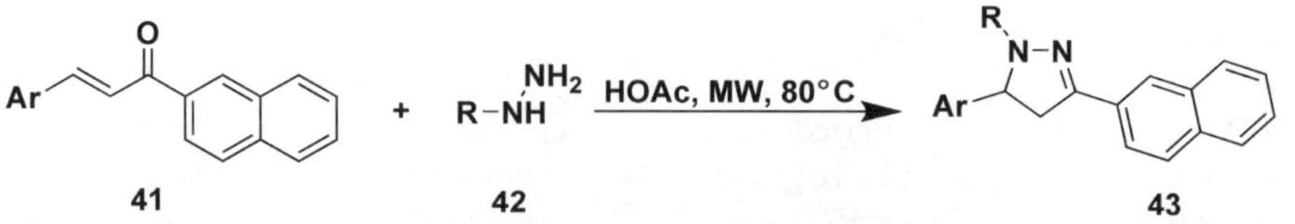


Scheme 2.10

In [57], different conditions for the reaction of chalcones and phenylhydrazine—thermal activation, microwave heating and ultrasonic irradiation—were compared. The authors used a set of solvents (acetonitrile, tetrahydrofuran, dioxane, dichloromethane) to carry out reactions using sonication and conventional heating. For microwave experiments, under solvent-free conditions different absorbents (silica gel, neutral alumina, bentonite) were used. The best results (yields up to 98% in 1 min) were observed in the case of the microwave activation on silica gel when $KHSO_4\ xH_2O$ was impregnated on the absorbent. Ultrasonic irradiation can achieve similar yields but requires a significantly longer time period (2.5–3 h).

Synthesis of pyrazolines can be achieved in a one-pot process without preliminary isolation of the α,β-unsaturated ketone. For example, the sequential treatment of aryl halogenides **44** containing strong electron-withdrawing substituents with 1-phenylprop-2-ynole **45** and then with methylhydrazine gives pyrazolines **48** in high yields (up to 90%) [58] (Scheme 2.11). It appears that the reaction initially forms the Sonogashira coupling intermediate **46**, which in tetrahydofuran in the presence of $NEt_3$ yields a chalcone **47**. Then this product undergoes the Michael-addition–cyclocondensation sequence with hydrazine, yielding pyrazoline **48**.

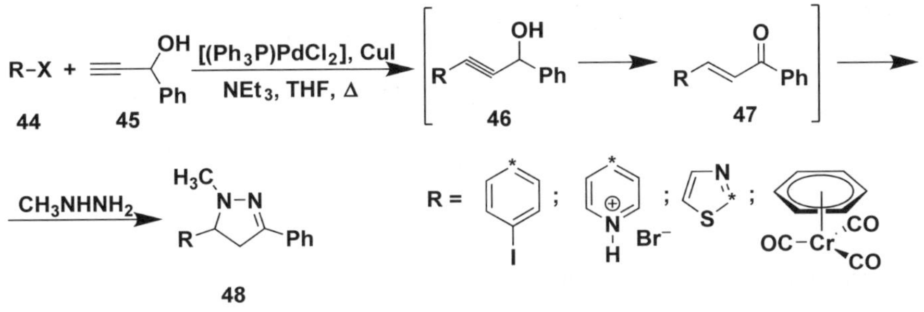


Scheme 2.11

Synthesis of oligo(1,4-phenylenepyrazole-3,5-diyl)s **50** from the reaction of oligochalcones **49** with methylhydrazine under alkaline catalysis was reported in [59] (Scheme 2.12). In this synthesis compound **50** is obtained when $n = 2$ in yields up to 12%, while in the case when $n = 3$, the target oligomer can be observed only in trace amounts.

Another method for the preparation of polymers containing pyrazoline fragments was suggested in [18]. Poly{(9*H*-fluoren-2,7-ylene)-*alt*-[3,5-bis(1,4-phenylene)-4,5-dihydro-1-phenylpyrazole]} **53** was synthesized by the reaction of pyrazoline **51** and fluorene-2,7-bis(trimethylene boronate) **52** in toluene in the presence of a catalyst, $Pd(PPh_3)_4$, with 85% yield.

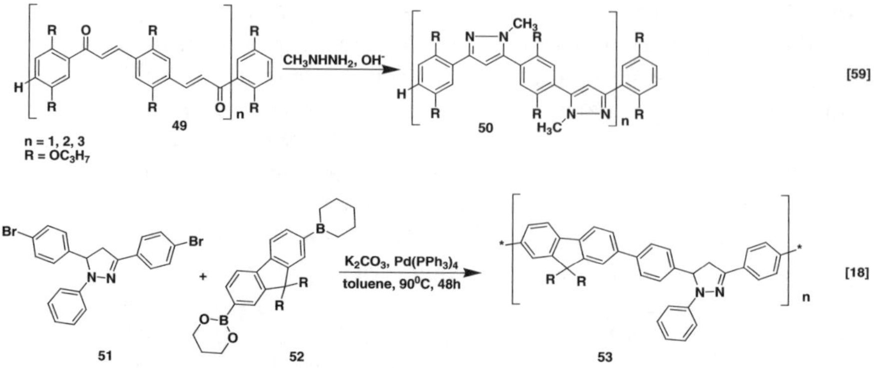


Scheme 2.12

The radical reaction of hydrazine and methylhydrazine with some α,β-unsaturated ketones **54** initiated by 2,2-dipenyl-1-picrylhydrazyl (DPPH) is described in [60]. At −78°C the treatment is practically regioselective and yields pyrazoles **55**, while regioisomeric heterocycles **56** are observed in trace amounts (Scheme 2.13).

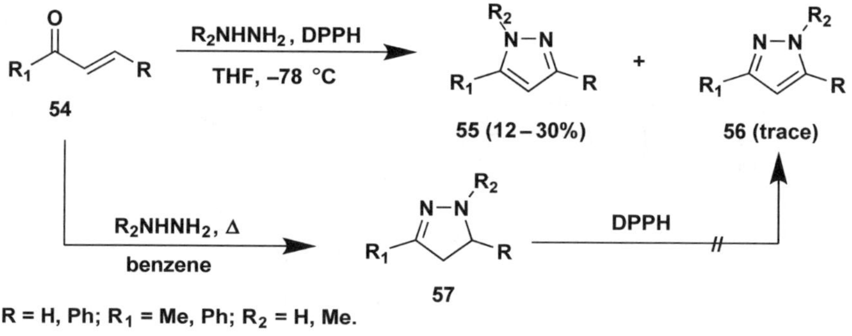


Scheme 2.13

The regioselectivity of pyrazole **55** formation is completely the reverse of that of pyrazoline **57** in the reaction of ketones **54** with hydrazines without DPPH. In addition, heterocycles **56** cannot be obtained from **57** by the action of DPPH. These facts show that the radical-initiated treatment of α,β-unsaturated ketones with hydrazines does not occur via the initial pyrazoline **57** formation.

Another situation is observed in the case of oxidative cyclization of unsaturated ketones **58** with phenylhydrazine **33** in the presence of iodine [61]. The reaction passes through the initial formation of pyrazoline **59**, which is oxidized in the presence of iodine to pyrazole **60** (Scheme 2.14).

*N*-Thiocarbamoyl derivatives of pyrazoline can be synthesized by the reaction of α,β-unsaturated carbonyl compounds with thiosemicarbazides under basic conditions [14, 62, 63, 64, 65].

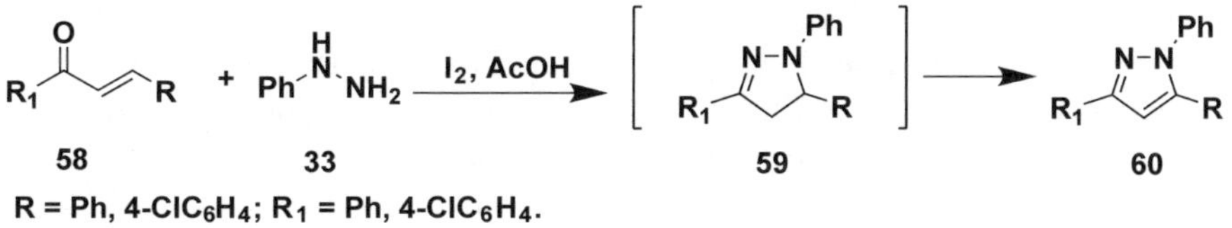


**Scheme 2.14**

The mechanisms of the reaction involving α,β-unsaturated ketones and hydrazines were studied in several publications [30, 66, 67, 68, 69, 70, 71, 72, 73, 74]. The first stage of the reaction is the addition of hydrazine to the carbonyl group of the ketone (compound **61**, Scheme 2.15). Subsequent cyclization by addition of the second nucleophilic center to the ethylene bond under acidic conditions is the rate-determining stage—its rate significantly depends on the stereochemistry and electronic structure of the intermediate hydrazone **62** (Scheme 2.15).

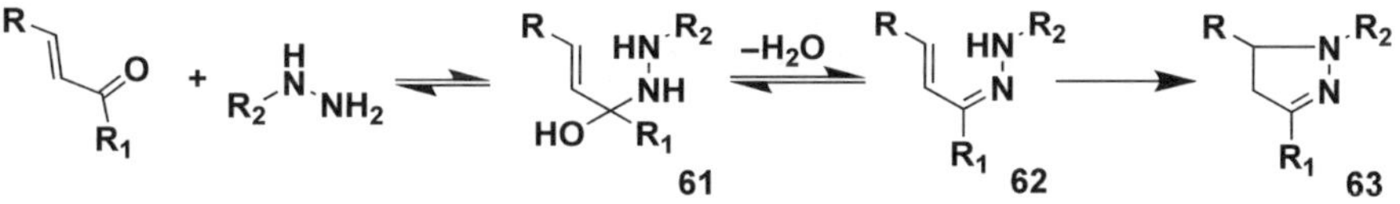


**Scheme 2.15**

Though intermediates **62** are usually not isolated from the reaction mixture, a conclusion about the formation of hydrazone (at least under acidic conditions) in most cases is based on numerous experimental data [30, 67, 68, 69] and is indisputable. However, the dependence of the direction of the first stage of the reaction of hydrazine derivatives and unsaturated ketones on acidity and other factors has been described [60, 75].

In addition, the kinetics and the mechanism of the pyrazoline formation in the reaction of chalcones with phenylhydrazine was studied in [70, 74] by polarography. It was definitely shown [74] that this reaction involves a three-stage process. The first stage, as already pointed out above, is the addition of the more nucleophilic β-nitrogen atom of phenylhydrazine to the carbon of the carbonyl group on the unsaturated ketone. The next step, consisting of dehydration of adduct **61**, is a rate-determining stage [70, 74], while subsequent cyclization in **63** occurs instantaneously (Scheme 2.15). Bezuglyi et al. [70] noted that for chalcones, the rate constants of these reactions with phenylhydrazine under acidic conditions depend on the substituents on the benzoyl fragment in an inverse manner, and this is similar for basic conditions. It is probably caused by the change in the ratio of the rates of the addition and dehydration stages. The influence of the substituents of the cinnamoyl moiety on reactivity does not depend on the acidity of the medium [71, 72, 73], although electron donors increase the reaction rate. For example, the reaction of diarylideneacetones $RC_6H_4CH{=}CHCOCH{=}CHC_6H_4R'$ when R is not the same as R′ involves the enone fragment containing the electron-donor substituent [71, 72].

The fixed geometry of α,β-unsaturated ketones in the *S-cis* form, as observed in arylidene alkanones, influences their reactivity. For example, the reaction of arylhydrazines with 3,5-diarylidene-4-piperidones, 2-arylidene-1-tetralones, 2,6-diarylidenecyclohexanones and 2-arylideneindan-1,3-diones requires stronger conditions than for the case of their noncyclic analogues [76, 77, 78, 79, 80, 81, 82, 83, 84]. However, arylidene derivatives of cyclohexanone, 1-indanone, 4-chromanone, 4-thiochromanone and *N*-methyl-4-piperidone hydrochloride react with hydrazines more easily [85, 86, 87, 88]. It is interesting to note that the more complicated and sterically hindered unsaturated ketones of the spiro type **64** react with hydrazine and phenylhydrazine very easily in the presence of piperidine, leading to pyrazoles **65** in high yields [89] (Scheme 2.16).

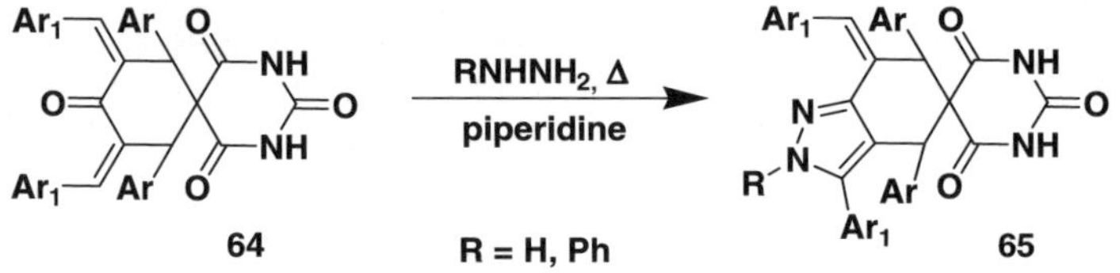


**Scheme 2.16**

The reaction of arylidenecyclonones with hydrazines proceeds with the formation of regio- and stereoisomers. Thus, treatment of bisbenzylidene-4-thiopyranones **66** with propylhydrazine **67** in boiling methanol yields mixtures of two isomeric pyrazolines **68** and **69**, with **68** predominating [90] (Scheme 2.17). The amount of isomer **69** increases when X is S to SO and $SO_2$.

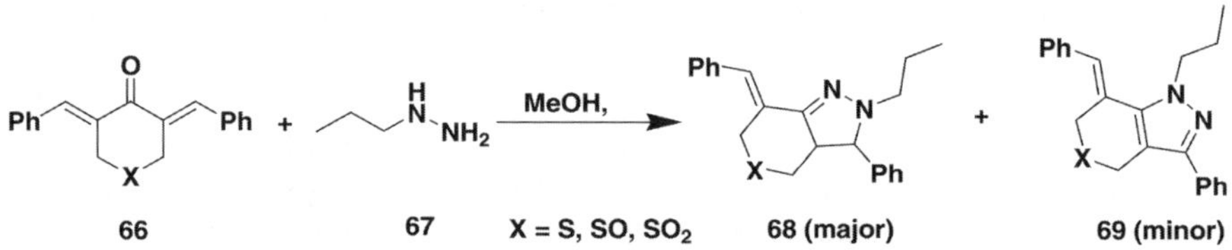


**Scheme 2.17**

The stereochemistry of the reaction of arylidenecycloalkanones with hydrazines has been discussed in several publications. It is established that in most cases during the formation of *trans*-pyrazolines or their mixtures with *cis* isomers, the predominant *trans* form is observed [84, 85, 91, 92, 93, 94]. For example, the reaction of 2-benzylidenetetralone **70** with hydrazines in boiling pyridine yields solely *trans*-pyrazoline **71** [85] (Scheme 2.18).

The reaction of arylidenechromanones with hydrazine in methanol also yields only the *trans* isomer of the appropriate pyrazoline [91]. Cyclization of diarylidenecyclonones **72** with numerous hydrazines in boiling methanol according to [17] yields a mixture of *trans*-pyrazoline **73** and its regioisomer **74** (Scheme 2.18).

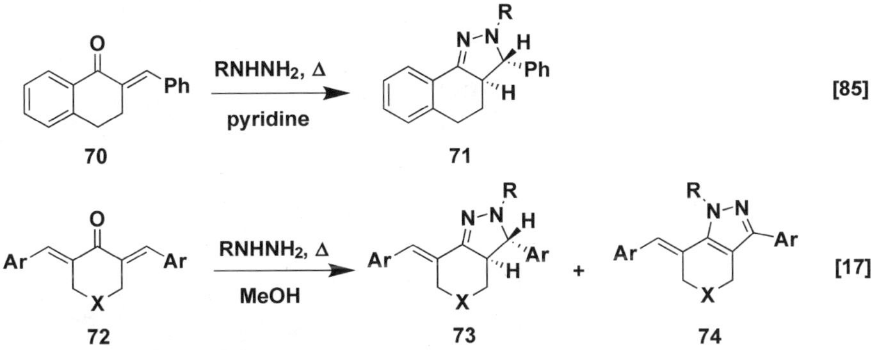


X = S, SO, $SO_2$, O, $NCH_3$;
R = H, n-$C_3H_7$, C(O)$CH_3$, $CH_2CH_2OH$, $CH_2CCH$, $CH_2CF_3$.

**Scheme 2.18**

Gella et al. [95] described the dependence of the direction of the reaction of diarylidenecycohexanones **75** with hydrazones. It was shown that treatment involving phenylhydrazine **33** yields *trans*-phenylhexahydro[2*H*]indazoles **77**, while in the case of 4-nitrophenylhydrazine **76** only *cis* isomers **79** were isolated from the reaction mixture (Scheme 2.19). The reaction of **75** with phenylhydrazine sometimes gives 1-phenyltetrahydro[1*H*]indazoles **78** as coproducts.

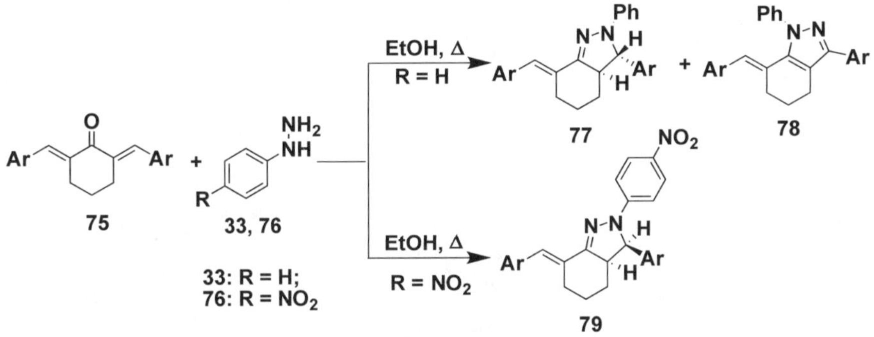


**Scheme 2.19**

Reactions of arylidenecycloalkanones with thiosemicarbazide in contrast to hydrazines usually yield *cis* isomers of the appropriate *N*-thiocarbamoyl derivatives of pyrazolines. For instance, treatment of 3,5-diarylidene-1-methyl-4-piperidones **80** and some other cyclic unsaturated ketones with thiosemicarbazide **81** under acidic conditions leads to *cis*-pyrazolines **82** in high yields [65] (Scheme 2.20).

However, reactions involving semicarbazide **83** under the same conditions in most cases give the *trans* isomers **84** (Scheme 2.20). The formation of only *cis*-pyrazolines has also been described for the reaction of thiosemicabazide with diarylidenecyclohexanone [64].

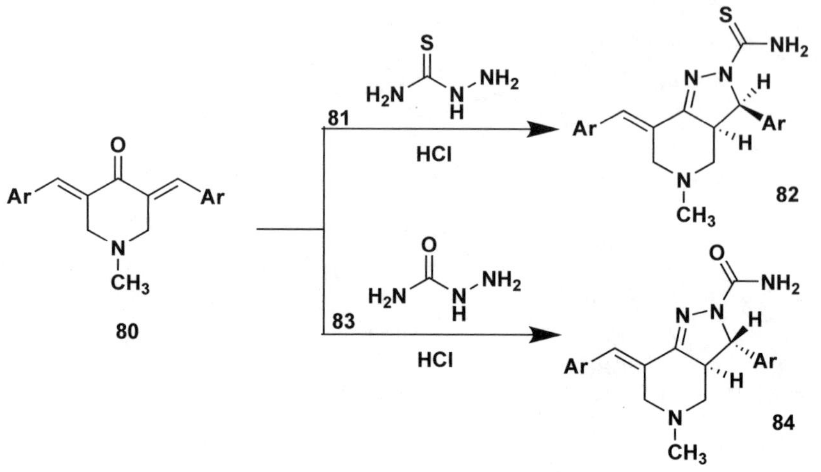


Scheme 2.20

Easily separated mixtures of stereoisomeric 2-pyrazolines (*cis*-**86** and *trans*-**86**) are isolated when cyclic unsaturated ketones **85** and semicarbazone **83** or thiosemicarbazone **81** react [96] (Scheme 2.21).

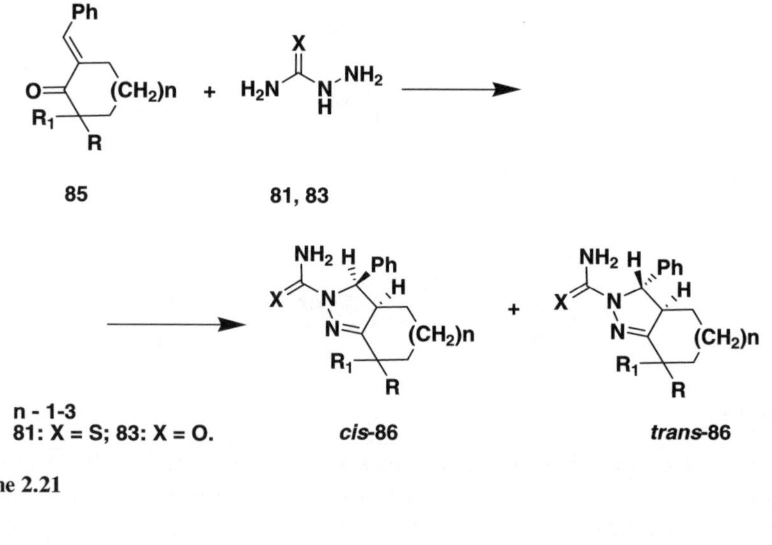


Scheme 2.21

The stereoselective synthesis of both *trans*-**88** and *cis* isomers **89** of pyrazolines was reported in [62] (Scheme 2.22). The configuration of the reaction products depends on the conditions of the cyclocondensation involving unsaturated ketones **87** and methyl(phenyl)hydrazines or (thio)semicarbazide. In boiling pyridine the treatment yields only *trans*-pyrazolines, while reaction in ethanol with catalytic amounts of hydrochloric acid leads mainly to the *cis*

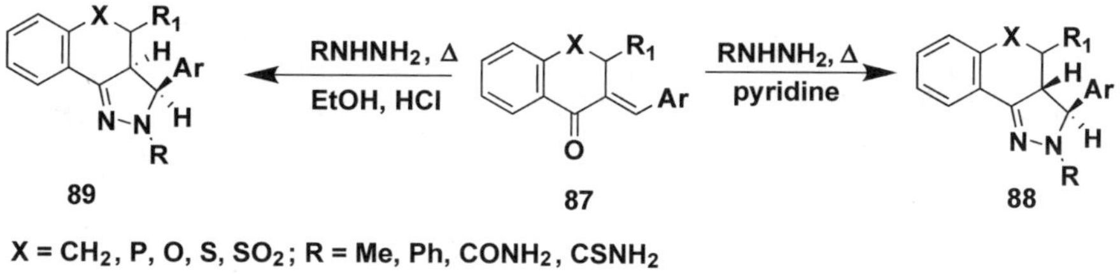

Scheme 2.22

isomers, with some impurities of the *trans* form. The mechanism of the reaction and its dependence on the conditions were not discussed in [62].

The mechanism of isoxazoline formation in the reaction of α,β-unsaturated carbonyls with hydroxylamines has no principal differences from that of the pyrazoline ring generation. However, we would like to note the reduced ability of oximes forming in the first stage, in comparison with the appropriate hydrazones, prior to cyclization. It initiates conditions for the presence of oximes as impurities in the reaction products [97, 98]. A change in the sequence of the heterocyclization is observed for hydroxyurea **90**, which is capable in some cases of reacting with α,β-unsaturated ketones **91** as a substituted hydroxylamine [99] (Scheme 2.23). This treatment, however, proceeds via a β-addition of the hydroxyl group to the carbon–carbon double bond of the enone. The β-adduct **92** formed in the first stage depending on the nature of the ketone and conditions of the reaction may eliminate water, yielding 1,2,4-oxadiazepin-3-one **93** [100], or may via 3-hydroxyisoxazolidine-2-carboxamide **94** form 4,5-dihydroisoxazole **95** [99]. It is clear that in the first case hydroxyurea acts as a 1,2-binucleophile, while in the second case its acts as a 1,4-binucleophile.

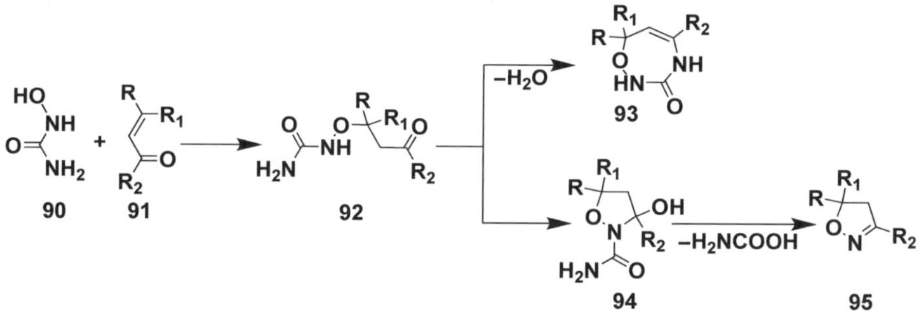

Scheme 2.23

An unexpected result is observed when 2,6-dimethylhepta-2,5-dien-4-one **96** reacts with hydroxylamine (Scheme 2.24). Instead of forming 1-hydroxy-4-piperidone **98** or isoxazole **99**, a double Michael-type addition of the OH and NH groups of hydroxylamine across the double bonds of **96** with formation of hexahydro1,2-oxazepine **97** occurs [101].

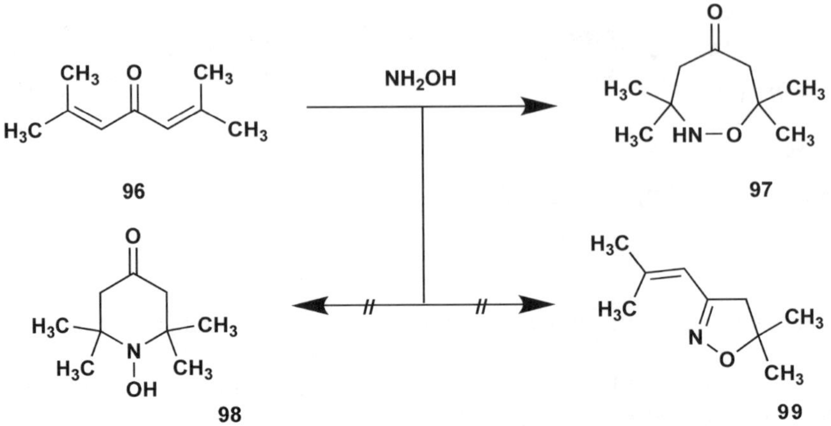


Scheme 2.24

## 2.3 Structure, Chemical and Physicochemical Properties of Pyrazolines

Most chemical and physicochemical properties of aromatic substituted pyrazolines as well as isoxazolines are formed by the $\pi,p,\pi$-system of the diarylhydrazone (oxime) moiety.

Some results of the $\pi$-electron structure analysis of pyrazolines were systematized in [102]. According these data, several main conclusions can be drawn:

- The diarylhydrazine $\pi$-system formally consists of two cross-conjugated fragments (ellipses $A_1$ and $A_2$, Scheme 2.25), but most of the molecular orbitals have well-defined delocalized character and these parts of the molecule cannot be considered as quasiautonomous.

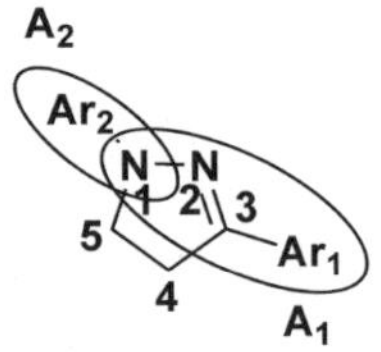


Scheme 2.25

- A strong interaction of fragments $A_1$ and $A_2$ through a “bridge” amine nitrogen in position 1 is apparent from the influence of substituents on the aromatic rings $Ar_1$ and $Ar_2$ on the electron density both in substituted and in unsubstituted aryl cycles.
- Both the general $\pi$-electron density and the density on the highest occupied molecular orbital in the *para* position of the *N*-phenyl substituent are much greater than in the phenyl at position 3 of the pyrazoline; in the case of the

Ph–N–N=$CH_2$ $\pi$-system, simulating pyrazolines unsubstituted at position 3, another situation is observed—although the $\pi$-electron density on the *ortho* and *para* atoms of the phenyl is greater than on $C_3$, the boundary electron density has a maximum value exactly at position 3.

The features of the electronic structure of aryl-substituted pyrazolines influence their chemical properties. For example, in the case of 3-substituted *N*-phenylpyrazolines **100** reactions of formylation, acylation, nitration, sulfonation, azocoupling and other electrophilic processes involve the *para* position of the *N*-phenyl ring, with formation of compounds **101** [103, 104, 105, 106, 107, 108, 109, 110, 111, 112, 113]. On the other hand, some electrophilic reactions, including nitration, bromination, chlorination, formylation and azocoupling, for 3-unsubstituted pyrazolines **102** occur at position 3, yielding heterocycles **103** and in some cases as a mixture with **104** [108, 114, 115] (Scheme 2.26). This fact provides evidence for orbital control of these reactions.

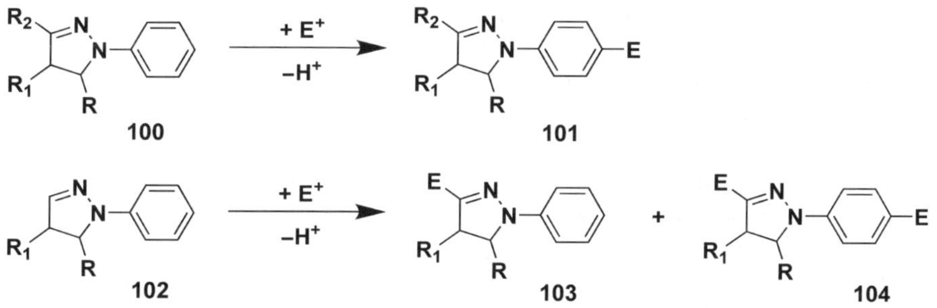


**Scheme 2.26**

Pyrazolines are the intermediates in one of the synthetic pathways to cyclopropanone and this fact has defined the development of pyrazoline chemistry for a long period. This aspect of their synthetic application is well described in the literature, including textbooks on general organic chemistry, and therefore we do not consider it necessary to cover it in this book.

Oxidative processes are very characteristic for pyrazoline derivatives such as dihydroheteroaromatic compounds. However, oxidation of pyrazolines is not a widely used preparative method for the synthesis of the appropriate pyrazoles owing to numerous side reactions following heteroaromatization, and other pathways are usually applied to obtain pyrazoles [66]. It should be noted that pyrazolines unsubstituted at positions 1 and 3 are very unstable compounds and easily decompose in air, with nitrogen elimination. Introducing alkyls or aryls in these positions leads to an increase of their stability but oxidative destruction is also possible, for example, under the action of nitrous acid [116, 117].

The most interesting direction of the oxidation of *N*-phenyl-substituted pyrazolines **105** is their dimerization to compounds like **107** under the action of $Fe^{3+}$, $K_2Cr_2O_7$, $Br_2$ or $SbCl_5$ or electrochemically at an anode (Scheme 2.27). This process was first observed by Knorr [117] and was then verified by Raiford and Peterson [51] and others [118, 119].

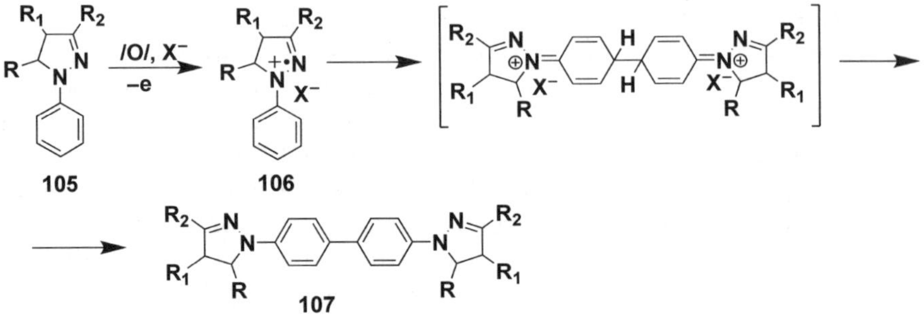

Scheme 2.27

The process is followed by the appearance of a bright blue, violet or green color of the reaction mixture and can be applied for qualitative analysis of pyrazolines. At the present time, it is proven that this dimerization passes via the preliminary formation of cation-radical particles **106**, which are responsible for the deep color of pyrazolines under acidic conditions in the presence of oxidizing agents.

Pyrazolines unsubstituted at position 1 are able to react as secondary amines. They undergo nitrozation with sodium nitrite in acids [48, 120, 121, 122, 123, 124, 125, 126, 127, 128, 129, 130], alkylation [48, 124, 131, 132, 133, 134, 135, 136, 137, 138, 139], acylation with acids or their derivatives [131, 140, 141, 142, 143, 144, 145, 146, 147, 148, 149, 150], isocyanates or isothiocyanates [127, 151, 152, 153, 154, 155, 156, 157, 158, 159], cyanoethylation [160, 161, 162, 163, 164, 165] or other numerous reactions (Scheme 2.28)

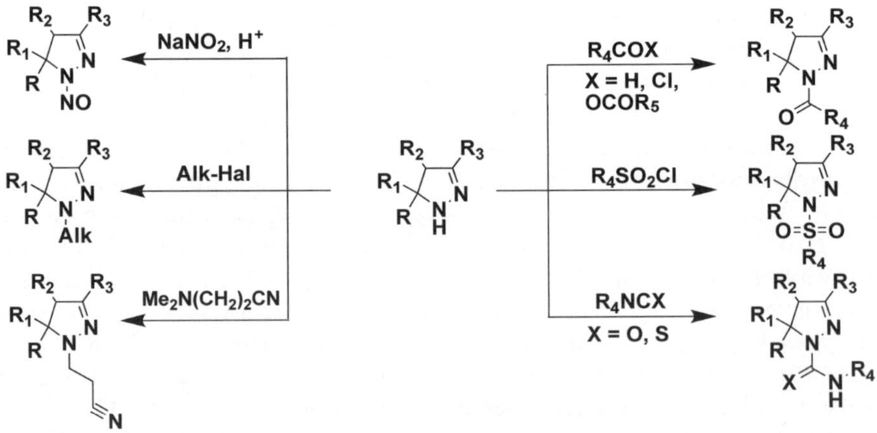


Scheme 2.28

However, acylation reactions of N-unsubstituted pyrazolines with halogen anhydrides of acids are very often followed by the destruction of the heterocyclic ring owing to the ease of breaking the $C_{(5)}-N_{(1)}$ bond located at the β-position of the azomethine group [140, 141, 142, 166]. For example, pyrazoline **108** treated with halogen anhydrides of acids yields α,β-unsaturated ketone **109** and diacylhydrazine **110** [166] (Scheme 2.29).

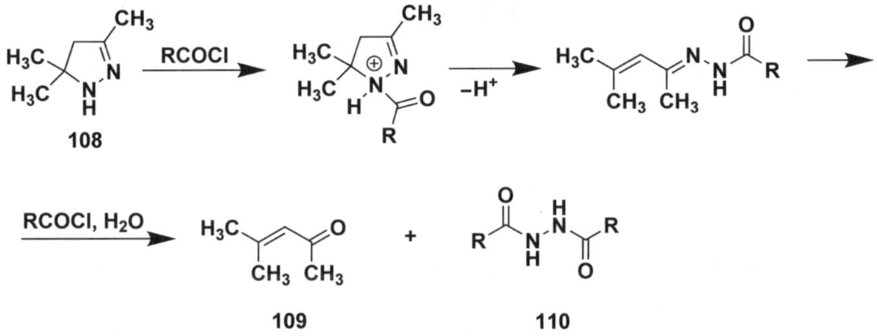

Scheme 2.29

Thus, several general conclusions about the reactions of pyrazolines with electrophilic reagents can be made:

- In the case of 3,5-disubstituted and 3,4,5-trisubstituted pyrazolines, electrophilic attack is directed to position 1 of the heterocycle (NH group).
- In the case of 1,5-disubstituted pyrazolines, electrophilic reaction involves position 3.
- 1-Phenyl-3-substituted and 1-phenyl-3,5-disubstituted pyrazolines react with electrophiles with participation of the *para* position of the *N*-phenyl substituent.
- Reactions of pyrazoline derivatives with electrophilic reagents very often become complicated owing to the oxidative destruction and/or processes related to the $C_5-N_1$ bond breaking.

Pyrazolines containing aryl substituents with some functional groups can be used in the synthesis of new heterocycles. For example, the reaction of 5-(2-hydrohyphenyl)pyrazolines **111** with noncyclic **112** or cyclic carbonyl compounds under acidic conditions yields dihydropyrazolo[1,5-*c*]-1,3-benzoxazines **114** and **115**, respectively [167, 168, 169] (Scheme 2.30).

Compounds like **114** and **115** have properties which are typical for semiaminales. In the presence of aqueous acids they are hydrolyzed and the starting pyrazolines **111** and carbonyls **112 (113)** are formed. Under the action of even trace amounts of acids (for instance, HCl forming in chloroform)

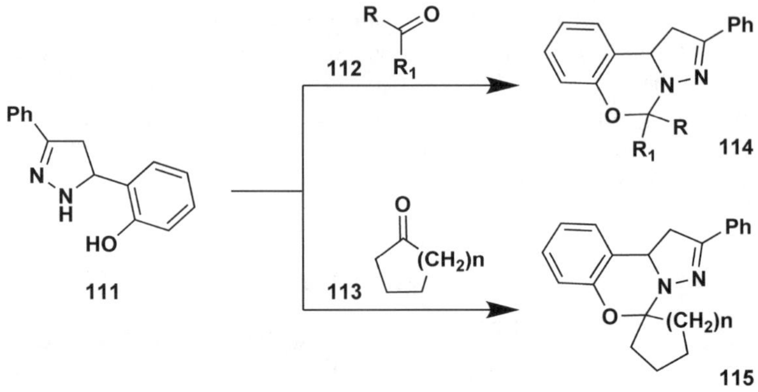


Scheme 2.30

pyrazolobenzoxazines easily form an equilibrium mixture of two diastereomers **A** and **B** [168] (Scheme 2.31). The diastereomerization process happens very fast, i.e., $^1$H NMR spectra for each diastereomer **A** and **B** in the individual state can be recorded only after thorough removal from chloroform -d6 an impurity of DCl.

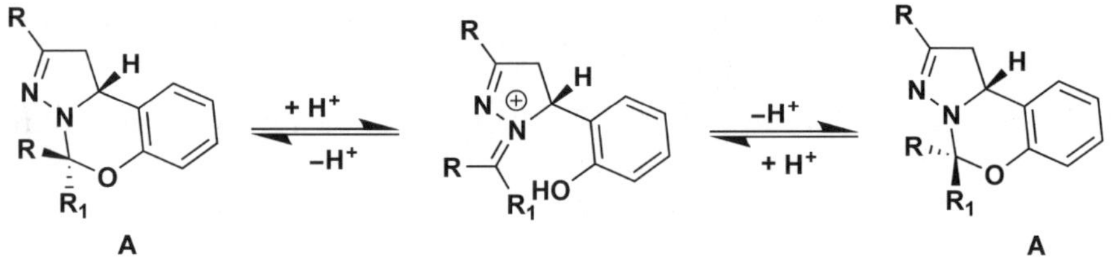


Scheme 2.31

Another interesting reaction of benzoxazines **114** (**115**) is the reductive opening of the oxazine ring with simultaneous dehydrogenation of the pyrazoline moiety [170]. This process is carried out in a KOH suspension of a mixture of dimethyl sulfoxide and dimethylformamide. For instance, this treatment involving 2,5-diaryl-1,10b-dihydro-1*H*-pyrazolo[1,5-*c*]benzo[*e*]-1,3-oxazines **116** leads to the formation of pyrazoles **117** (Scheme 2.32). Similar disproportionation reactions have also been described for some bezopiranes, for example, pyrazole derivative **118** [91, 170].

Some aryl halide derivatives of pyrazoline can undergo photocyclization. Thus, 5-(2-halogenphenyl)-1,3-diphenyl-$\Delta^2$-pyrazoline **120** proceeds by a simple bond homolysis from the $S_1^*$ state to give 2-phenylpyrazolo[1,5-*f*]phenathridine **121** [171] (Scheme 2.33).

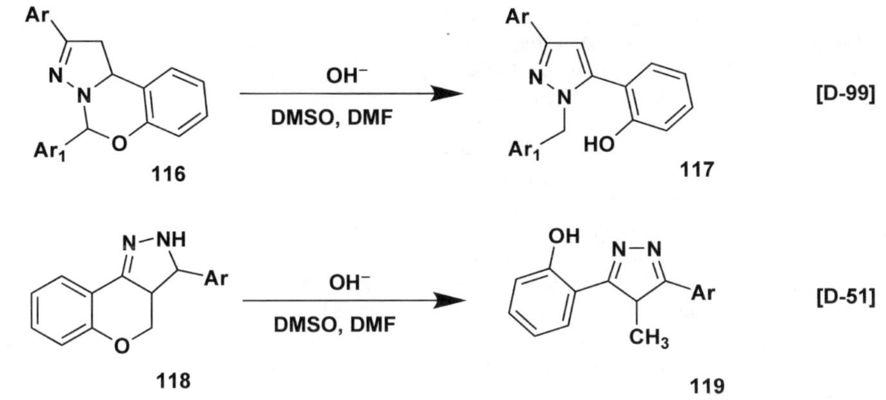


Scheme 2.32

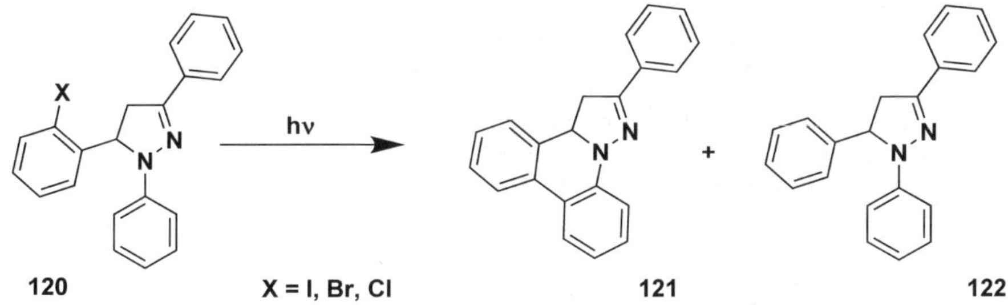


Scheme 2.33

Grimshaw and de Silva [171] described the influence of the type of halogen on the rate of the reaction. Iodine-substituted pyrazolines react at the highest rate, while in the case of bromo derivatives, the cyclization is slower and when X is Cl, the speed of transformation is negligible.

## References


1. Bilgin, A. A.; Palaska, E.; Sunal, R.; Guemuesel, D. *Pharmazie,* 1994, *49*, 67.
2. Abbady, M. A.; Hebbachy, R. *Indian J. Chem. Sect. B*, 1993, *32*, 1119.
3. Attia, A.; Michael, M. *Acta Chim. Hung*., 1983, *114*, 337.
4. Descacq, P.; Nuhrich, F.; Varache-Beranger, V.; Capdepuy, V.; Devaux, D. *Eur. J. Med. Chem. Chim. Ther*., 1990, *25*, 285.
5. Ankhiwala, M. D. *J. Indian Chem. Soc*.,1990, *67*, 514.
6. Fahmy, A. M.; Hassan, K. M.; Khalaf, A. A.; Ahmed, R. A. *Indian J. Chem. Sect. B*, 1987, *26*, 884.
7. Elguero, J.; Goya, P.; Jagerovic, N.; Silva, A. M. S. *Targets Heterocycl. Syst*. 2002, *6*, 52.

8. Penning, T. D.; Talley, J. J.; Bertenshaw, S. R.; Carter, J. S.; Collins, P. W.; Docter, S.; Graneto, M. J.; Lee, L. F.; Malecha, J. W.; Miyashiro, J. M.; Rogers, R. S.; Rogier, D. J.; Yu, S. S.; Anderson, G. D.; Burton, E. G.; Cogburn, J. N.; Gregory, S. A.; Koboldt, C M.; Perkins, W. E.; Seibert, K.; Veenhuizen, A. W.; Zhang, Y. Y.; Isakson, P. C.; *J. Med. Chem.* 1997, *40*, 1347.
9. Terrett, N. K.; Bell, A. S.; Brown, D.; Ellis, P.; *Bioorg. Med. Chem. Lett.* 1996, *6*, 1819.
10. Elguero, J. In *Comprehensive Heterocyclic Chemistry II*; Katritzky, A. R.; Rees, C. W.; Scriven, E. F. V. (Eds.); Pergamon Elsevier: Oxford, 1996, *6*, 1.
11. Singh, S. K.; Reddy, P. G.; Rao, K. S.; Lohray, B. B.; Misra, P.; Rajjak, S. A.; Rao, Y. K.; Venkatewarlu, A. *Bioorg. Med. Chem. Lett.* 2004, *14*, 499.
12. Palaska, E.; Erol, D.; Demirdamar, R. *Eur. J. Med. Chem. Chim. Ther.*,1996, *31*, 43.
13. Cox, C. D.; Breslin, M. J.; Mariano, B. J.; Coleman, P. J.; Buser, C. A.; Walsh, E. S.; Hamilton, K.; Huber, H. E.; Kohl, N. E.; Torrent, M.; Yan, Y.; Kuo, L. C.; Hartman, G. D. *Bioorg. Med. Chem. Lett.*, 2005, *15*, 2041.
14. Chimenti, F.; Maccioni, E.; Secci, D.; Bolasco, A.; Chimenti, P.; Granese, A.; Befani, O.; Turini, P.; Alcaro, S.; Ortuso, F.; Cirilli, R.; Torre, F. L.; Cardia, M. C.; Distinto, S. *J. Med. Chem.*, 2005, *48*, 7113.
15. Shaharyar, M.; Siddiqui, A. A.; Ali, M. A.; Sriram, D.; Yogeeswari, P. *Bioorg. Med. Chem. Lett.*, 2006, *16*, 3947.
16. Shahar, M. Y.; Siddiqui, A. A.; Ali, M. A. *Bioorg. Med. Chem. Lett.*, 2006, *16*, 4571.
17. Rovnyak, G. C.; Millonig, R. C.; Schwartz, J.; Shu, V. *J. Med. Chem.*, 1982, *25*, 1482.
18. Peng, Q.; Lu, Z.-Y.; Huang, Y.; Xie, M.-G.; Xiao, D.; Han, S.-H.; Peng, J.-B.; Cao, Y. *J. Mater. Chem.*, 2004, *14*, 396.
19. Krasovitskii, B. M.; Bolotin, B. M. *Organic Luminescent Materials*, Wiley-VCH: Weinheim, 1988.
20. Krasovitskii, B. M.; Zubanova, T. P.; Vinetskaya, Y. M. *Khim. Geterotsikl. Soedin.*, 1982, 1248.
21. Fahrni, C. J.; Yang, L.; VanDerveer, D. G. *J. Am. Chem. Soc.*, 2003, *125*, 3799.
22. Wiley, R. H.; Jarboe, C. H.; Hayes, F. N.; Hansbury, E.; Nielsen, J. T.; Callahan, P. X.; Sellars, M. C. *J. Org. Chem.*, 1958, *23*, 732.
23. Azarello, J. *Gazz. Chim. Ital.*, 1906, *36*, 50.
24. Smith, L. I.; Pings, W. B. *J. Org. Chem.*, 1937, *2*, 23.
25. Levai, A.; Cziaky, Z.; Jeko, J.; Szabo, Z. *Indian J. Chem.*, 1996, *35B*, 1091.
26. Ghate, S. G.; Kaushal, R.; Deshapande, S. S. *J. Indian Chem. Soc.*, 1950, *27*, 633.
27. Mustafa, A.; Freifel, A. M. *J. Org. Chem.*, 1959, *24*, 1470.
28. Sayed, G. H.; Kjosen, N. *Indian J. Chem.*, 1980, *19B*, 980.
29. Aleksandrova, I. A.; Dorofeeva, N. A.; Chernova, A. V.; Kharullin, V. K. *Zh. Org. Khim.*, 1978, *14*, 1974.
30. Levai, A. *Khim. Geterotskl. Soedin.*, 1997, 579.
31. Tokes, A. L.; Szollosy, A.; Levai. A. *Acta Chem. Hung.*, 1983, *112*, 335.
32. Levai, A. *Monatsh. Chem.*, 1995, *126*, 1245.
33. Pijewska, L.; Kamecki, J.; Perka-Karolczak, W. *Pharmazie*, 1993, *48*, 254.
34. Kamecki, J.; Perka, W.; Pijewska, L. *Pol. J. Chem.*, 1985, *59*, 285.
35. Tőkés, A. L.; Géza, J. *Synth. Commun.*, 1989, *19*, 3159.
36. Padmavathi, V.; Reddy, B. J. M.; Subbaiah, D. R. C. V. *New J. Chem.*, 2004, *28*, 1479.
37. Sausins, A.; Duburs, G. *Khim. Geterotsikl. Soedin.*, 1992, *4*, 435.
38. Sausins, A.; Duburs, G. *Heterocycles*, 1988, *27*, 269.
39. Mustafa, A.; Hilmy, M. K. *J. Chem. Soc.*, 1951, 3254.
40. Neudeck, H. K. *Monatsh. Chem.*, 1996, *127*, 417.
41. Toth, G.; Szollosy, A.; Levai, A.; Kotovich, G. *J. Chem. Soc. Perkin Trans. II*, 1986, 1895.
42. Toth, G.; Levai, A.; Duddeck, H. *Magn. Reson. Chem.*, 1992, *30*, 235.
43. Toth, G.; Levai, A.; Szollosy, A.; Duddeck, H. *Tetrahedron*, 1993, *49*, 863.
44. Molchanov, A. P.; Korotkov, V. S.; Kostikov, R. R. *Zh. Org. Khim.*, 2004, *40*, 501.

45. von Auwers, K.; Muller, K. *Chem. Ber.*, 1908, *41*, 4230.
46. von Auwers K.; Muller, K. *Chem. Ber.*, 1909, *42*, 4411.
47. von Auwers, K.; Kreuder, A. *Chem. Ber.*, 1925, *58*, 1974.
48. von Auwers, K.; Heimke, P. *Ann. Chem.*, 1927, *458*, 186.
49. Kishner, N. *J. Phys. Chem. Soc.*, 1913, *45*, 949.
50. Raiford, L. C.; Entrikin, J. B. *J. Am. Chem. Soc.*, 1933, *55*, 1125.
51. Raiford, L. C.; Peterson, W. J. *J. Org. Chem.*, 1936, *1*, 544.
52. Raiford, L. C.; Gundy, G. V. *J. Org. Chem.*, 1938. *3*, 265.
53. Nisbert, H. B. *J. Chem. Soc.*, 1938, 1237.
54. Fathi, T.; Dinh An, N.; Schmitt, G.; Cerutti, E.; Laude, B. *Tetrahedron*, 1988, *44*, 4527.
55. Kidwai, M.; Misra, P. *Synth. Commun.*, 1999, *29*, 3237.
56. Azarifar, D.; Ghasemnejad, H. *Molecules*, 2003, *8*, 642.
57. Kapoor, K. K.; Ganai, B. A.; Kumar, S.; Andotra, C. S. *Synth. Commun.*, 2006, *36*, 2727.
58. Müller, T. J. J.; Ansorge, M.; Aktah, D.; *Angew. Chem. Int. Ed.*, 2000, *39*, 1253.
59. Meier, H.; Hormaza, A. *Eur. J. Org. Chem.*, 2003, *17*, 3372.
60. Yoshihara, N.; Hasegawa, T.; Hasegawa, S. *Bull. Chem. Soc. Jpn.*, 1991, *64*, 719.
61. Ponnala, S.; Sahu, D. P. *Synth. Commun.*, 2006, *36*, 2189.
62. Toth, G.; Szollosy, A.; Lorand, T.; Szabo, D.; Foldesi, A.; Levai, A. *J. Chem. Soc. Perkin Trans. II*, 1989, 319.
63. Turan-Zitouni, G.; Chevallet, P.; Kilic, F. S.; Erol, K. *Eur. J. Med. Chem. Chim. Ther.*, 2000, *35*, 635.
64. Gella, I. M.; Yaya, A. R.; Cherkaev, G. V.; Fialkova, S. V.; Orlov, V. D. *Khim. Geterotsikl. Soedin.*, 1996, 1650.
65. Lóránd, T.; Kocsis, B.; Emôdy, L.; Sohár, P. *Eur. J. Med. Chem. Chim. Ther.*, 1999, *34*, 1009.
66. Barton D., Ollis W.D. (Eds.) *Comprehensive Organic Chemistry: The Synthesis and Reactions of Organic Compounds*, vol. 4. *Heterocyclic Compounds*, Pergamon: Oxford, 1979.
67. von Auwers, K.; Muller, K. *Chem. Ber.*, 1921, *54*, 1000.
68. Suchchar, S. P.; Singh, A. K. *J. Indian Chem. Soc.*, 1985, *62*, 142.
69. Ferres, H.; Hamdam, M. S.; Jackson, W. R. *J. Chem. Soc. B*, 1971, 1892.
70. Bezuglyi, V. D.; Kotok, L. A.; Shimanskaya, N. P.; Bondarenko, V. E. *Zh. Obsch. Khim.*, 1969, *39*, 2167.
71. Raiford, L. C.; Manley, R. H. *J. Org. Chem.*, 1940, *5*, 345.
72. Bayer, H.; Dieterle, H. *Chem. Ber.*, 1911, *44*, 2697.
73. Tsukerman, S. V.; Danil'chenko, E. V.; Lavrushin, V. F. *Ukr. Khim. Zh.*, 1976, *42*, 383.
74. Lavrushin, V. F.; Bezugly, V. D.; Belous, G. G.; Tishchenko, V. G. *Zh. Obshch. Khim.*, 1964, *34*, 7.
75. Al-Farkh, Y. A.; Al-Hajjar, F. H.; Al-Shamali, F. S.; Hamoud, H. S. *Chem. Pharm. Bull.*, 1979, *27*, 257.
76. Award, W. I.; Ismail, M. F.; Moustafa, A. H. *U. A. R. J. Chem.*, 1971, 561.
77. Orlov, V. D.; Tishchenko, V. G.; Lavrushin, V. F. *Ukr. Khim. Zh.*, 1975, *41*, 862.
78. Roberman, A. I.; Madraso, K.; Orlov, V. D. *Vestnik Khark. Univer. Ser. Khim.*, 1977, 100.
79. Orlov, V. D.; Roberman, A. I.; Lavrushin, V. F. *Khim. Geterotsikl. Soedin.*, 1979, 130.
80. Turk, C. F. *Ger. Offen.* 2520171.
81. Krapcho, J.; Schwartz, J. US Patent 3,969,527.
82. El-Rayyes, N. R.; Al-Jawhary, A. *J. Heterocycl. Chem.*, 1986, 135.
83. El-Rayyes, N. R.; Bahtiti, N. H. *J. Heterocycl. Chem.*, 1989, 209.
84. El-Subbagh, H. I.; Abu-Zaid, S. M.; Mahraan, M. A.; Badria, F. A.; Al-Obaid, A. M. *J. Med. Chem.*, 2000, *43*, 2915.
85. Levai, A.; Szollosy, A.; Toth, G. *J. Chem. Res. Synop.*, 1985, 392.
86. Sangwan, N. K.; Rastogy, S. N. *Indian J. Chem.*, 1981, *20B*, 135.
87. Sammour, A.; Selim, M. I. B.; Elkasaby, M.; Sated, F. *J. Prakt. Chem.*, 1972, *314*, 941.
88. Krapcho, J.; Turk, C. F. *J. Med. Chem.*, 1979, *22*, 207.

89. Behera, R. K.; Behera, A. K.; Pradhan, R.; Pati, A.; Patra, M. *Synth. Commun.*, 2006, *36*, 3729.
90. Rovnyak, G. C.; Shu, V. *J. Org. Chem.*, 1979, *44*, 2518.
91. Rudenko, R. V.; Desenko, S. M.; Chebanov, V. A.; Chernenko, V. N.; Musatov, V. I. *Mendeleev Commun.*, 2005, *15*, 83.
92. Bugaev, A. A.; Golikov, A. G.; Kriven'ko, A. P. *Khim. Geterotsikl. Soedin.*, 2005, 986.
93. Szollosy, A.; Toth, G.; Konya, T.; Aradi, F.; Levai A. *J. Chem. Soc. Perkin Trans. II*, 1991, 489.
94. Lorand, T.; Aradi, F.; Szollosy, A.; Toth, G.; Konya, T. *Monatsh. Chem.*, 1966, *127*, 971.
95. Gella, I. M.; Yaya, A. R.; Orlov, V. D. *Kharkov University Bulletin*, 2001, 30, 103.
96. Lorand, T.; Szabo, D.; Foldesi, A.; Parkanyi, L.; Kalman, A.; Neszmelyi, A. *J. Chem. Soc. Perkin Trans. I*, 1985, 481.
97. Dhar D. N. *The Chemistry of Chalcones and Related Compounds*, Wiley-Interscience: New York, 1981.
98. Zelenin, K. N.; Ershov, A. Y.; Bezhan, I. P. *Khim. Geterotsikl. Soedin.*, 1990, 1559.
99. Jacquier, R.; Olive, J. L.; Petrus, C.; Petrus, F. *Tetrahedron Lett.*, 1975, *16*, 2337.
100. Jacquier, R.; Olive, J. L.; Petrus, C.; Petrus, F. *Tetrahedron Lett.*, 1975, *16*, 2979.
101. Rice, K. C.; Weiss, U. *Tetrahedron Lett.*, 1973, *14*, 1615.
102. Mitina, V. G.; Kutulya, L.A.; Sukhorukov, A. A. *Problems of Physical Organic Chemistry (Structure, Spectral, Luminescent, and Photochemical Properties of Unsaturated Ketones and 2-Pyrazolnes)*, Kharkov Sate University: Kharkov, 1989, p. 152.
103. Kutulya, L. A.; Shevchenko, A. E.; Surov, Y. N. *Khim. Geterotsikl. Soedin.*, 1975, 250.
104. Kutulya, L. A.; Shevchenko, A. E.; Pchelinova, L. D. *Khim. Geterotsikl. Soedin.*, 1982, 807.
105. Orlov, V. D.; Aziz, M. A. *Khim. Geterotsikl. Soedin.*, 1982, 88.
106. Orlov, V. D.; Aziz, M. A.; Savran, A. T.; Asoka, P. K. D. *Khim. Geterotsikl. Soedin.*, 1984, 965.
107. Orlov, V. D.; Aziz, M. A.; Mchedlov-Petrosyan., N. O.; Orlova, N. N. *Khim. Geterotsikl. Soedin.*, 1985, 1511.
108. Bouchet, P.; Elguero, J.; Jacquier, R. *Bull. Soc. Chim. Fr.*, 1967, *95*, 4716.
109. Dhar, D. N.; Raghunathan, R. *Indian J. Chem. Sect. B*, 1984, *23*, 1187.
110. Gella, I. M.; Vakula, V. N.; Orlov, V. D. *Khim. Geterotsikl. Soedin.*, 1981, 1245.
111. Simiti, I.; Zaharia, V.; Coman, M.; Demian, H.; Muresan, A. *Pharmazie*, 1988, *43*, 82.
112. Rivett, D. E.; Rosevear, J.; Wilshire, J. F. K. *Aust. J. Chem.*, 1983, *36*, 1649.
113. Ternon, M.; Outurquin, F.; Paulmier, C. *Tetrahedron*, 2001, *57*, 10259.
114. Duffin, G. F.; Kendall, J. D. *J. Chem. Soc.*, 1954, 408.
115. Pillai, P. M.; Bhat, V. S. *Indian J. Chem. Sect. B*, 1989, *28*, 1026.
116. Locquin, R.; Heilmann, R. *Compt. Rend.*, 1925. 1787.
117. Knorr, L. *Chem. Ber.*, 1893, *26*, 100.
118. Pragst, F. *J. Prakt. Chem.*, 1973, *315*, 549.
119. Morkovnik, A. S.; Okhlobystin, O. Y. *Khim. Geterotsikl. Soedin.*, 1985, 551.
120. von Auwers, K.; Cauer, E. *Liebigs Ann. Chem.*, 1929, *470*, 284.
121. Bech, S. G.; Turnbell, J. H.; Wison, W. *J. Chem. Soc.*, 1952, 4686.
122. Jorlander, H. *Chem. Ber.*, 1916, *49*, 2782.
123. Michaelis, A. *Liebigs Ann. Chem.*, 1905, *338*, 183.
124. Gorpinchenko, V. A.; Petrov, D. V.; Spirikhin, L. V.; Dokichev, V. A.; Tomilov, Y. V. *Zh. Org. Khim.*, 2006, *42*, 1718.
125. Mohamed, S. M.; Khalile, M.; Ismail, A. A.; Kadh, M. S. *J. Indian Chem. Soc.*, 2005, *82*, 833.
126. Molchanov, A. P.; Stepakov, A. V.; Kostikov, R. R. *Zh. Org. Khim.*, 2004, *40*, 1561.
127. Bekhit, A. A.; Abdel-Aziem, T. *Bioorg. Med. Chem.*, 2004, *12*, 1935.
128. Reddy, D. B.; Sarma, M. R.; Padmaja, A.; Padmavathi, V. *Phosphorus Sulfur Silicon Relat. Elem.*, 2000, *164*, 23.

129. Abdel-Rahman, T. M. *Boll. Chim. Farm.*, 1998, *137*, 43.
130. Faidallah, H. M.; Makki, M. S. I.; ElMassry, A. H. I.; Hassan, S. Y. *Pharmazie*, 1997, *52*, 101.
131. Sayed, G. H.; Kjosen, H. *J. Prakt. Chem.*, 1980, *322*, 716.
132. Trener, G. *Monatsch. Chem.*, 1900, 1111.
133. Braendvang, M.; Gundersen, L.-L. *Tetrahedron Lett.*, 2007, *48*, 3057.
134. Ballell, L.; Field, R. A.; Chung, G. A. C.; Young, R. J. *Bioorg. Med. Chem. Lett.*, 2007, *17*, 1736.
135. Huisgen, R.; Mloston, G.; Poechlauer, P.; Fisera, L.; Giera, H. *Helv. Chim. Acta*, 2007, *90*, 1.
136. Gamage, S. A.; Spicer, J. A.; Rewcastle, G. W.; Milton, J.; Sohal, S.; Dangerfield, W.; Mistry, P.; Vicker, N.; Charlton, P. A.; Denny, W. A. *J. Med. Chem.*, 2002, *45*, 740.
137. Mojzych, M.; Rykowski, A.; Wierzchowski, J. *Molecules*, 2005, *10*, 1298.
138. Bouissane, L.; Kazzouli, S. El; Leonce, S.; Pfeiffer, B.; Rakib, E. M.; Khouili, M.; Guillaumet, G. *Bioorg. Med. Chem.*, 2006, *14*, 1078.
139. Semple, G.; Skinner, P. J.; Cherrier, M. C.; Webb, P. J.; Sage, C. R.; Tamura, S. Y.; Chen, R.; Richman, J. G.; Connolly, D. T. *J. Med. Chem.*, 2006, *49*, 1227.
140. Mowry, D.; Renoll, M.; Huber, W. *J. Am. Chem. Soc.*, 1946, *68*, 1105.
141. Locquin, R.; Heilmann, R. *Bull. Soc. Chim. Fr.*, 1929, *45*, 541.
142. Sayed, G. H. *Indian J. Chem.*, 1980, *19B*, 364.
143. Abass, M.; Abdel-Megid, M.; Hassan, M. *Synth. Commun.*, 2007, *37*, 329.
144. Goodell, J. R.; Puig-Basagoiti, F.; Forshey, B. M.; Shi, P.-Yo.; Ferguson, D. M. *J. Med. Chem.*, 2006, *49*, 2127.
145. Kourafalos, V. N.; Marakos, P.; Mikros, E.; Pouli, N.; Marek, J.; Marek, R. *Tetrahedron*, 2006, *62*, 11987.
146. Turan-Zitouni, G.; Oezdemir, A.; Kaplancikli, Z. A.; Chevallet, P.; Tunali, Y. *Phosphorus Sulfur Silicon Relat. Elem.*, 2005, *180*, 2717.
147. Guerra, F. M.; Mish, M. R.; Carreira, E. M. *Org. Lett.*, 2000, *2*, 4265.
148. Curtis, M. D.; Hayes, N. C.; Matson, P. A. *J. Org. Chem.*, 2006, *71*, 5035.
149. Amr, A. E.-G. E.; Abdel-Latif, N. A.; Abdalla, M. M. *Bioorg. Med. Chem.*, 2006, *14*, 373.
150. Heydenreich, M.; Koch, A.; Sarodnick, G.; Kleinpeter, E. *Tetrahedron*, 2005, *61*, 2373.
151. Maire, M. *Bull. Soc. Chim. Fr.*, 1908, *3*, 272.
152. Freudenberg, K.; Stoll, W. *Ann. Chem.*, 1924, *440*, 38.
153. Smith, L.; Showell, J. *J. Org. Chem.*, 1952, *17*, 827.
154. Lange, J. H. M.; Stuivenberg, H. H. van; Veerman, W.; Wals, H. C.; Stork, B.; Coolen, H. K. A. C.; McCreary, A. C.; Adolfs, T. J. P.; Kruse, C. G. *Bioorg. Med. Chem. Lett.*, 2005, *15*, 4794.
155. Ruhoglu, O.; Oezdemir, Z.; Calis, U.; Guemuesel, B.; Bilgin, A. A. *Arzneim. Forsch.*, 2005, *55*, 431.
156. Cox, C. D.; Breslin, M. J.; Mariano, B. J. *Tetrahedron Lett.*, 2004, *45*, 1489.
157. Wang, S.; Shi, B.; Li, Y.; Wang, Q.; Huang, R., *Synth. Commun.*, 2003, *33*, 1449.
158. Malhotra, V.; Pathak, S.; Nath, R.; Mukerjee, D.; Shanker, K. *Indian J. Chem. Sect. B*, 2002, *41*, 1310.
159. Metwally, M. A.; Abdel-Latif, E.; Amer, F. A. *Boll. Chim. Farm.*, 2000, *139*, 213.
160. Kost, A. N.; Ershov, V. V.; Suminov, S. I. *Zh. Obshch. Khim.*, 1957, *27*, 25142.
161. Rousseau, V.; Lindwall, H. G. *J. Am. Chem. Soc.*, 1950, *72*, 3047.
162. Shao; S. *Zh. Obshch. Khim.*, 1959, *29*, 3971.
163. Elgemeie, G. H.; Elghandour, A. H.; Elshimy, H. M. *J. Prakt. Chem.*, 1989, *331*, 466.
164. Holla, D. C.; Seshadri, S. *Bull. Chem. Soc. Jpn.*, 1984, *57*, 2984.
165. Batt, D. G.; Petraitis, J. J.; Houghton, G. C.; Modi, D. P.; Cain, G. A.; Corjay, M. H.; Mousa, S. A.; Bouchard, P. J.; Forsythe, M. S.; Harlow, P. P.; Barbera, F. A.; Spitz, S. M.; Wexler, R. R.; Jadhav, P. K. *J. Med. Chem.*, 2000, *43*, 41.

166. Ershov, V. V.; Kost A. N. Evreinova, E. V. *Dokl. Akad. Nauk USSR*, 1957, *113*, 813.
167. Orlov, V. D.; Getmanskii, N. V.; Oksenich, I. A.; Iksanova, S. V. *Khim. Geterotsikl. Soedin.*, 1991, 1131.
168. Desenko, S. M.; Getmanskii, N. V.; Chernenko, V. N.; Zemlin, I. M.; Shishkin, O. V.; Orlov, V. D. *Khim. Geterotsikl. Soedin.*, 1999, 805.
169. Desenko, S. M.; Chernenko, V. N.; Orlov, V. D.; Musatov, V. I. *Kharkov Univ. Bull.*, 2000, *29*, 46.
170. Desenko, S. M.; Chernenko, V. N.; Orlov, V. D.; Musatov, V. I. *Khim. Geterotsikl. Soedin.*, 2001, 1427.
171. Grimshaw, J.; de Silva, A. P., *J. Chem. Soc. Perkin Trans. 2*, 1983, 1679.

# Chapter 3
# Six-Membered Azaheterocycles Based on 1,3-Binucleophiles

Heterocyclizations of α,β-unsaturated ketones are the most widely investigated reactions leading to six-membered heterocycles. For convenience these processes can be separated into three large groups (Scheme 3.1):

1. Synthesis of dihydropyridines through preliminary transformation of unsaturated ketones into 1,5-dicarbonyl compounds (Scheme 3.1, reaction a—variant of Hantzch synthesis)
2. Cyclocondensations of unsaturated carbonyls with nitrogen-containing 1,3-binucleophiles (Scheme 3.1, reactions b–g)
3. Heterocyclization with participation of additional functional groups of enons

Hantzch synthesis is a well-known pathway to dihydro derivatives of azines. In some of the variations of this synthesis, the electrophilic reagents can be unsaturated ketones (Scheme 3.1, reactions a and b). Hantzch syntheses, including reaction b in Scheme 3.1 in which enamines are formed in situ from ammonia and the corresponding carbonyls, are not a topic of this book. However, synthetic applications of such condensations were described in detail in some reviews [1, 2].

In addition, heterocyclizations based on enamines (Scheme 3.1, reaction b) are an example of other general synthetic approaches to dihydroazine systems—condensations of nitrogen-containing 1,3-binucleophiles. Among the numerous variations, aromatic and heterocyclic amines can be considered as cyclic enamines (Scheme 3.1, reactions f and g). 1,3-Binucleophilic components of the cyclocondensations, in addition to enamines, may be used in these reactions. Thus, substituted pyridines can be easily obtained by condensation of α,β-unsaturated ketones with 1,3-binucleophiles, like derivatives of acetamide, malonamide and malonodinitrile containing an CH-acidic nucleophilic center (Scheme 3.1, reaction e). Reactions of α,β-unsaturated ketones with urea and its analogues—amidines, thiourea, guanidine and its derivatives—yielded a pyrimidine ring compound (Scheme 3.1, reactions c and d).

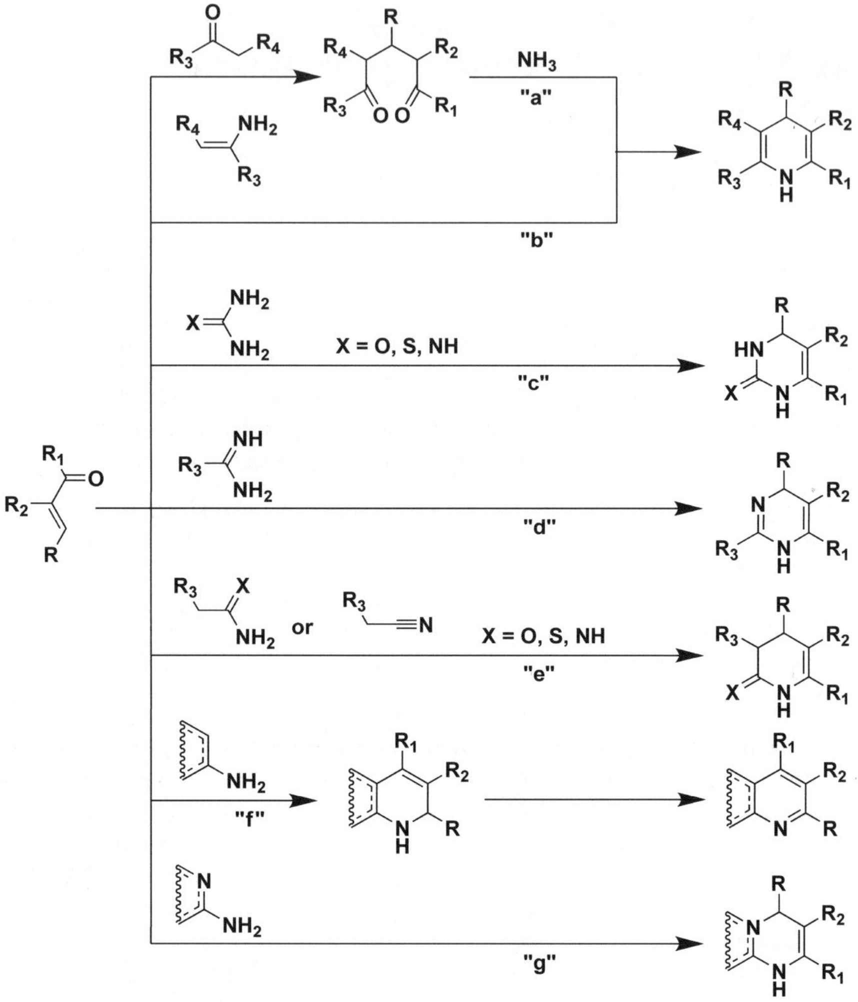


Scheme 3.1

## 3.1 Reactions of Enamines

As mentioned already, a reaction of enamines with α,β-unsaturated ketones may be one of the stages of Hantzch synthesis. But in the literature there are a lot of examples of independent applications of a broad set of enamines in dihydropyrimidine syntheses. For example, in [3, 4], a reaction of 3-aminocyclohex-2-enones **1** with an unsaturated ketone **2** was described which results in a dehydrogenation leading to the formation of a quinoline **3** (Scheme 3.2).

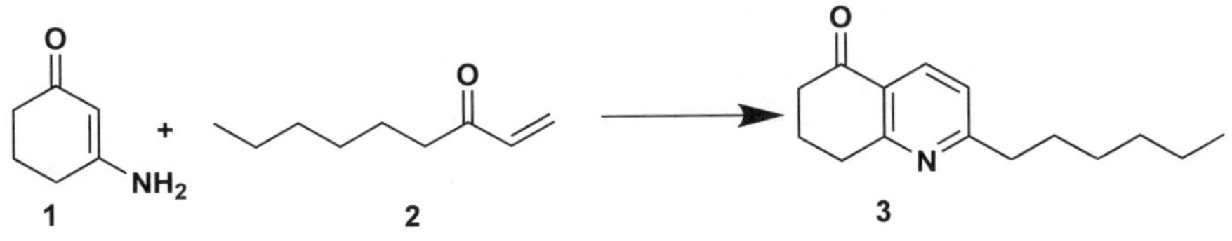


Scheme 3.2

In [5, 6] the emphasis of the authors was on the reactions of cyclic emanine **4** with some α,β-unsaturated carbonyl compounds. These reactions involve both noncyclic chalcone **5** and arylidenecyclohexanones **7** (Scheme 3.3). The solvent used was acetonitrile and the duration of the reaction varied from 15 to 48 h. Zhang et al. [6] also showed that at room temperature the reaction of enamine **4** with unsaturated ketone **9** led to the isolation of 5-hydroxyhexahydroimidazo[1,2-*a*]pyridine **10** without further dehydration. Subsequently, compound **10** reacted with another equivalent of unsaturated ketone to yield a heterocyclic salt **11** (Scheme 3.4).

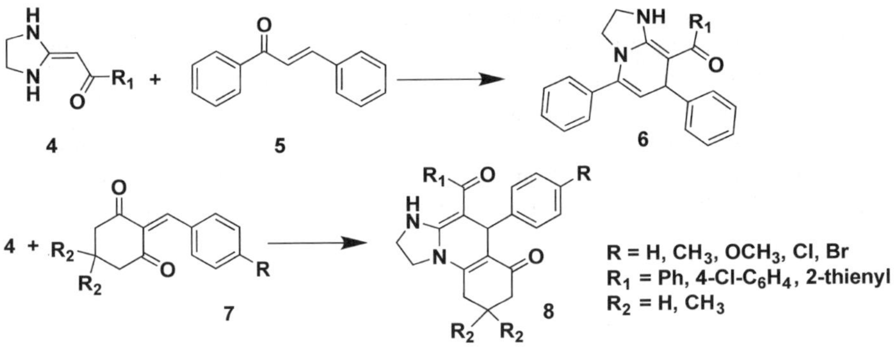


Scheme 3.3

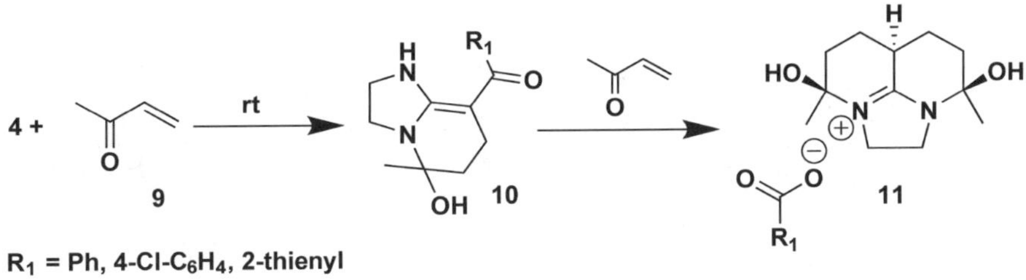


Scheme 3.4

In [7, 8] the reactivity of nitroethanediamines **12** and **14** with unsaturated ketone **9** was investigated. In this case, as expected, the reaction products were the respective dihydropyridines **13** and **15** (Scheme 3.5).

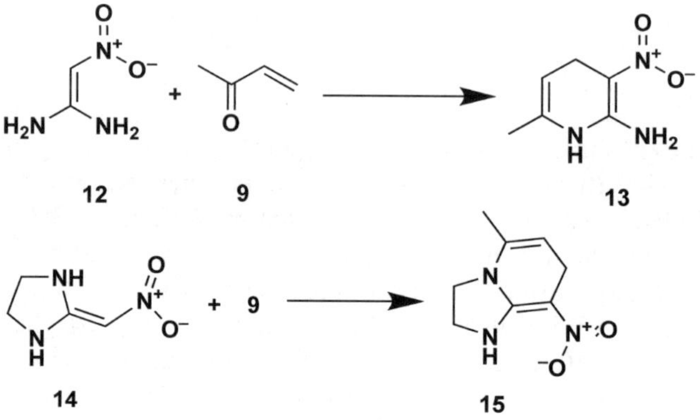


**Scheme 3.5**

There are several examples of similar reactions of enaminonitriles [9, 10, 11, 12, 13, 14]. Katsuyama et al. [9] determined that refluxing 3-aminobut-2-enenitrile **16** with fluorinated unsaturated ketone **17** in ethanol leads to high yields of dihydropyridines **18** (Scheme 3.6). The reaction of **16** with fluorinated cyclic ketone **19** leading to heterocycle **20** can be carried out in glacial acetic acid and does not need any additional catalyst or reagent [14] (Scheme 3.6). On the other hand, treatment of aminobutenenitrile **16** with ketones without strong

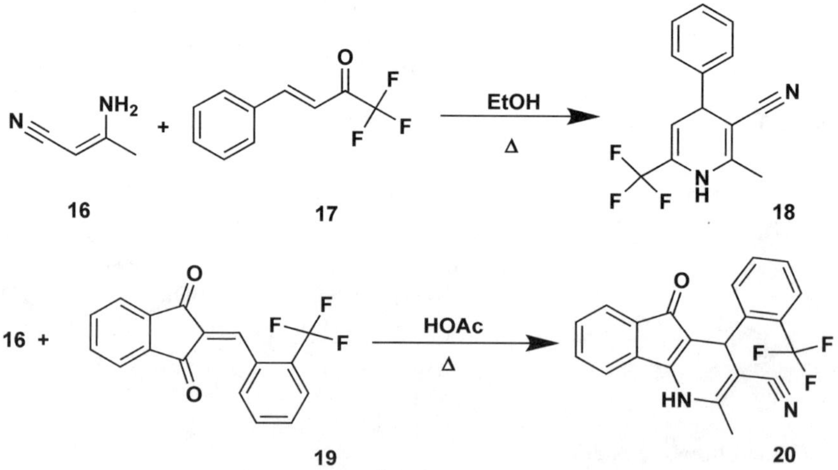


**Scheme 3.6**

electron-withdrawing substituents requires stronger conditions, i.e., additional reagents such as potassium *tert*-butoxide (under UV irradiation) [11], sodium ethoxide in ethanol [12] or sodium methoxide in methanol [13].

It is interesting to note that the reaction of **16** with *ortho*-hydroxybenzylideneacetone **21** is followed by intramolecular addition of a hydroxyl group to the double bond of the dihydropyridine. The final isolated product in this case is the azatricyclic heterocyclic system **22** (Scheme 3.7). Intramolecular addition of a hydroxyl group to a carbon–carbon double bond in the dihydropyridine ring also occurs in the reaction of enaminoester **23** with benzylideneacetone **24** [15] (Scheme 3.8).

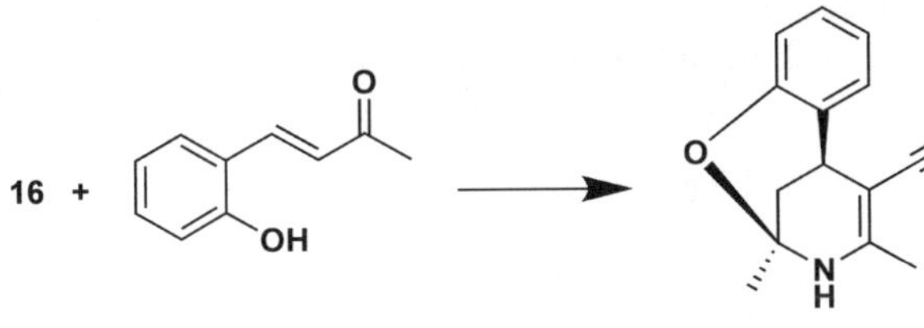


21 22

Scheme 3.7

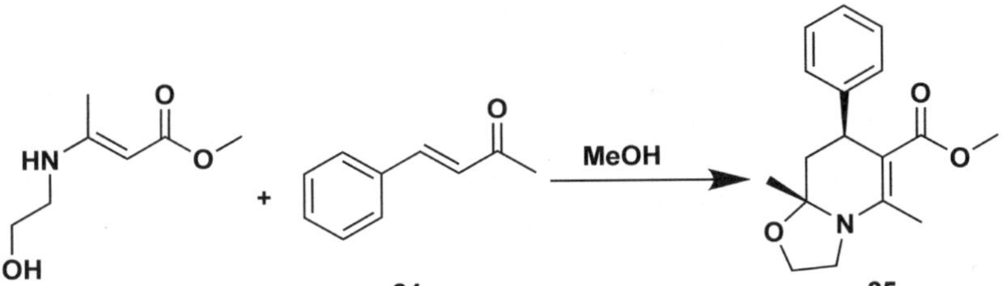


23 24 25

Scheme 3.8

## 3.2 Reactions of Urea and Its Analogues

As already mentioned at the beginning of this chapter, one of the facile methods for the synthesis of dihydropyrimidine derivatives is the treatment of α,β-unsaturated carbonyls with urea and its analogues—thiourea, guanidine and amidines. However, the majority of the publications have dealt with syntheses involving thiourea. Most likely is the possibility of the modification of 3,4-dihydropyrimidine-2-thiones or their heteroaromatized analogues, which produces a diverse class of heterocycles. The reagent involved in this modification process can act like α,β-unsaturated carbonyls [16] (Scheme 3.9).

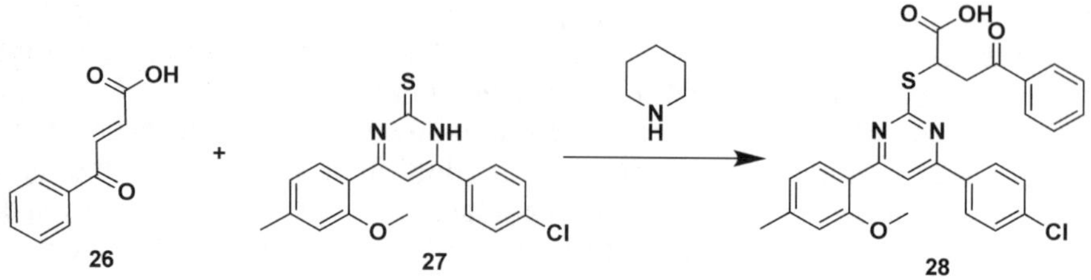


**Scheme 3.9**

The conditions of the heterocyclizations of thiourea **28** with unsaturated ketones can be very different and usually a basic catalyst is used. The most common catalysts are sodium hydroxide [17, 18, 19, 20, 21, 22, 23, 24, 25, 26, 27, 28, 29, 30, 31, 32, 33, 34, 35, 36, 37, 38, 39, 40, 41, 42], potassium methoxide and ethoxide [23, 29, 43, 44, 45, 46, 47, 48] or the more gentle sodium carbonate [49, 50] (Scheme 3.10).

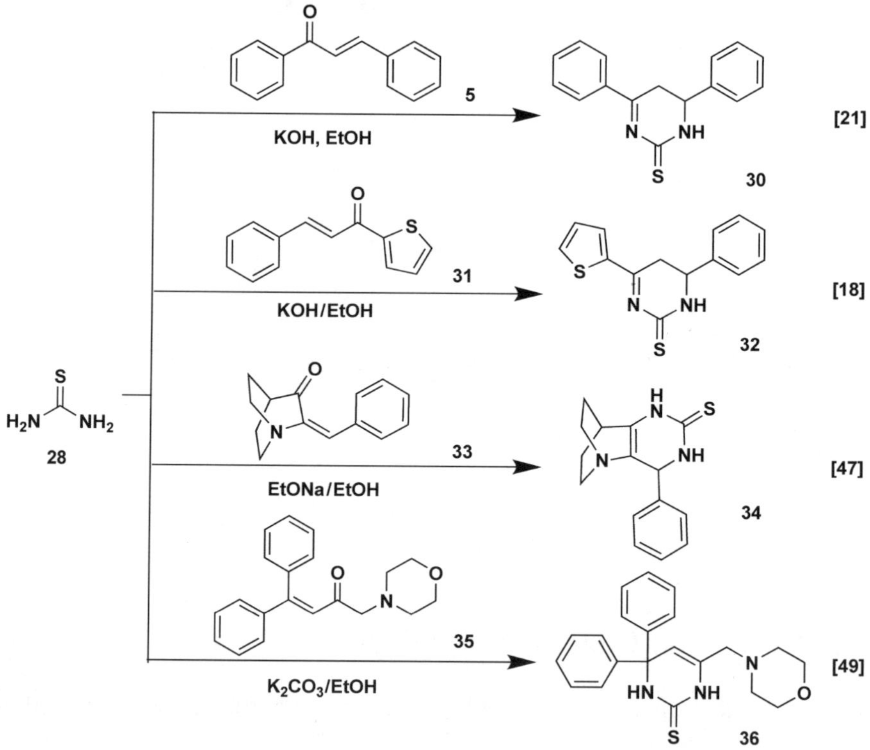


**Scheme 3.10**

But acidic catalysis is also possible. Some examples include the application of sulfuric [51], hydrochloric [52, 53, 54, 55, 56] and acetic [56, 57, 58] acids. Usually acidic catalysis is used in the case of α,β-unsaturated carbonyl compounds with heterocyclic nature or ones containing one or more heterocyclic substituents.

For example, sulfuric acid was used to catalyze a reaction of thiourea **28** with 2-(3-oxo-3-phenylpropenyl)-3-phenyl-3*H*-quinolizin-4-one **37,** which yielded pyrimidinethione **38** [51] (Scheme 3.11).

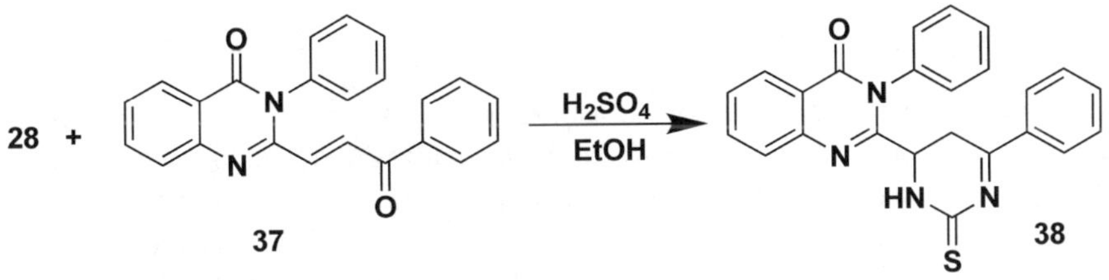


**Scheme 3.11**

An example of the use of hydrochloric acid in reactions of thiourea with unsaturated ketone **39**, based on 5-chloro-3-methyl-1-phenylpyrazole, leads to the formation of dihydropyrimidinethione **40** as mentioned in [51] (Scheme 3.12). Acetic acid, in particular, was used to carry out the reaction of thiourea **28** with 2-benzylideneindene-1,3-dione **41** to synthesize heterocycle **42** [58] (Scheme 3.12).

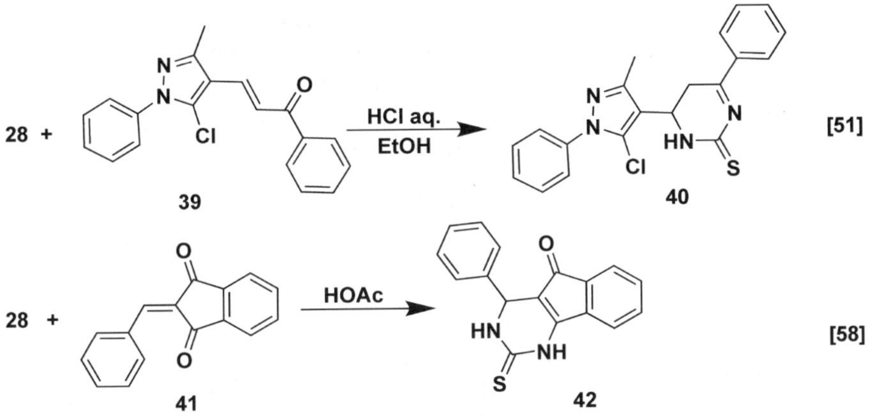


**Scheme 3.12**

The analysis of literature data shows that irrespective of the catalyst used the reaction time varies from 30 min to 12 h with yields of 40–80%. But under microwave irradiation the treatment of chalcone **43** with thiourea was completed in 4 min with 85% yield of dihydropyrimidine-2-thione **44** [59] (Scheme 3.13).

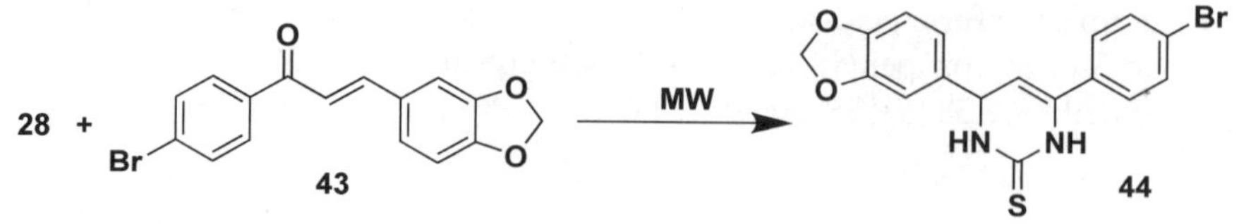


Scheme 3.13

In some cases the application of a basic catalyst can be followed by further oxidation of dihydropyrimidene-2-thiones, yielding heteroaromatized heterocycles [60, 61]. For example (Scheme 3.14), it was shown that the reaction of 4-oxo-4-arylbut-2-enoic acids **45** with thiourea **28** in the presence of sodium hydroxide led to 2-mercapto-6-arylpyrimidine-4-carboxylic acids **46** [60], while their dihydro analogues were not isolated.

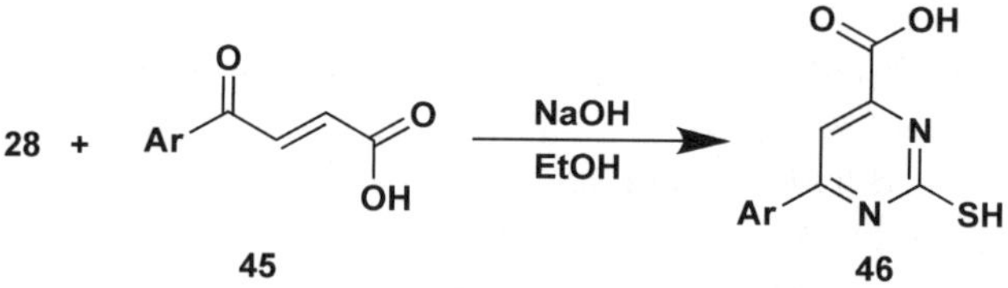


Scheme 3.14

Because of the presence of three nucleophilic centers in thiourea, an alternative direction of its reactions with α,β-unsaturated carbonyls with the formation of dihydrothiazines was found in several cases (Scheme 3.15). This trend is usually observed in cases of N-substituted thioureas or cyclic α,β-unsaturated ketones using an acidic catalyst [62, 63, 64, 65].

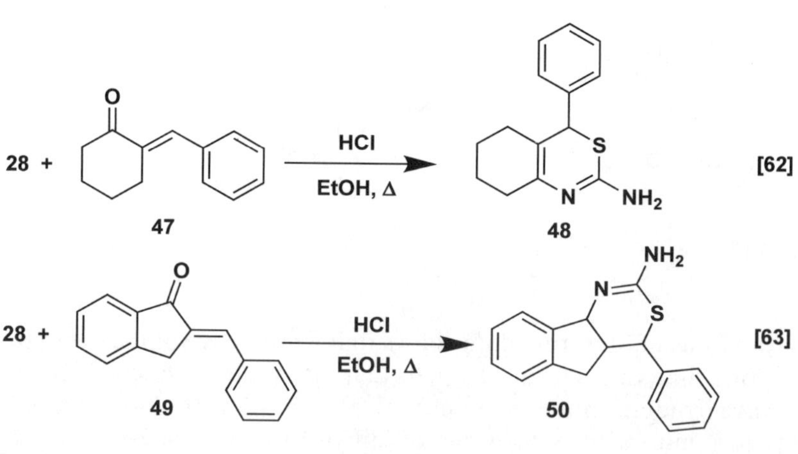


Scheme 3.15

*N*-Aryl ureas **51** and **54** react with mesityl oxide **52** [66] or 3-buten-2-one **9** [67] to form thiazines **53** and **55**, respectively. But in the second case, alkylation of the hydroxyl group with solvent molecules occurs without dehydration (Scheme 3.16).

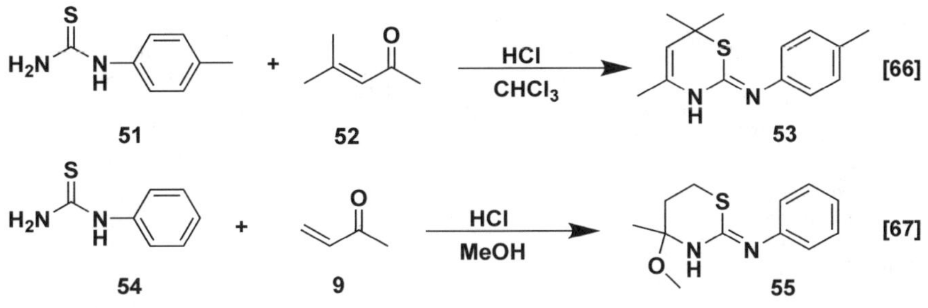


**Scheme 3.16**

Reaction of 4-benzylidene-1-phenylpyrrolidine-2,3,5-trione **56** with thiourea **28** occurs with hydrolysis of the amino group and leads to the formation of 1,4-dihydropyrrolo[3,4-*d*][1,3]thiazine-2,5,7-trione **57** [65] (Scheme 3.17). There is just one reference to the formation of a dihydrothiazine derivative **59** in the reaction of thiourea with an unsaturated ketone **58** under basic catalysis [68] (Scheme 3.18).

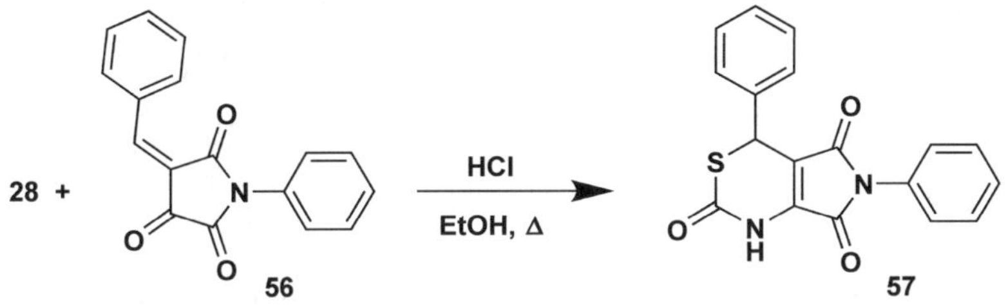


**Scheme 3.17**

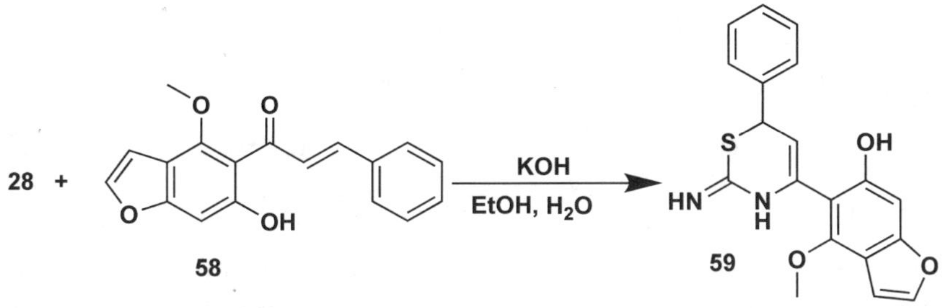


**Scheme 3.18**

Reactions of unsaturated ketones **62** containing a strong σ electron-withdrawing trifluoromethyl group with urea and its analogues **28**, **60** and **61** usually stop at the stage of the formation of hydroxytetrahydropyrimidines **63** [69] (Scheme 3.19).

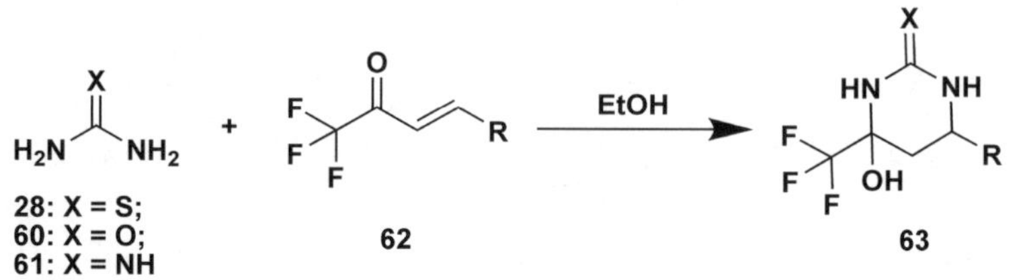


**Scheme 3.19**

There have been investigations of the reaction of cyclic thioureas with unsaturated carbonyls. For example, interactions involving imidazolidine-2-thione **64** ($n = 1$) or tetrahydropyrimidine-2-thione **64** ($n = 2$) with ketones **65** ($k = 1–3$) in the presence of the catalyst boron trifluoroetherate yielded only one stereoisomer **66** according to Perjesi et al. [70] (Scheme 3.20).

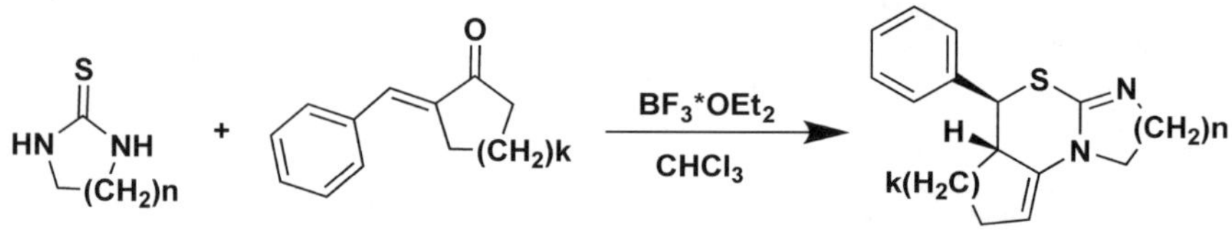


**Scheme 3.20**

Another cyclic thiourea—4,5-dihydrothiazole-2-ylamine **67**—reacted with unsaturated heterocyclic carbonyls **68**, yielding tricyclic compounds **69** [71] (Scheme 3.21).

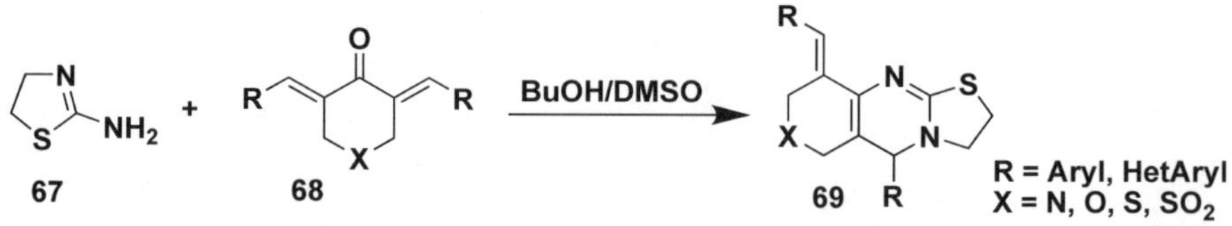


**Scheme 3.21**

Reactions of unsaturated ketones with urea **60**, as already mentioned, have been much less investigated and require stronger conditions. In such processes, instead of using alkaline metal hydroxides, a catalyst like sodium ethoxide

[72, 73, 74] is usually used. The use of acidic catalysis–concentrated sulfuric [75] and hydrochloric [76] acids—has also been also described. In this case, the yields of the target dihydropyrimidines were lower than those for thiourea reactions. As examples of reactions involving urea it should be noted that chalcone **70** [73] and cyclic unsaturated ketone **72** [74] produce pyrimidinones **71** and **73**, respectively (Scheme 3.22).

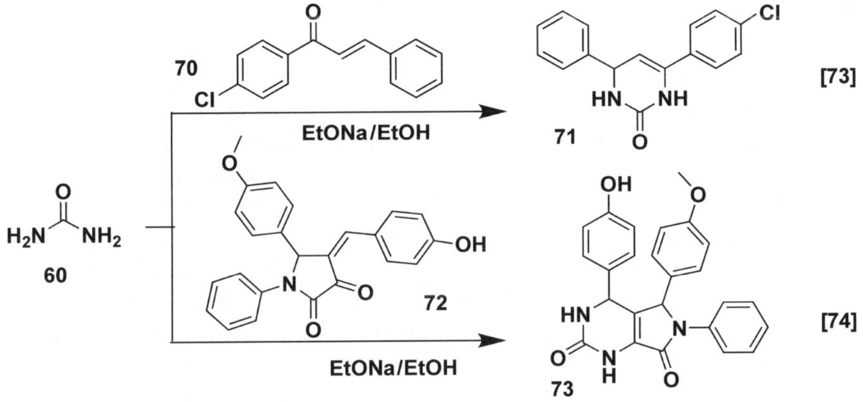


**Scheme 3.22**

Basic catalysis of the reactions of urea with unsaturated ketones, as well as thiourea, can lead to heteroaromatization with the formation of oxidized pyrimidinones instead of their dihydro analogues [77] (Scheme 3.23).

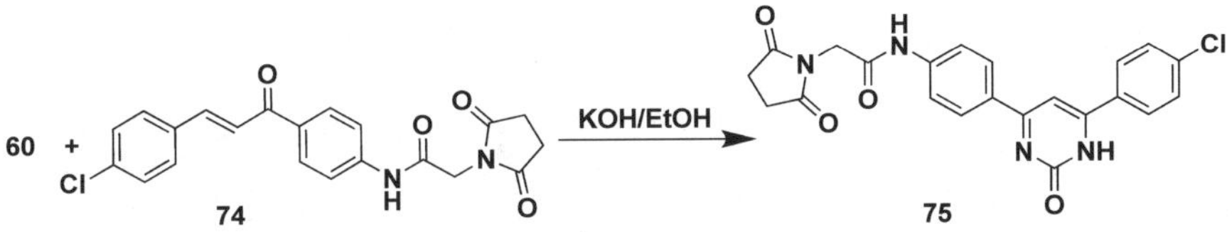


**Scheme 3.23**

Another representative of urea-like binucleophiles is guanidine and its derivatives. There are dozens of publications devoted to their reactions with α,β-unsaturated carbonyls, which are usually very simple processes. For example, routine magnetic stirring of guanidine **61** with chalcone **5** at room temperature in ethanol led to the formation of 1,6-dihydropyrimidin-2-ylamine **76** [78] with good yields (Scheme 3.24). But in several publications it was mentioned that such reactions may be complicated by side reactions and isolation of the target dihydropyrimidines is difficult. Very often among

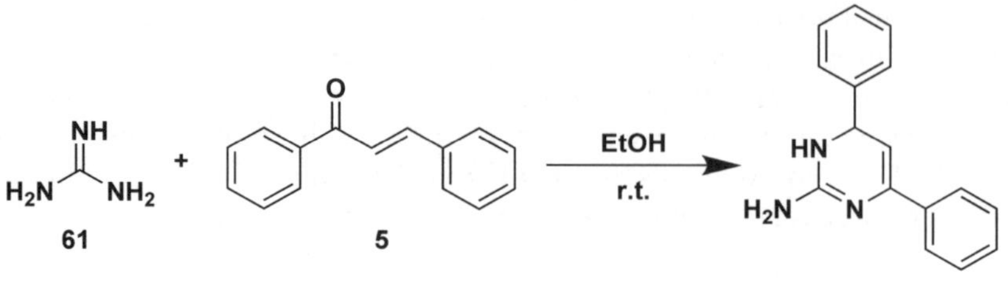


Scheme 3.24

competitive reactions oxidation by oxygen in air occurs [79, 80, 81, 82, 83]. The most detailed studies of this oxidation problem were carried out by El-Rayyes [79] and Wendelin and Schermanz [82]. In particular, in [82] it was shown that the reaction products can vary over a wide range. Treatment of guanidines **77** (Scheme 3.25) with benzylideneacetone **24** in benzene under a nitrogen atmosphere yielded 4-hydroxy derivative **78**, which rapidly oxidized to pyrimidine **79** in air. Stirring **78** in methanol at room temperature led to the formation of 6-methoxy-6-methyl-4-phenyl-1,4,5,6-tetrahydropyrimidine-2-amines **80**. Briefly heating (5–10 min) the starting materials in methanol

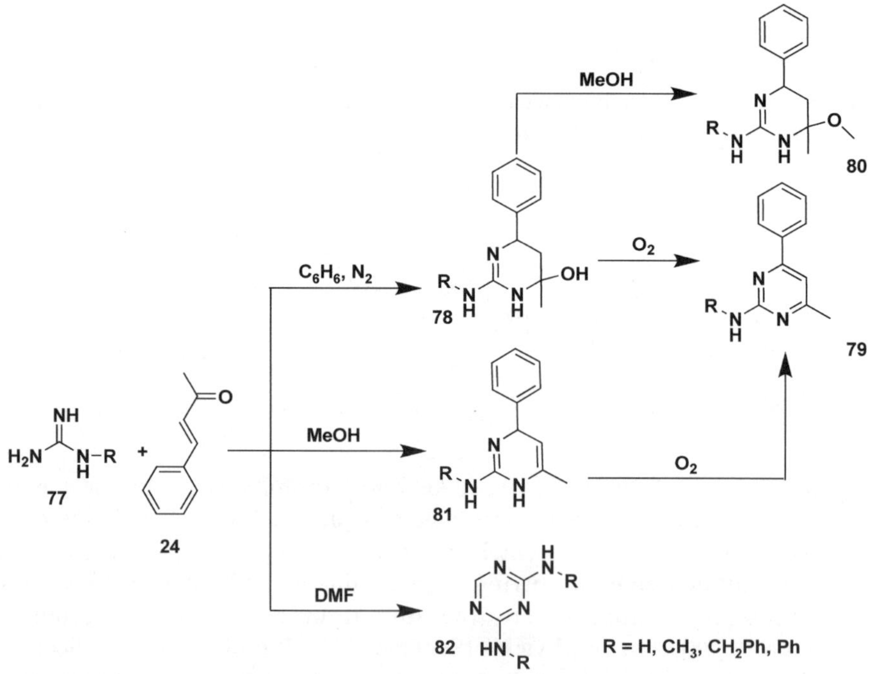


Scheme 3.25

resulted in the formation of 1,4-dihydropyrimidine-2-amines **81**, which were easily heteroaromatized by oxygen in air into **79**. At the same time, after compounds **77** and **24** had been refluxed in dimethylformamide (DMF), only 2,4-diamine-1,3,5-triazine **82** was generated from two molecules of guanidine and a product of DMF decomposition.

Dihydropyrimidines obtained in the reactions of guanidine and α,β-unsaturated carbonyl compounds can further react with the starting reagents [81, 84, 85]. For example (Scheme 3.26), Wendelin et al. [85] showed the possibility of several such processes: treatment of **83**, formed

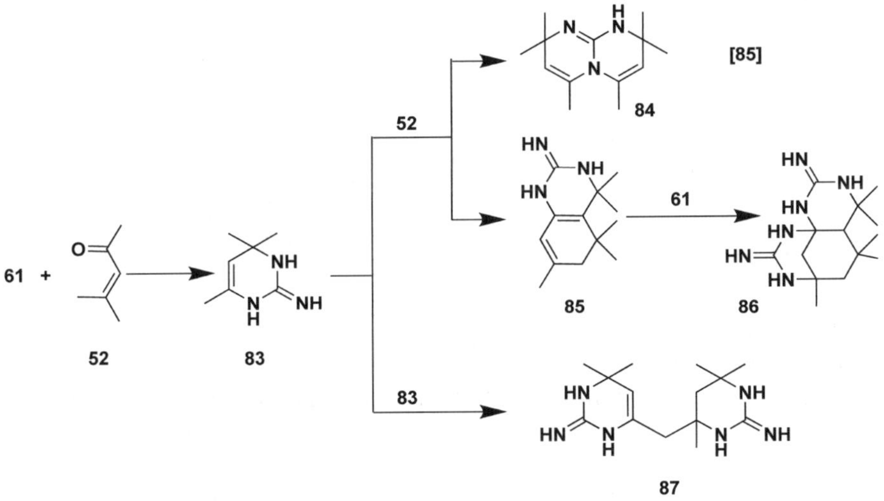


**Scheme 3.26**

from guanidine **61** and mesityl oxide **52**, with another molecule of unsaturated ketone yielded pyrimido[1,2-*a*]pyrimidine **84** or 3,4,5,6-tetrahydro-1*H*-quinazolin-2-ylideneamine **85**. The latter can further react with guanidine and leads to a tricyclic compound **86**. It is also possible to add a methyl group to one molecule of **83** and to a carbon–carbon double bond of another to form compound **87**.

Amidines and thioamidines can also be viewed as urea-like binucleophilic compounds. There are a series of publications about their treatment with unsaturated carbonyls [7, 86, 87, 88, 89, 90, 91, 92]. The most general reaction products in this case are 1,4-dihydropyrimidines or 1,6-dihydropyrimidines. Interaction of acetamidine **88** with mesityl oxide **52** [86] and benzamidine **90** with arylidenetetraline **91** [91] led to the corresponding 1,6-dihydropyrimidines **89** and **92**, while the reaction of bezamidine and benzylideneacetone **24** yielded 1,4-dihydropyrimidine **93** [87] (Scheme 3.27).

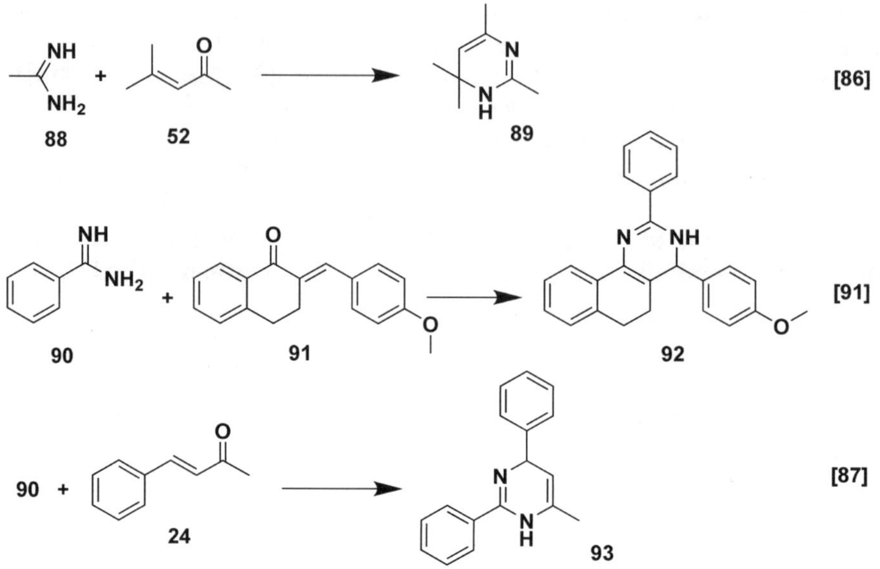


Scheme 3.27

The presence of an *ortho*-hydroxyphenyl substituent in an α,β-unsaturated ketone (e.g., **94**) can complicate the reaction with intramolecular addition of a hydroxyl group to the double bond of dihydripyrimidine [93] (Scheme 3.28).

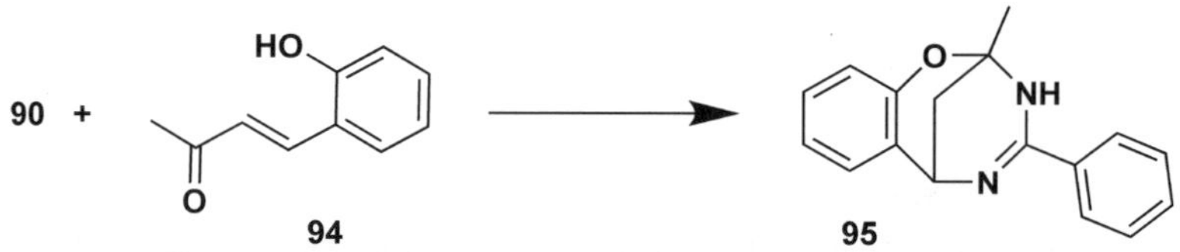


Scheme 3.28

Unlike amidines, the multicomponent reaction of α,β-unsaturated ketones **96** (aliphatic [94] or aromatic [95, 96]) with carbonyl compounds **97** and ammonia, which are the synthetic precursors of amidines, yielded 1,2,5,6-tetrahydropyrimidines **98** instead of dihydroheterocycles. When $R_3$ is not the same as $R_4$ tetrahydropyrimidines **98** were mixtures of diastereomers **A** and **B**, in which the relative configurations of stereogenic centers were also established [95, 96]. In contrast to conventional mechanical shaking requiring about 48 h [95], sonicated reactions were completed within 90 min at room temperature and provided the target heterocycles in high yields and purities [96]. Ultrasonic irradiation also significantly expanded the possibilities of such three-component reactions (Scheme 3.29).

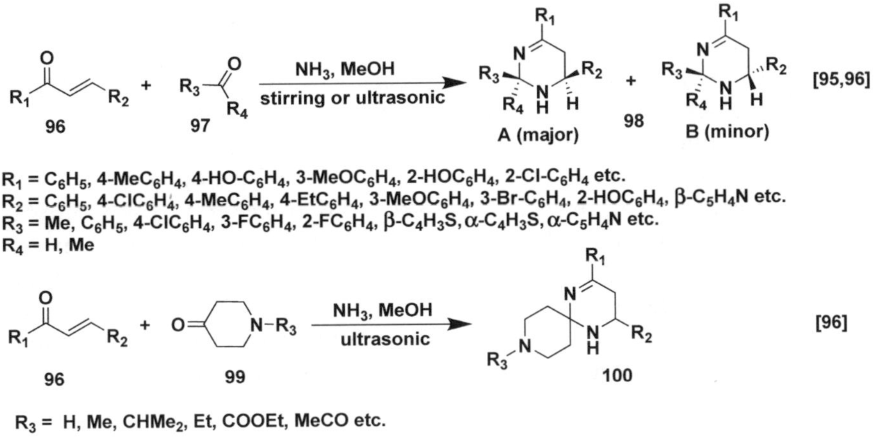


**Scheme 3.29**

Reactions of unsaturated ketones with thioamidines **101** were usually carried out with difficulty and to form thiazines requires the presence of dehydrating agents in the reaction mixture [88, 90]. Without a dehydrating agent, water is not eliminated and the reaction products are 5,6-dihydro-1,3-thiazin-4-oles **102** [89] (Scheme 3.30).

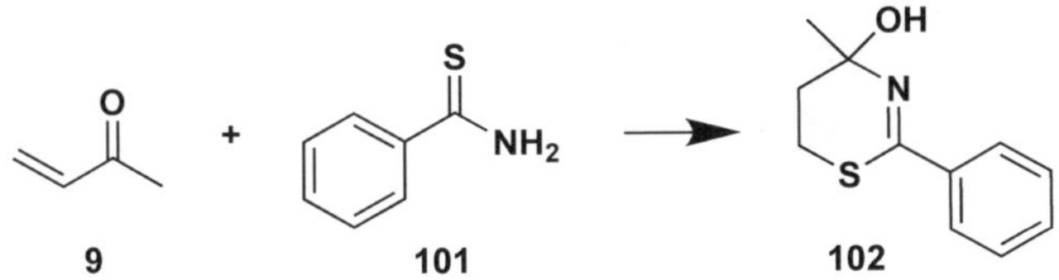


**Scheme 3.30**

The mechanism of the reaction of urea and its analogues with unsaturated carbonyls is described in several publications [97, 98, 99, 100]. The first stage of the reaction mechanisms discussed is the β-addition of one of the nucleophilic centers of the urea-like compound to the enone system irrespective of catalyst type (Scheme 3.31). A sequence of stages of 1,2- or 1,4-addition was not rigorously proven, but the low reactivity of urea and its analogues towards carbonyl compounds indirectly confirms a preference for the β-addition stage. Cyclocondensation of the β-adduct involving the carbonyl group is an intramolecular process and is easier as the reaction centers become closer.

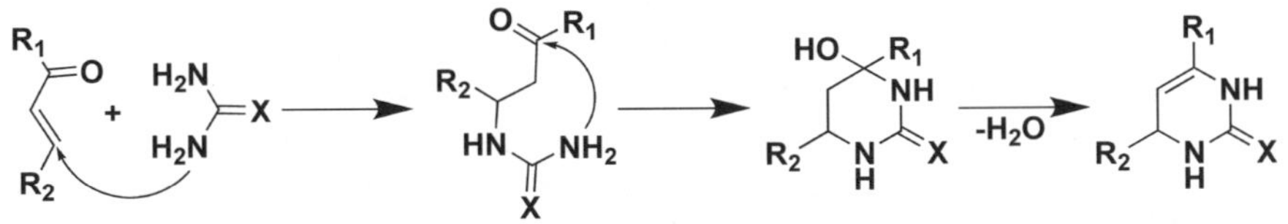


**Scheme 3.31**

There are a series of communications about the formation of dihydroazines by direct reaction of urea-like compounds with synthetic precursors of unsaturated carbonyls—ketones, containing an activated methyl or methylene group. The reaction products formed in this case are usually identical to the heterocycles obtained in reactions of the same binuclephiles with α,β-unsaturated ketones. For example, interaction of 2 equiv of acetophenone **103** with urea under acidic catalysis yielded 6-methyl-4,6-diphenyl-2-oxi-1,6-dihydropyrimidine **106** and two products of the self-condensation of acetophenone—dipnone **104** and 1,3,5-triphenylbenzene **105** [100] (Scheme 3.32). When urea was absent from the reaction mixture or substituted with 1,3-dimethylurea, the only isolated product was dipnon **104**. In addition, ketone **104** and urea in a multicomponent reaction form the same pyrimidine derivative **106**. All these facts suggest mechanism for the heterocyclization shown in Scheme 3.32.

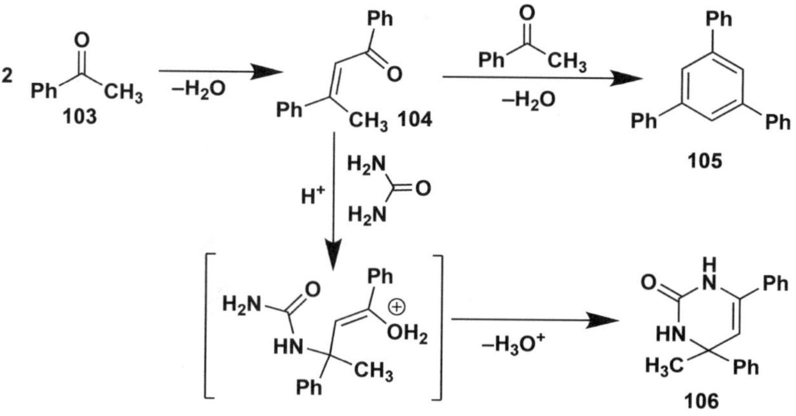


**Scheme 3.32**

The reactions of thiourea and guanidine with cycloalkanones are carried out in the same manner [101]. But it is worthwhile noting that muticomponent reactions of urea-like compounds with ketones, containing an activated methyl or methylenegroup, often do not stop after the formation of a pyrimidine ring. In fact numerous derived condensation processes can lead to more complicated polycyclic compounds, which are especially typical for the reaction of cycloalkanones [101].

## 3.3 Reactions of Malonic Acid Derivatives

The 1,3-binucleophilic components of cyclocondensations can react with α,β-unsaturated carbonyl compounds containing a CH-acidic reaction center (Scheme 3.1, reaction e). Examples of these are derivatives of malonic acid (malononitrile, malonoamide, cyanoacetic acid, ethyl cyanoacetate, cyanoacetamide, cyanothioacetamide etc,), cyclic monoketones and diketones, and barbituric acid derivatives.

Most of the publications on the reactions of malonic acid derivatives are devoted to the reaction of α,β-unsaturated carbonyls with malononitrile **107**. In this book we do not describe the results of all known publications but only the most characteristic or interesting ones in our opinion. The reactions of unsaturated ketones with malononitrile are usually carried out in methanol or ethanol in the presence of ammonium acetate [102, 103, 104, 105, 106, 107, 108, 109, 110, 111, 112, 113, 114, 115, 116, 117]. Treatment of malononitrile **107** with chalcone **5** [115] and its heterocyclic analogue **109** [104] yielded 3-cyanopyridines **108** and **110** (Scheme 3.33).

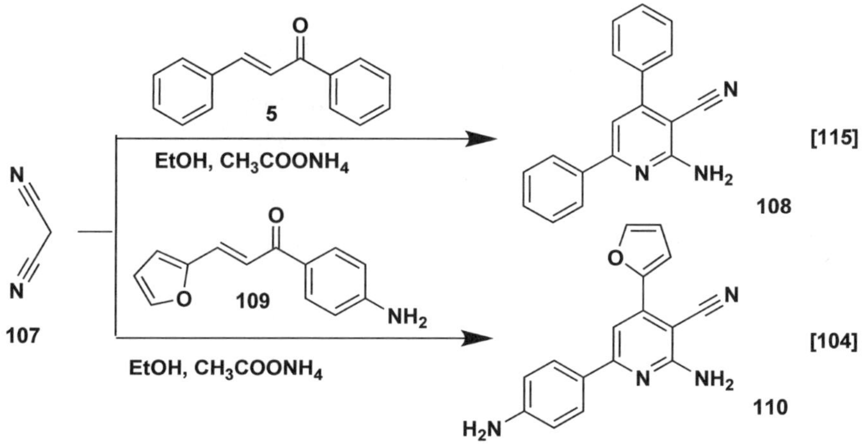


**Scheme 3.33**

In several cases the application of other basic catalysts is described, but such reactions often are followed by numerous side reactions or proceed in another direction. For example, in the reaction of malononitrile **107** with dimethylaminochalcone **111** in the presence of pyrrolidine as the catalyst, instead of the expected cyanopyridines being formed the heterocycle 5-amino-7-(pyrrolidin-1-yl)-1,6-naphthyridine-8-carbonitrile **112** [118] is produced (Scheme 3.34).

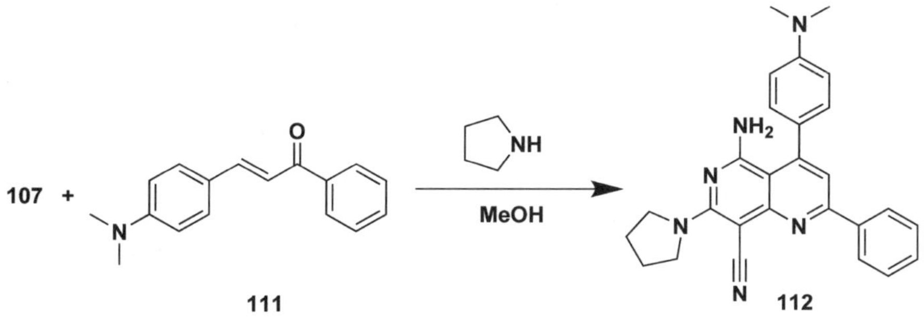


**Scheme 3.34**

This reaction involves 1 equiv of chalcone and pyrrolidine and two molecules of malononitrile. The structure of compound **112** was unambiguously established only with help of X-ray analysis [119].

The reactions using some strong bases (sodium hydroxide [120, 121, 122, 123, 124] or sodium ethoxide [125]) are accompanied by nucleophilic substitution of the amino group at the 2 position of the cyanopyridine (Scheme 3.35).

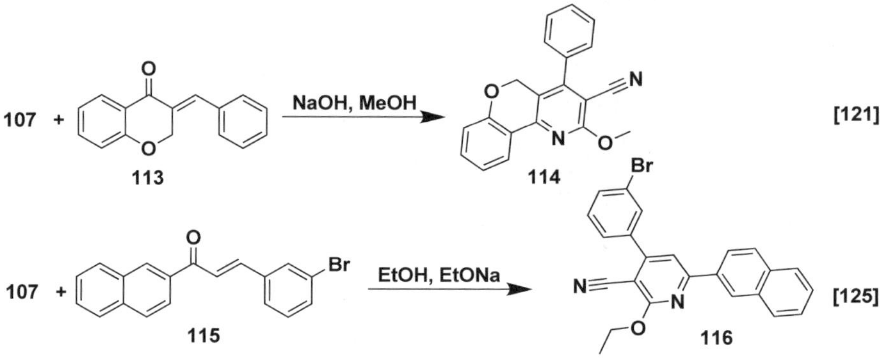


**Scheme 3.35**

The nitrile group in cyanopyridine can participate in intramolecular reactions. The presence in chalcone **118** of an *ortho*-hydroxyl substituent can lead to its addition to a carbon–nitrogen triple bond, with the formation of the 2*H*-pyran-2-iminic moiety (compound **119**) [126, 127] (Scheme 3.36).

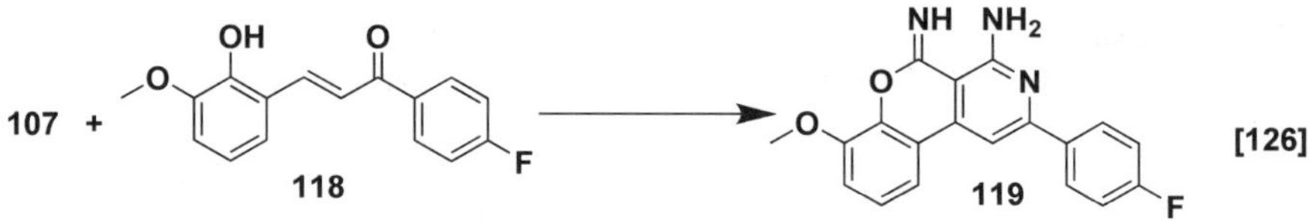


**Scheme 3.36**

The reaction products of the treatment of α,β-unsaturated ketones with malononitrile in most cases are the expected derivatives of 2-amino-3-cyanopyridine. But an alternate direction with the formation of carbocyclic compounds **120** was described [128]. It is clear from structure **120** that two molecules of unsaturated ketone **5** and 1 equiv of malononitrile **107** are involved in the reaction (Scheme 3.37).

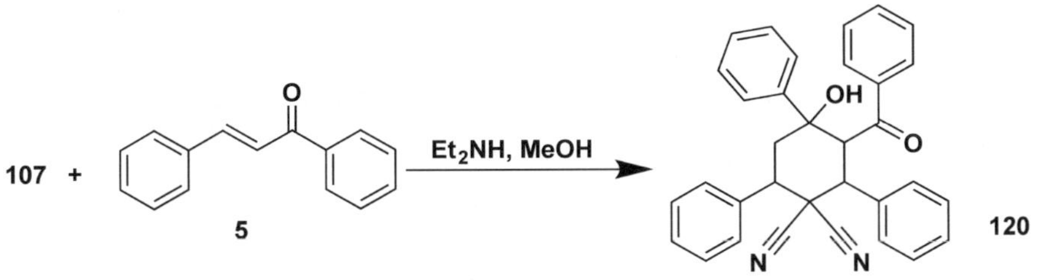


**Scheme 3.37**

The reaction of α,β-unsaturated carbonyls with other derivatives of malonic acid, for example, with malonoamide **121** [129, 130] and malonothioamide **124** [131, 132], were also described.

Treatments of malonoamide **121** with unsaturated ketones are carried out using a basic catalyst (sodium methoxide and ethoxide). At room temperature these reactions lead to 2-oxo-1,2,3,4-tetrahydro-3-pyridinecarboxamide like **122** (Scheme 3.38). Heating of the reaction mixture is followed by further addition of the carboxamide group to the carbon–carbon double bond, resulting in the formation of 2,7-diazabicycle [2.2.2]octan-3,8-dione **123** [130].

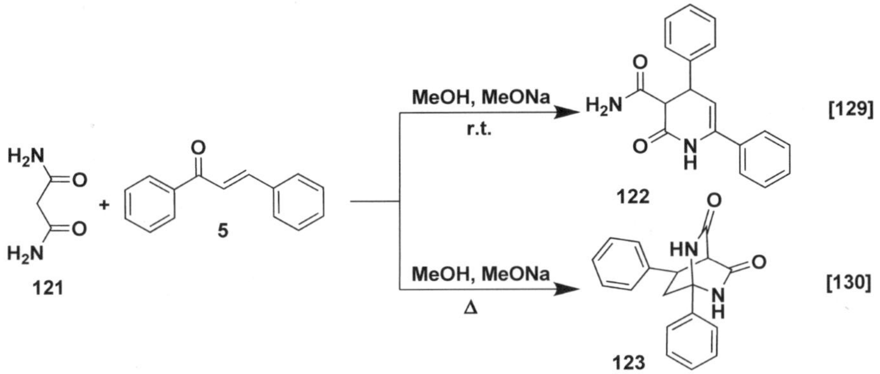


**Scheme 3.38**

Under sodium methoxide catalysis the reactions of α,β-unsaturated ketones with malonothioamide **124** are chemoselective and yielded the expected 2-oxo-1,2,3,4-tetrahydropyridine-3-carbothioamide **126**, while 2-thioxo-1,2,3,4-tetrahydropyridine-3-carboxamides **127** were not observed [131,132] (Scheme 3.39). For reaction in the presence of piperidine, Krauze et al. [132] isolated and described an intermediate of the reaction—6-hydroxy-2-oxopiperidine-3-carbothioamide **125** (Scheme 3.39).

A lot of publications are devoted to the reactions of α,β-unsaturated carbonyls with derivatives of cyanoacetic acid: cyanoacetamide **128**, cyanothioacetamide

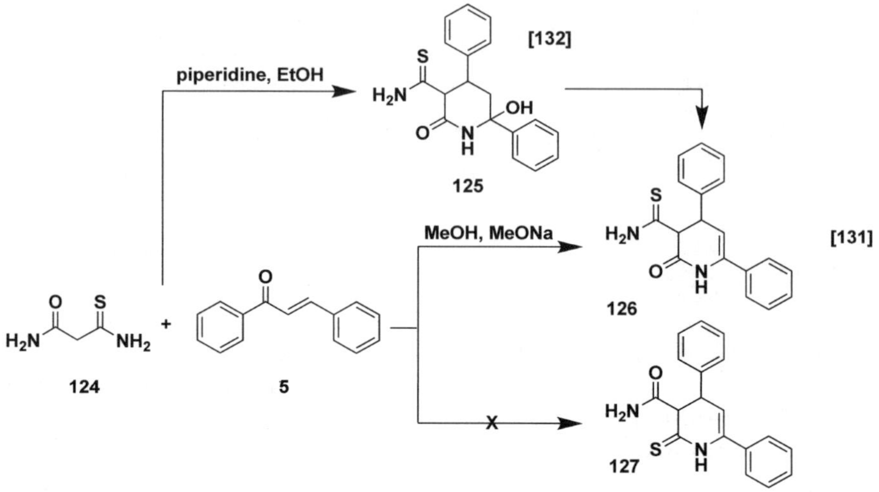


Scheme 3.39

**130** and ethyl cyanoacetate **133**. Also, all these CH-acids can be considered as derivatives of malonic acids. The reactions of **128**, **130** and **133** with unsaturated ketones **5**, **31** and **134**, as expected, yielded 2-oxo-1,2,3,4-tetrahydropyridine-3-carbonitriles or 2-thioxo-1,2,3,4-tetrahydropyridine-3-carbonitriles **129, 132** and **135**, respectively (Scheme 3.40). Usually the catalysts used are sodium methoxide or ethoxide [133, 134, 135, 136, 137, 138, 139, 140, 141, 142], ammonium acetate, [114, 143, 144], triethylamine or piperidine [145, 146, 147, 148, 149].

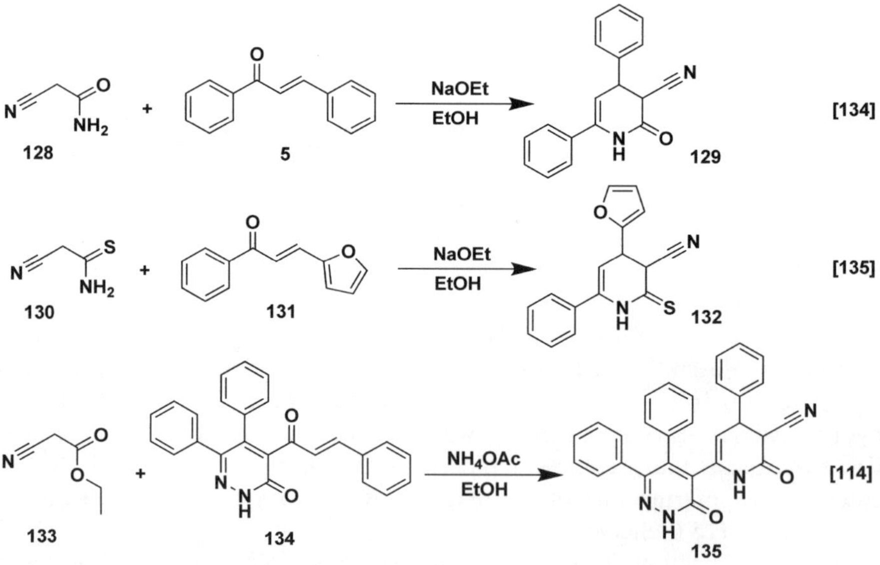


Scheme 3.40

Sometimes the reaction is followed by oxidation and leads to 1,2-dihydropyridines instead of their 1,2,3,4-tetrahydro analogues. As often as not, such a situation is observed in the case of cyclic unsaturated ketones in the presence of triethylamine or piperidine [137, 138, 139, 140, 141, 142, 144, 147, 148, 149, 150, 151] (Scheme 3.41). The possibility of such reactions occurring instead of α,β-

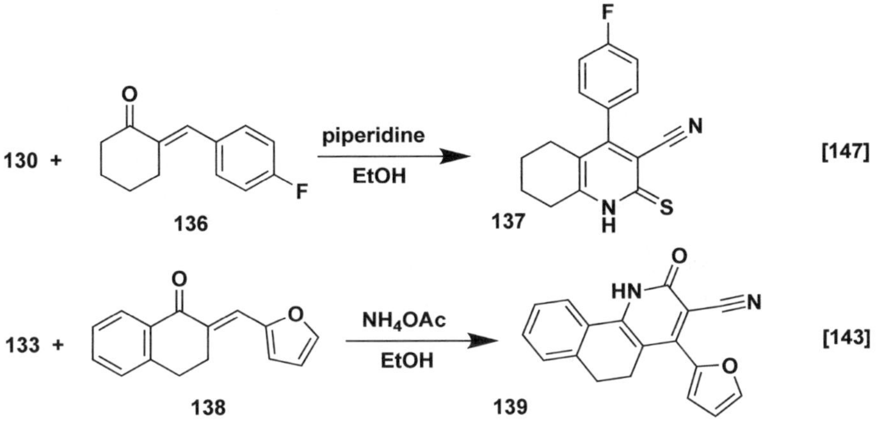


Scheme 3.41

unsaturated ketones and their synthetic precursors was shown in [152]. The authors described a multicomponent protocol involving aldehydes **131**, indan-1,3-diaones **132** and cyanothioacetamide **130** leading to 5-oxo-4,5-dihydro-1*H*-indeno[1,2-*b*]pyridine-3-carbonitriles **133** in high yields (Scheme 3.42).

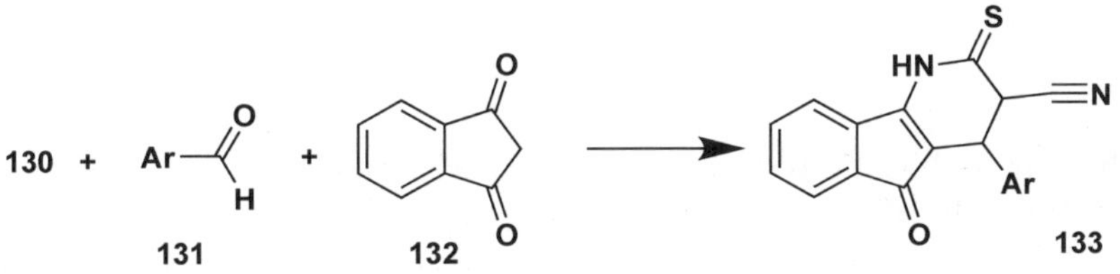


Scheme 3.42

Some assumptions about a sequence of reactions of unsaturated ketones and cyanoacetic acid derivatives can be drawn from the results of [146], which described several reaction products **134**–**140** and their isolation (Scheme 3.43). In the same publication it was also established that the presence of an excess of α,β-unsaturated ketone **24** in the reaction mixture led to the addition of the initially formed tetrahydropyridine to the ethylene bond of the enone system.

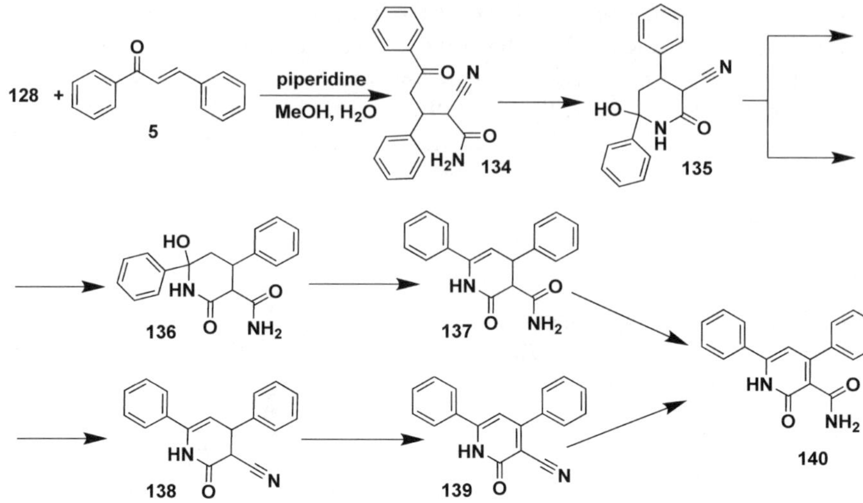


Scheme 3.43

Further, adduct **141** can intramolecularly react in the presence of sodium ethoxide to form the bicyclic product **142** (Scheme 3.44).

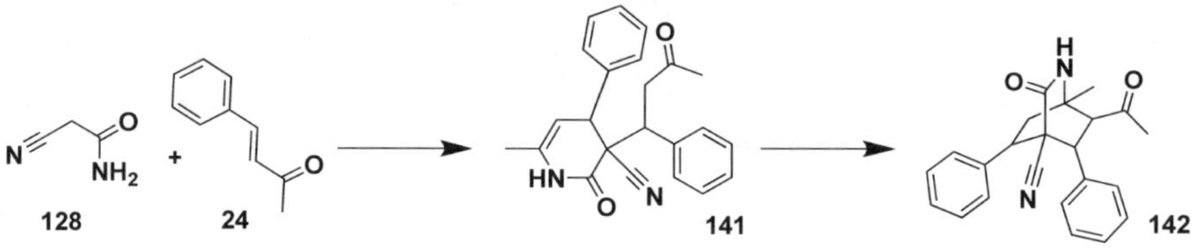


Scheme 3.44

There are publications devoted to a study of the mechanism of the treatment of unsaturated carbonyls with cyanoacetic acid derivatives [134, 153]. According to these communications, the sequences of such reactions are similar to those of urea-like binucleophiles. Using piperidine as the catalyst, Al-Hajjar and Jarrar [134] and Otto [153] isolated a series of known intermediates.

The mechanism of the reactions of aromatic unsaturated ketones with malonamide was described in [154]. The expected product of β-addition—δ-ketamide—was isolated and fully characterized. The proposed mechanism of the formation of the terahydropyridone cycle in [154] was similar to one suggested in [134, 153] for cyanamide. The first stage of the reaction is an addition of the CH-acid to the ethylene bond of the unsaturated ketone and generation of the β-adduct, which can easily undergo a cyclization leading to dihydro derivatives of 2-pyridone **146** (Scheme 3.45).

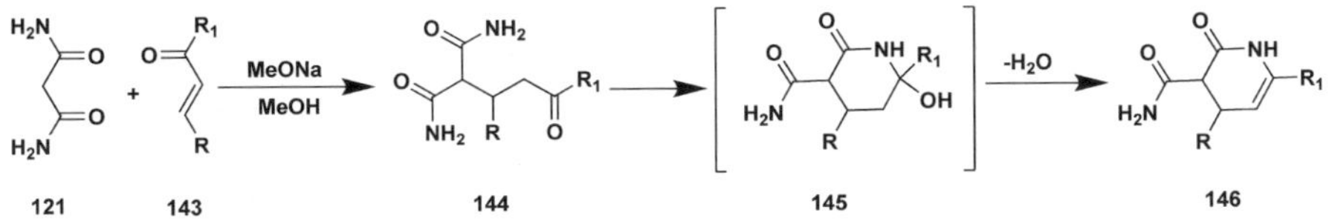


**Scheme 3.45**

## 3.4 Reactions of Aminoazoles

Cyclocondensation of aminoazoles and α,β-unsaturated carbonyl compounds or Mannich bases is the most common method for the synthesis of dihydroazolopyrimidines [99, 155, 156, 157]. Various alkyl- and aryl-substituted dihydropyrimidines were prepared in this way. For example, cyclocondensation of 3-amino-1,2,4-triazole **147** with chalcones **148** leads to 5,7-diaryl-4,7-dihydro-1,2,4-triazolo[1,5-*a*]pyrimidines **149** [158], while reaction with hydrochlorides of Mannich bases **150** leads to heterocycles **151** (Scheme 3.46).

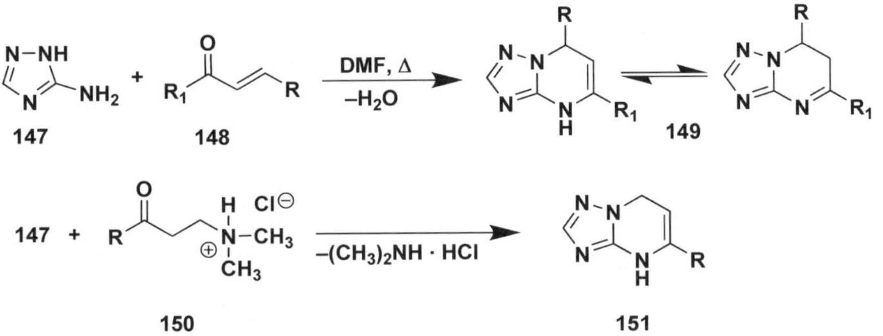


R = $C_6H_5$, $R_1$ = $C_6H_5$; R = 4-$CH_3C_6H_4$; $R_1$ = $C_6H_5$; R = $C_6H_5$, 4-$CH_3OC_6H_4$;
R = $C_6H_5$, 4-$(CH_3)_2NC_6H_4$; R = 4-$ClC_6H_5$, 4-$ClC_6H_4$; R = $C_6H_5$, 4-$NO_2C_6H_4$;

**Scheme 3.46**

The common character of cyclocondensations involving α,β-unsaturated carbonyls and their synthetic equivalents with aminoazoles was described in several publications. The reaction of chalcones **152** and the corresponding Mannich bases **153** with 2-aminobenzimidazole **154** yields 1,4-dihydrobenzo[4,5]imidazo[1,2-*a*]pyrimidines **158** [159], with 5-aminopyrazoles **155** yields 4,7-dihydropyrazolo[1,5-*a*]pyrimidines **159** [160, 161], with 5-aminotetrazole **156** yields 4,7-dihydrotetrazolo[1,5-*a*]pyrimidines **160** [162] and with 2-aminoimidazoles **157** yields 5,8-dihydroimidazo[1,2-*a*]pyrimidines **161** [163] (Scheme 3.47).

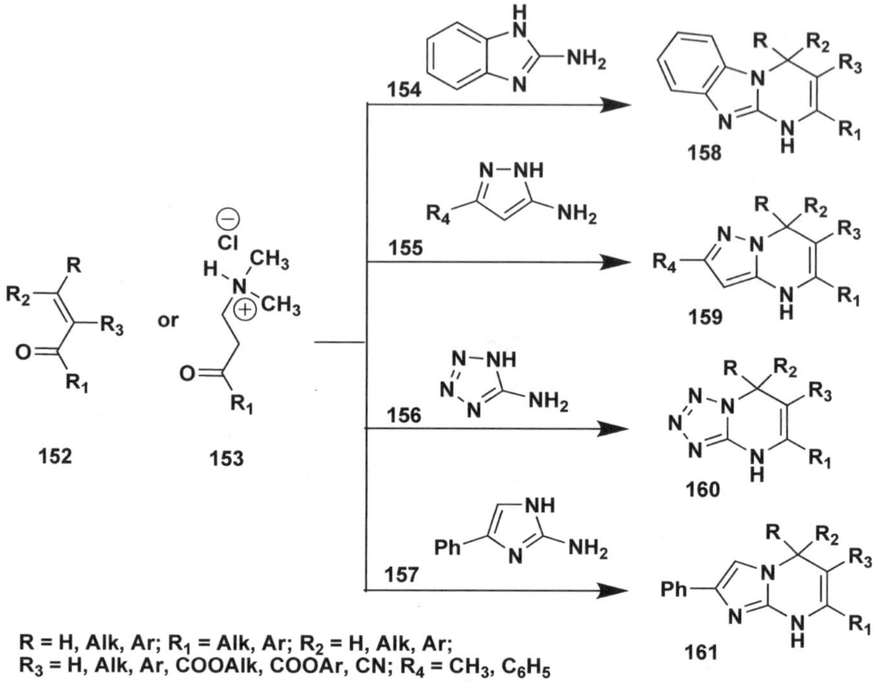


**Scheme 3.47**

In contrast to the analogous reaction of amines with β-diketones [164], cyclocondensations involving unsaturated ketones are characterized by high regioselectivity, which makes it possible to obtain compounds where R is not the same as $R^1$ even with a low degree of differentiation of the electronic properties of these substituents. The formation of pyrimidine rings corresponds to the reaction of a β-carbon of the enone with the endocyclic nitrogen of the aminoazole and a carbonyl group with an exocyclic amino group ("anti-Scraup" direction).

If there are several nonequivalent endocyclic reaction centers in the aminoazole molecule, a question arises about regioselectivity of the condensation. Some difficulties in the determination of the isomer structures in these cases are illustrated, for example, in [165, 166]: the product of the reaction of 2-aminoindole **162** with chalcone initially was assigned the structure of a pyrimidoindole **163** [165], but in a subsequent publication another arrangement of α-carboline **165** was proven (Scheme 3.48).

This particular experimental fact allowed Kost [164] to speculate that carbon atoms were more preferable endocyclic reaction centers in reactions with unsaturated ketones than nitrogens. But applying this generalization to other aminoazoles is not correct because of a dramatic difference of the nucleophilicity of the pyrrole-type nitrogen of indole and pyrimidine-type polyazoles. Indeed, the

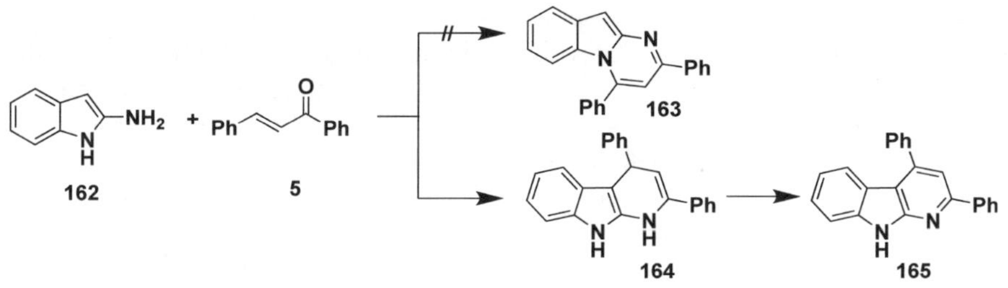


**Scheme 3.48**

reactions of N-unsubstituted derivatives of pyrazole **166** with unsaturated ketones **167** yield pyrimidine **169** but not the pyridine derivatives **168** [160, 161, 167, 168] (Scheme 3.49).

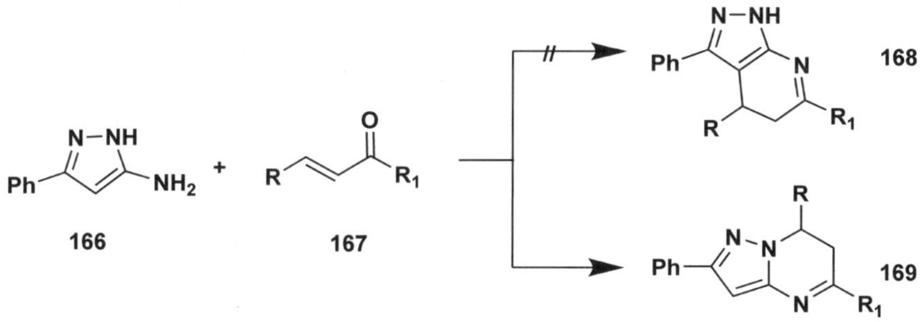


**Scheme 3.49**

Thus, if there are several nonequivalent endocyclic reaction centers in the aminoazole molecule (Scheme 3.50), then usually more nucleophilic ones participate in the interaction with the carbonyl compound (for example, $N_{(2)}$ in 3-amino-1,2,4-triazole or 5-aminopyrazole [158, 160]).

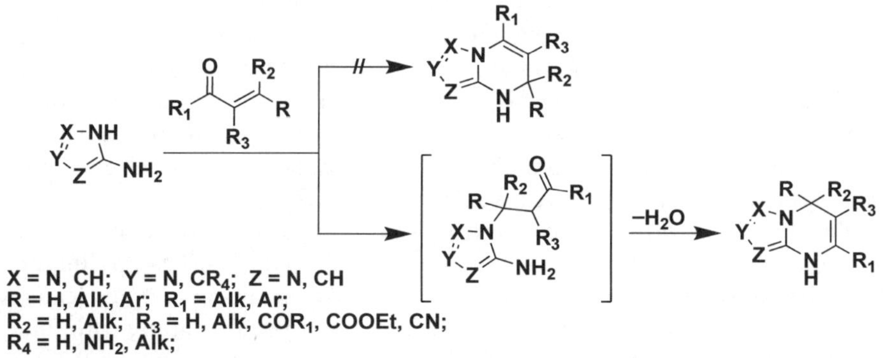


**Scheme 3.50**

There are several exceptions in the literature. One of them is a reaction of 3-amino-1,2,4-triazole **147** with 1,4-diphenylbut-2-ene-1,4-dione (dibenzoylethylene) **170** [169]: the first stage—alkylation of the amine with α,β-unsaturated ketone—is carried out with participation of all possible endocyclic reaction centers of the aminoazole (Scheme 3.51).

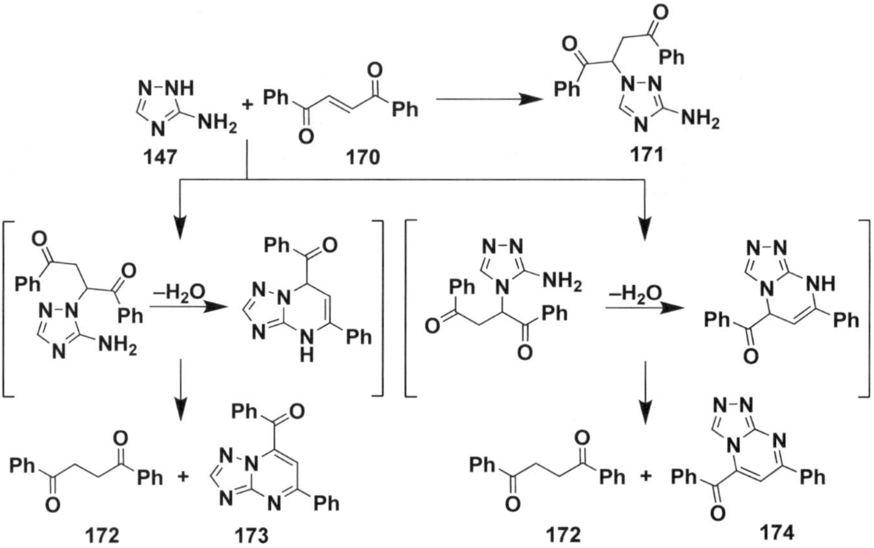


**Scheme 3.51**

In addition, the ability of dibenzoylethylene to act as a hydride ion acceptor became apparent in the formation β-adduct **171**, dibenzoylethane **172** and triazolopyrimidines **173** and **174**, while their dihydro analogues were not observed as reaction products. Another example, involving pyruvic acid derivatives and 3-amino-1,2,4-triazole, will be shown below.

A process of heteroaromatization was also observed in reactions of some iminoazoles with chalcones containing a nitro substituent [159], arylidenecycloalkanones [170] and during treatment of α,β-unsaturated ketones with 3,5-diaminotriazoles [155]. In some cases reactions carried out under ambient conditions (in air) become complicated by the oxidation of dihydroazolopyrimidines to their hydroxy derivatives [168].

This ability to form hydroxy derivatives is more characteristic for dihydropyrimido[1,2-*a*]benzimidazoles **158** and dihydropyrazolo[1,5-*a*]pyrimidines **159**. Storing these solutions in chloroform, DMF or alcohols in air leads to the appropriate hydroxy derivatives **176** as a major product along with minor compounds **177** and **178** [158, 171, 172, 173, 174, 175] (Scheme 3.52). Often heterocycles like **176** were the only products of oxidation in air. Surprisingly,

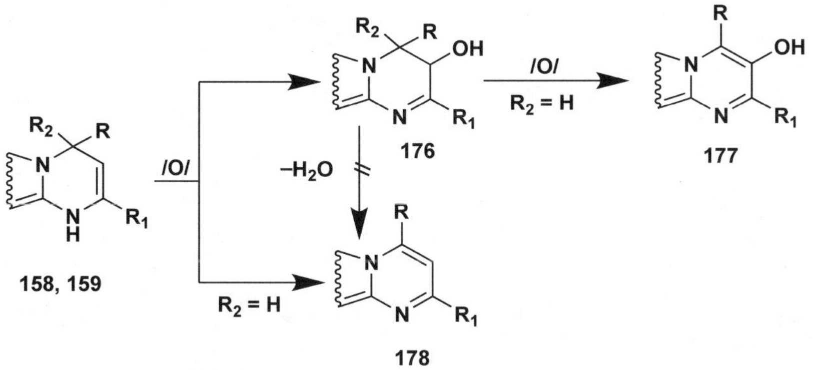


Scheme 3.52

compounds **176** showed high stability towards dehydration agents when $R_2$ was H [158, 171, 172, 173, 174, 175]. This interesting fact was explained by specific geometry of the CH–COH fragment of the dihydropyrimidine ring, which is closest to the *cis* orientation and thereby obstructs the dehydration process.

A possibility of formation of 5,7-diaryl-4,7-dihydro-1,2,4-triazolo[1,5-*a*]pyrimidines **182** in reactions of 3-amino-l,2.4-triazole **147** with synthetic precursors of unsaturated ketones—aldehydes **179** and ketones **180**—or in condensation of acetophenones **180** and azomethynes **183** was established in [176, 177]. It was shown that in the first case the multicomponent procedure was not an independent method but resulted in the initial formation of unsaturated ketones **181** themselves and their further reaction with aminoazol (Scheme 3.53).

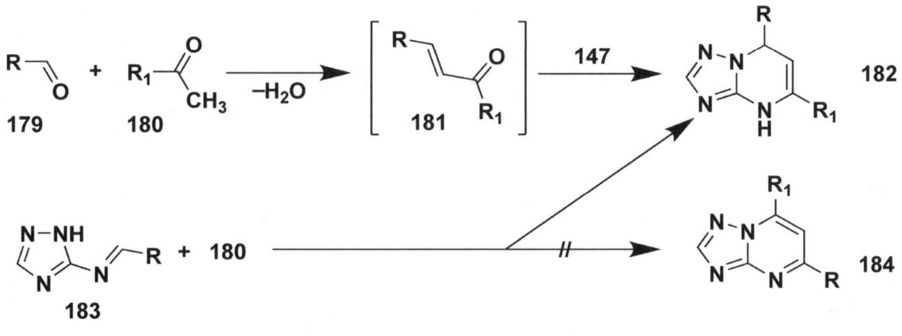


Scheme 3.53

Another situation was observed in the reaction of aminotriazole **147** with benzaldehyde **185** and benzocycloalkanones **186** [176] (Scheme 3.54): in addition to the major product 4,7-dihydro-l,2,4-triazolo[l,5-*a*]pyrimidines **187** in yields of 15–25%, isomeric compounds **188** and/or products of their

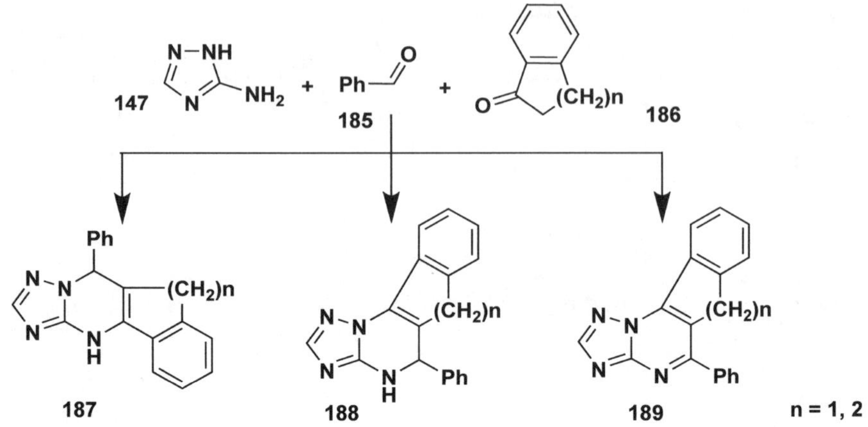


Scheme 3.54

dehydrogenation **189** were also isolated. It is clear that structures **187** and **188** (**189**) demonstrate the different potential directions of cyclocondensation.

The formation of mixtures of 4,5-dihydro and 4,7-dihydro isomers **191** and **192** was observed in the reaction of 3-amino-1,2,4-triazole **147** with methyl aryl ketone **190** in the presence of $ZnCl_2$ as a catalyst [178] (Scheme 3.55). But the

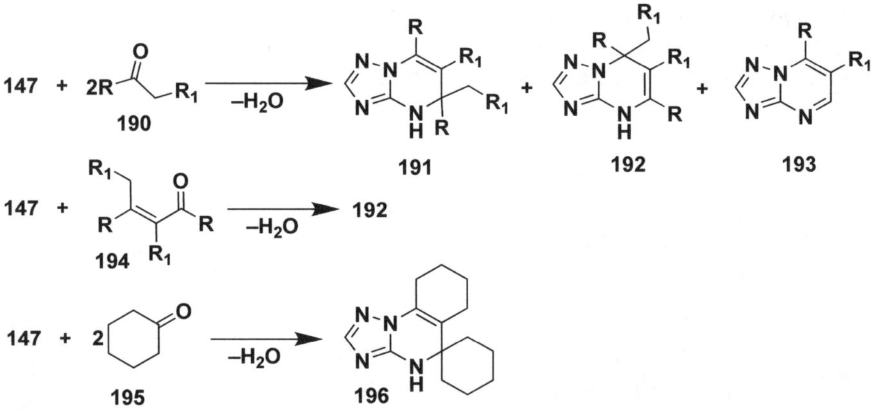


Scheme 3.55

interaction of **147** with two molecules of acetophenone under acidic catalysis (acetic or mineral acids) yielded only 4,5-dihydro derivatives **191** and heterocycles **193** [179]. In the latter case, the third component of the multicomponent condensation was the solvent—DMF.

The multicomponent reaction of **147** with 2 mol of cycloalkanones **195** follows in a very similar manner [180, 181]. Subject to the nature of the carbonyl component, either derivatives of 4,5-dihydro-1,2,4-triazolo[1,5-*a*]pyrimidine **196** (in the reaction with cyclohexanone [180]) or heteroaromatized compounds **193** [R and $R_1$ are $(CH_2)_4$] were obtained in reactions with benzocycloalkanones in DMF [181].

Taking into account that self-condensation of two molecules of ketones cannot be the first step of the formation of dihydrotriazolopyrimidines **191** and **196** (treatment of unsaturated ketones **190** with aminoazoles leads exclusively to heterocycles **192**) led to two alternative mechanisms (A and B) being suggested in [178, 179] (Scheme 3.56). In the eyes of Desenko et al. [178], the key intermediate in the formation of the dihydropyrimidine ring is azomethines (pathway A), while in the work reported in [179] enamines were formed (pathway B).

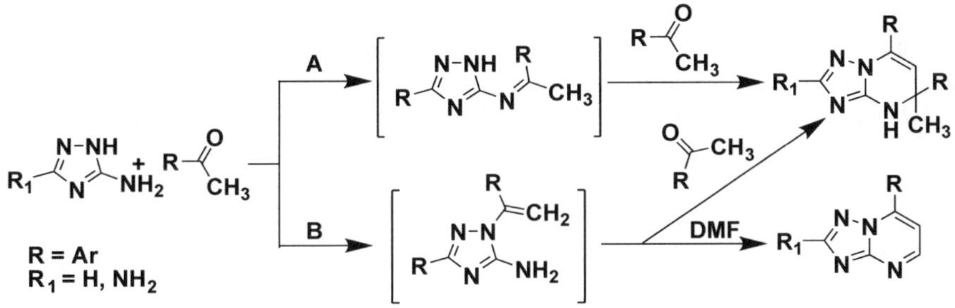


**Scheme 3.56**

In our opinion, mechanism B is more probable for the formation of compound **193** because the reaction product corresponds to the cyclization of the enemine intermediate but not azomethine. Indirect evidence against mechanism A is the fact that azomethines like **183**, as mentioned above, in the reaction with acetophenones act as benzaldehyde "producers" and do not directly undergo cyclization.

Thus, in three-component reactions of aminoazoles with synthetic precursors of α,β-unsaturated ketones, two contrary directions of the dihydrocycle formation corresponding to different sequences of cyclocondensation stages are possible and can be realized (Scheme 3.57, reactions I and II). It is worth noting

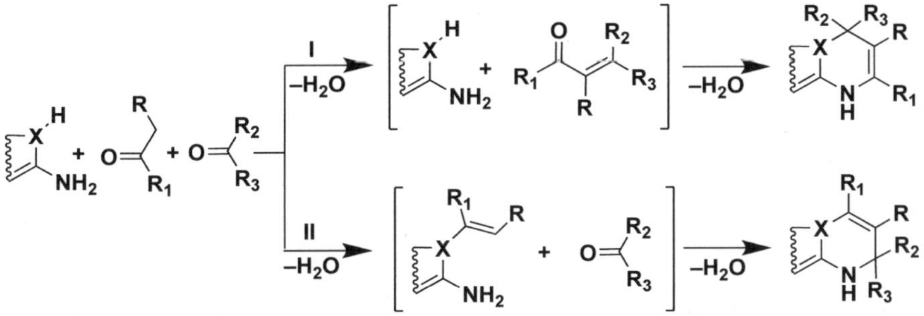


**Scheme 3.57**

again that the reaction carried out according to pathway I is not an independent method for the formation of the dihydroazine system and corresponds to the normal treatment of α,β-unsaturated carbonyls because the generation of the latter occurs in situ. In contrast, reaction II follows a different mechanism, leading to 1,2-dihydroisomers of azines, which are hard to synthesize by other methods.

Reactions of 3-amino-1,2,4-triazole **147** [182], 3-amino-5-methylthio-1,2,4-triazole **197** [183], 3,5-diamino-1,2,4-triazole **198** [184], 2-aminobenzimidazole **154** [185]and 4-substituted 5-aminopyrazoles **206–208** [186, 187] with arylidene-1,3-cyclonones **199** or their synthetic precursors— dimedone **200** and aromatic aldehydes **201**—have general characteristics that lead to the formation of fused pyrimidine rings. According to [182, 183, 184, 185, 186, 187] a positional direction of these interactions is always predictable, i.e., aldehyde reacts with an endocyclic nitrogen atom, but not with the amino group, yielding compounds **202–204**, **205** and **209–211**, while formation of angular heterocycles like **212** is not observed (Scheme 3.58).

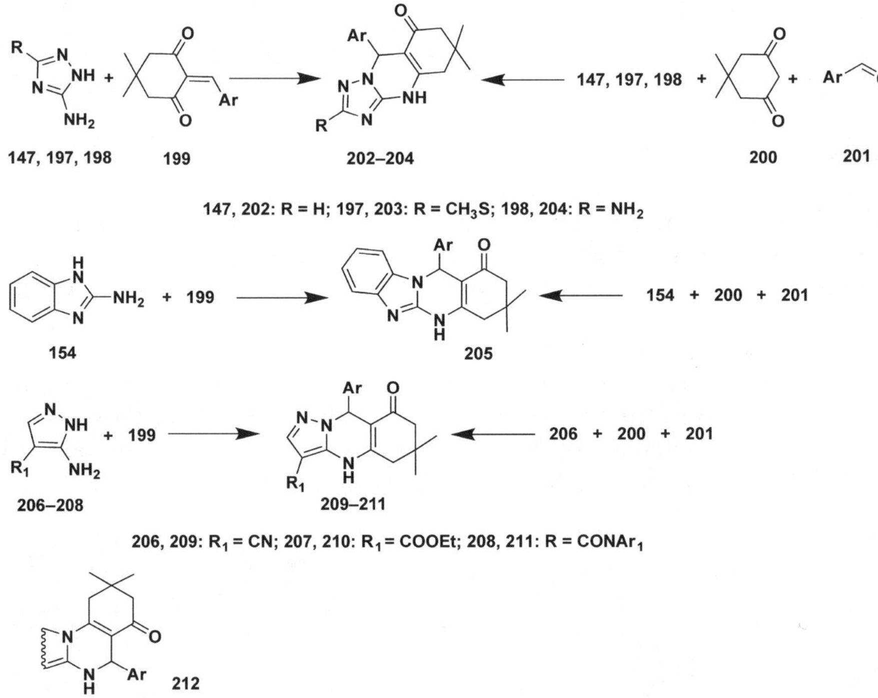


**Scheme 3.58**

More recently this last conclusion was disproved in [188], where the formation of the angular compound **215** (minor) as well as the linear one **214** (major) in the reaction of 5-amino-*N*-phenyl-1*H*-pyrazole-4-carboxamide **213** with

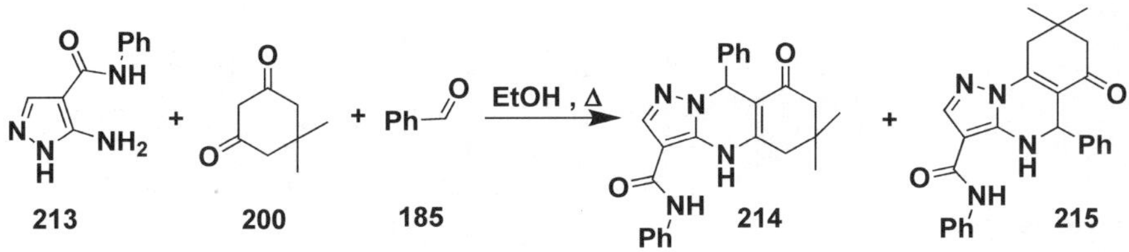


**Scheme 3.59**

dimedone **200** and benzaldehyde **185** (Scheme 3.59) was unambiguously established by means of X-ray analysis and NMR spectroscopy.

When one of the components of such multicomponent reactions was a molecule of DMF, the formation of angular dihydropyrimidines was also observed. For example, amines **197, 198** and **208** in boiling DMF reacted with dimedone **200**, leading to azolopyrimidines **217**–**219** [184, 186, 187]. These treatments, in the opinion of Lipson et al. [184], resulted in the reaction passing through an enamine intermediate **216** formed at the endocyclic nitrogen atom (Scheme 3.60).

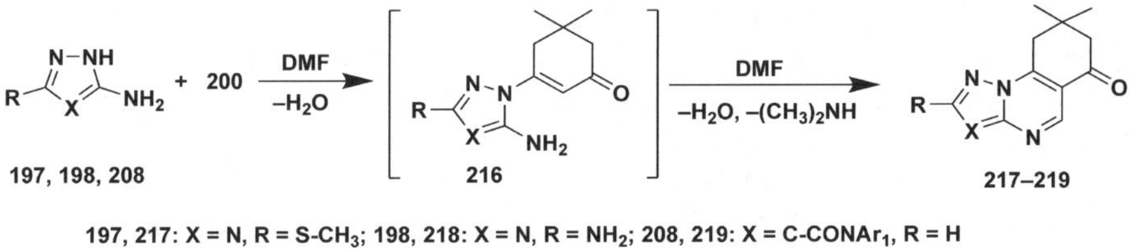


**Scheme 3.60**

The presence in the aminoazole molecule of nonequivalent endocyclic reaction centers can lead to several directions of the interaction as is observed in the case of multicomponent treatment of 5-aminopyrazoles **220**–**222** with dimedone **200** and aldehydes **201**. In [189, 190, 191] it was shown that reactions of amines **220** and **221** with 1,3-diketone **200** and aromatic aldehydes in boiling ethanol or DMF yielded fused pyridines **225** and **226** when R is $CH_3$ irrespective of the electronic nature of the substituent Ar (Scheme 3.61).

Quiroga et al. [189] suggested that the preliminary reactions generated the appropriate arylidenecyclodiones, which, in turn, reacted with aminopyrazole, forming the intermediates **223** and **224**. The cyclization of the latter with water elimination led to pyridine moiety formation and isolation of compounds **225** and **226** as reaction products (Scheme 3.61).

For 3-aryl-5-aminopyrazoles **222** the direction of the multicomponent reaction with cyclic β-diketones and aromatic aldehydes is not so unequivocal. Quiroga et al. [190] found that during refluxing of these starting materials in ethanol, the reaction products were only derivatives of pyrazoloquinoline **227**,

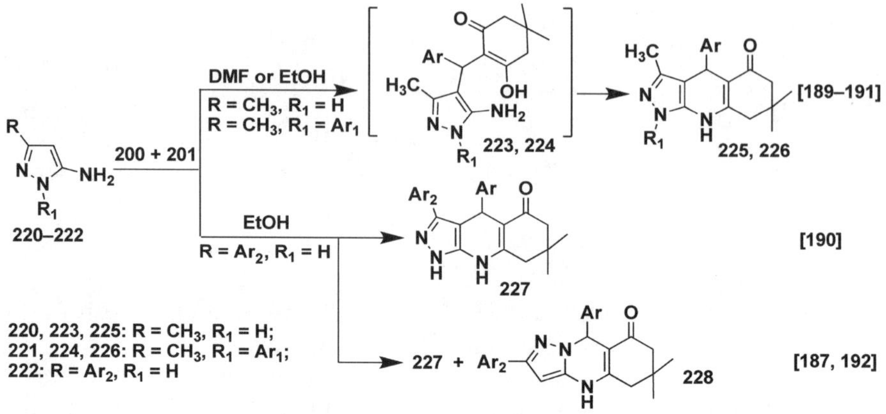


Scheme 3.61

while in [187, 192] under the same reaction conditions the formation of a mixture containing **227** and pyrazoloquinazoline **228** was established. The fused pyrimidines **228** were either major regioisomers or the only products.

A regioselective synthesis of pyrazoloquinolines **225** by the multicomponent reaction of 3-aryl-5-aminopyrazole **222** with 1,3-diketones and aromatic aldehydes in ethanol in the presence of $Et_3N$ under microwave irradiation at 150 °C (Scheme 3.62, reaction i) was described in [193, 194]. In the presence of a strong base such as EtONa (KOH), the three-component treatment proceeded via a different pathway and led to novel and unusual reaction products—quinolizinones **226** (Scheme 3.62, reaction ii) [193]. The formation of only one diastereomeric pair from two possible ones with *trans* relative configuration of the

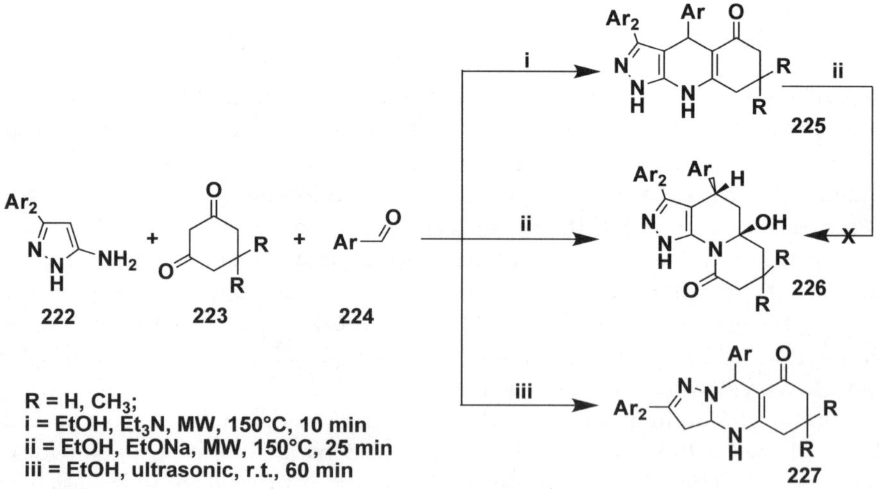


Scheme 3.62

stereogenic centers was observed. But it was shown that compounds **226** are not products of the rearrangement of Hantzsch-type pyridines **225**.

Although the exact reaction mechanism for this three-component condensation reaction was not confirmed in [193, 194], the hypothesized mechanism is likely to involve the initial base-catalyzed formation of the Michael adduct, and its subsequent reaction with the aminopyrazole component to furnish the tricyclic intermediate (Scheme 3.63). Elimination of water from this intermediate leads to the formation of the classic Hantzsch-type dihydropyridine

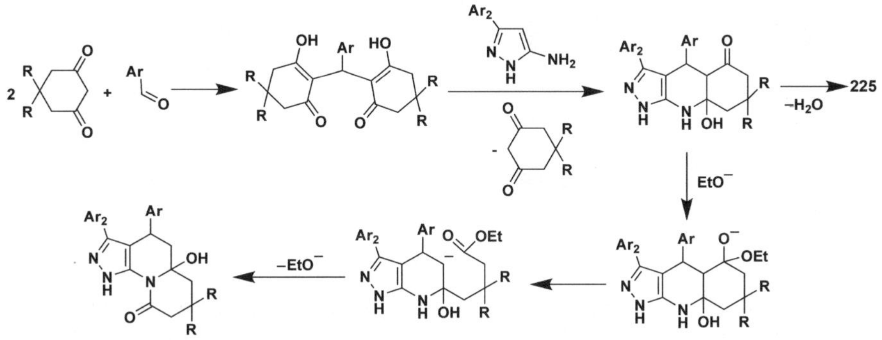


**Scheme 3.63**

derivatives **225** (an observed byproduct), while after nuceolphilic attack of the ethoxide (or hydroxide) ion on the carbonyl group, the intermediate apparently undergoes ring opening and recyclization to the quinolizinones **226**.

A regioselective synthesis of pyrazoloquinazoline **227** was also described [189]. The reaction was carried out under sonication of an equimolar mixture of starting materials in ethanol at room temperature (Scheme 3.62, reaction iii). Thus, compounds **227** are the kinetically controlled reaction products, while pyridines **225** are the thermodynamically preffered ones.

Three-component reactions of aromatic aldehydes with a series of cyclic CH-acids, for example, with **230**, and 5-amino-3-pyrazolones **228** have a high regioselectivity involving the carbon atom at position 4 and yield compounds like **231** [195] (Scheme 3.64).

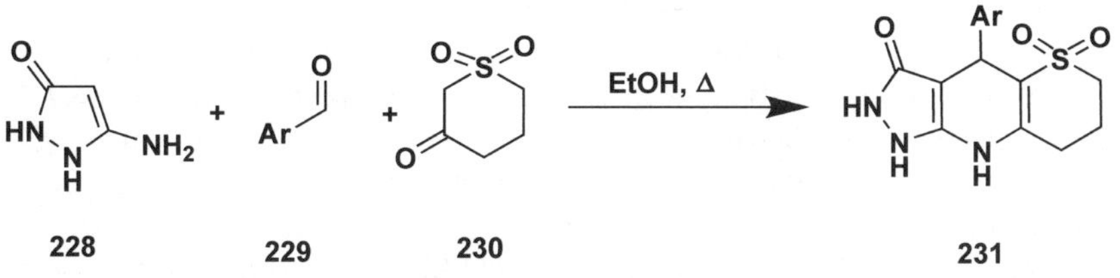


**Scheme 3.64**

The multicomponent reactions of noncyclic ketosulfones with 3-amino-1,2,4-triazole and 2-aminobenzoimidazole were described in [196]. It was found (Scheme 3.65) that the microwave-assisted three-component condensation

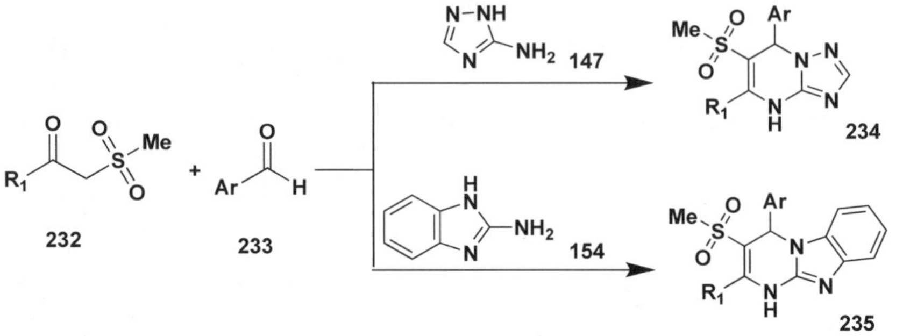


$R_1$ = Me, Ph
Ar = Ph, 4-$MeC_6H_4$, 4-$MeOC_6H_4$, 4-$FC_6H_4$, 4-$ClC_6H_4$, 4-$BrC_6H_4$, 3-$MeOC_6H_4$, 2-$MeOC_6H_4$

**Scheme 3.65**

of **147** and **154** with methylsulfonylacetone **232** ($R_1$ is Me) or $\alpha$-methylsulfonylacetophenone **232** ($R_1$ is Ph), and the appropriate aromatic aldehydes **233** in DMF at 135 °C for 30 min led to the formation of the corresponding 4,7-dihydro[1,2,4]triazolo[1,5-*a*]pyrimidines **234** or 5,8-dihydroimidazolo[1,2-*a*]-pyrimidines **235**, respectively, while the same reaction with 5-aminotetrazole was unsuccessful. An explanation for this unreactivity according to [196] could be due to the decrease in nucleophilicity of 3-amino-1,2,4-triazole and 2-aminobenzimidazole compared with that of 5-aminotetrazole. The same observations regarding the reactivity characteristics for these aminoazoles had already been reported [197].

Thus, reactions of α,β-unsaturated ketones and their synthetic precursors with aminoazoles can occur in several directions and lead to different reaction products. The reaction direction and the structure of the compounds obtained depends on five factors—solvent and catalyst used, electronic nature of the substituents, the reaction activation type (conventional heating, microwave field, ultrasonic irradiation) and temperature.

An interesting example for achieving different pathways in reactions of aminoazoles with unsaturated carbonyls is heterocyclization involving pyruvic acids. The use of pyruvic acid and its derivatives in the synthesis of nitrogen-containing heterocycles was first described at the beginning of the twentieth century in [198]. For a long time such heterocyclizations were limited to the synthesis of quinoline carboxylic acids based on aniline reactions (see Sect. 3.6), while reactions with other binucleophiles were not found in the literature. The first reaction of arylidenepyruvic acid with aminoazole was described in [199]. The authors showed that treatment of unsaturated ketoacids **236** with 5-aminotetrazole **156** or 2-aminobenzoimidazole **154** in boiling DMF led to yields of 23–62% for 5-aryl-5,8-dihydrotetrazolo[1,5-*a*]pyrimidine-7-carboxylic acid **237** and 4-aryl-1,4-dihydro[1,2-*a*]benzimidazol-2-carboxylic acid **238**, respectively (Scheme 3.66).

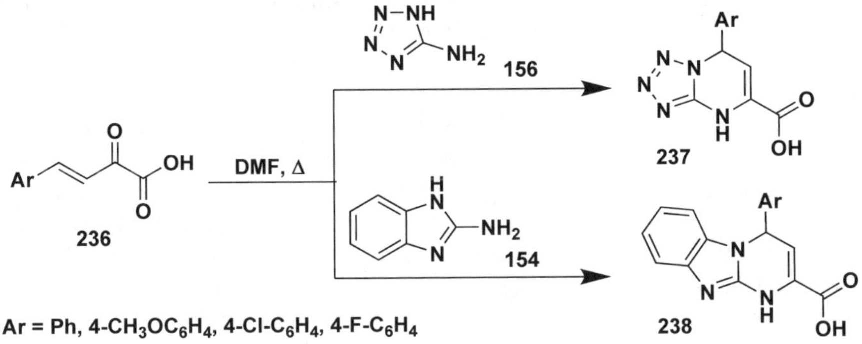


**Scheme 3.66**

However, the key disadvantage of the reaction described in [199] is observed in the first step—synthesis of unsaturated ketoacids **236** with very low yields (20–30%) [200]. The same authors [201] proposed a multicomponent approach to azolopyrimidinecarboxylic acids. The reactions of pyruvic acid **239** and aromatic aldehydes **240** with 2-aminobenzimidazole, 5-aminotetrazole **156** and 3-amino-1,2,4-triazole **147** in alcohols, DMF and acetic acid were studied (Scheme 3.67).

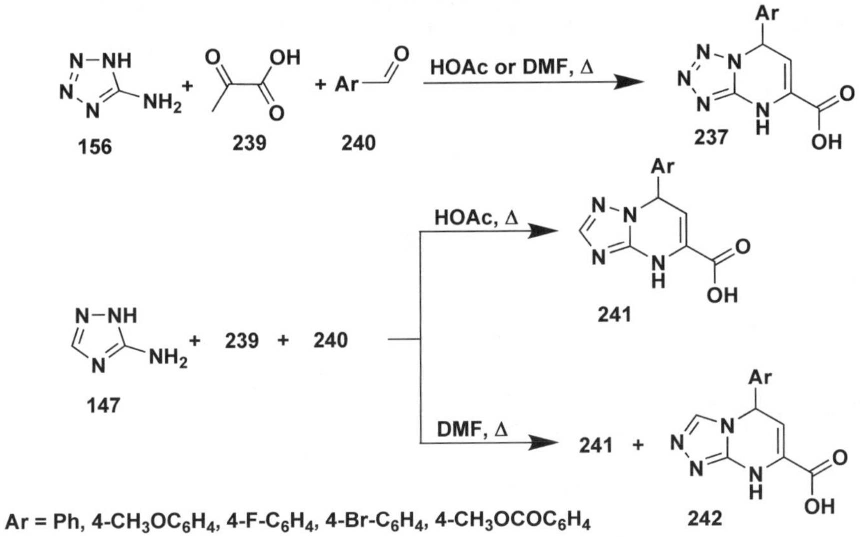


**Scheme 3.67**

It was found that the reaction of 2-aminobenzimidazole led to a hard-to-separate resinous mixture under all reaction conditions. Chebanov et al. [201] connected this fact with the inclination of 4-aryl-1,4-dihydropyrimido[1,2-*a*] benzoimidazole-2-carboxylic acids **238** to undergo oxidation.

But refluxing pyruvic acid **239** and aldehydes with 5-aminotetrazole **156** in DMF or glacial acetic acid yields high-purity carboxylic acids **237** in 54–65% yield. Three-component reactions involving 3-amino-1,2,4-triazole **147** in glacial acetic acid allowed the isolation from the reaction mixture of the pure acids **241** (51–65%), while refluxing of the starting compounds in DMF was not a regioselective process and led to mixtures of **241** and **242** (Scheme 3.67).

Reactions of pyruvic acids with 3-methyl-5-aminopyrazoles described in [202] are also interesting from the regioselectivity point of view (Scheme 3.68). The authors showed that reaction of arylidenepyruvic acids **236** with amine **220** (R is $CH_3$, $R_1$ is H) in DMF or in acetic acid led to the formation of a pyridine ring and allowed the isolation of 4-aryl-3-methylpyrazolo[3,4-*b*]pyridine-6-carboxylic acids **243** in satisfactory yields. However, heterocyclic products in these reactions were not isolable as dihydro derivatives, which indicates, in the opinion of Chebanov et al. [202], the high propensity of the expected 4,7-dihydropyrazolo[3,4-*b*]pyridine intermediates to undergo oxidation. Refluxing of the amine **221** (R is $CH_3$, $R_1$ is Ph) with arylidenepyruvic acids **236** in DMF also yielded heteroaromatized pyrazolo[3,4-*b*]pyridine-6-carboxylic acids **244**. However, treatment in boiling acetic acid led to the corresponding dihydro analogues **245** accompanied by only small amounts of the aromatized heterocycles **244**.

Treatment of arylidenepyruvic acids **236** with 5-amino-3-arylpyrazoles **222** (R is Ar, $R_1$ is H) in most cases was not regioselective and yielded mixtures of several regioisomers and products of their heteroaromatization: 2,7-diaryl-4,7-dihydropyrazolo[1,5-*a*]pyrimidine-5-carboxylic acids **246** (main product), 3,4-diaryl-4,7-dihydropyrazolo[3,4-*b*]pyridine-6-carboxylic acids **247** and their heteroaromatized derivatives **248** (Scheme 3.68).

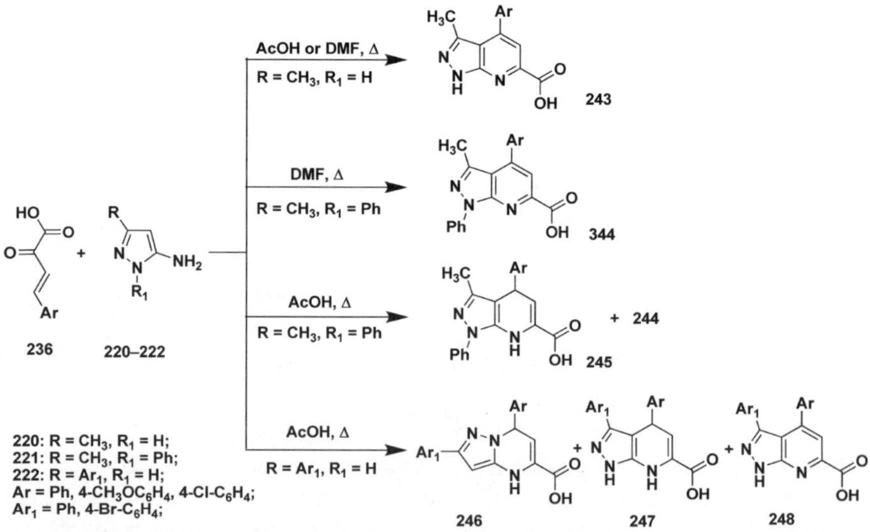


**Scheme 3.68**

Chebanov et al. [202] noted that condensation of the unsaturated acids **236** with 5-aminopyrazoles **220–222** never yielded isomers with opposite location of the aryl and carboxyl groups on the pyridine or pyrimidine rings, respectively. In the case of the multicomponent reaction of aminopyrazoles **220–222** with pyruvic acid **239** and aromatic aldehydes a different direction was observed. Refluxing of the starting materials in acetic acid led exclusively to pyrazolo[3,4-*b*]pyridine-4-carboxylic acids **249–251** instead of the anticipated carboxylic acids **243–248** (Scheme 3.69). The three-component procedures led only to the formation of heteroaromatized compounds even under a nitrogen atmosphere [202].

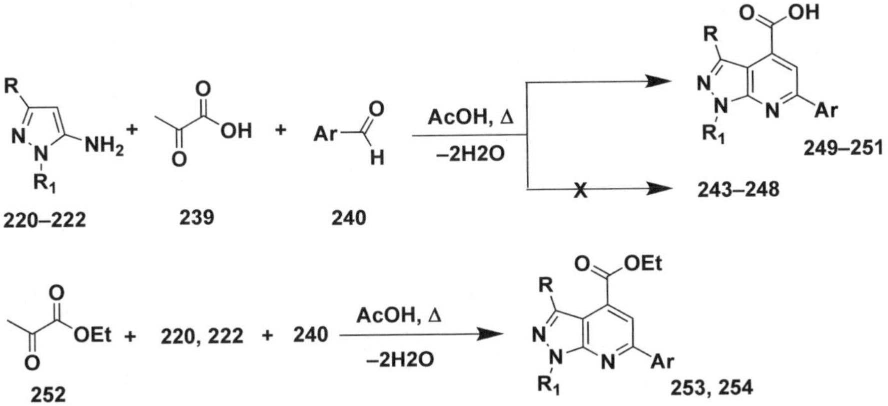


**220, 249, 253:** R = $CH_3$, $R_1$ = H;
**221, 250:** R = $CH_3$, $R_1$ = Ph;
**222, 252, 254:** R = $Ar_1$, $R_1$ = H;
Ar = Ph, 4-$CH_3$-$C_6H_4$, 4-$CH_3OC_6H_4$, 4-Cl-$C_6H_4$, 4-F-$C_6H_4$;
$Ar_1$ = Ph, 4-Br-$C_6H_4$;

**Scheme 3.69**

Thus, it is clear that in the case of 3-amino-1,2,4-triazole and 5-aminotetrazole, reactions with arylidenepyruvic acids and their synthetic precursors lead to identical reaction products [199, 201], while applying sequential or multicomponent procedures for interaction with 3-substituted 5-aminopyrazoles allows for the isolation of heterocycles of different structures [202]. However, it is interesting to note that, according to publication [202], reactions of 5-amino-*N*-arylpyrazolo-4-carboxamides **213** with arylidenepyruvic acids **236** or with pyruvic acid **239** and aldehydes **240** also yielded identical reaction products—pyrimidine heterocycles **255** (Scheme 3.70).

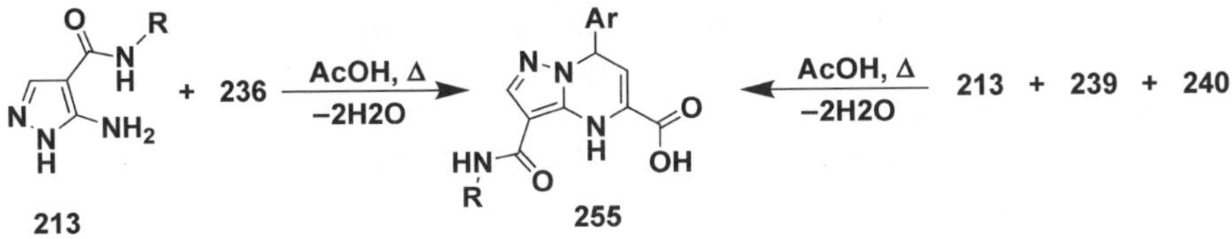


**Scheme 3.70**

The different reaction pathways described in [202] were rationalized from a mechanistic point of view. The reactions of arylidenepyruvic acids **236** with 3-substituted 5-aminopyrazoles most likely start with nucleophilic attack of the enone system of the unsaturated ketoacid at the β-position to the carbonyl group (Scheme 3.71). In the case of amines **220** and **221**, the nucleophilic center CH group at position 4 of aminopyrazole causes the formation of

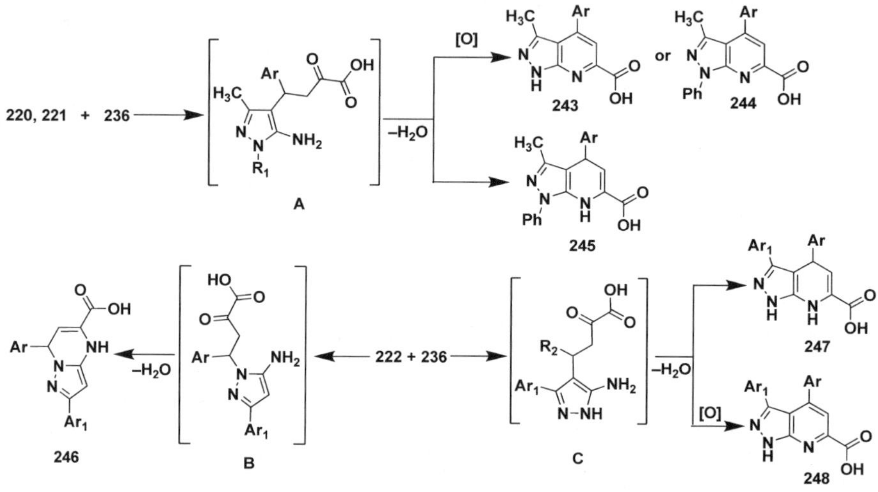


**Scheme 3.71**

intermediate A. On the other hand, in the case of 3-aryl-5-aminopyrazoles **222** the enone carbon–carbon double bond of the unsaturated ketoacid may be attacked both by the endocyclic nitrogen atom (intermediate **B**) and less likely by the sterically hindered CH group (intermediate **C**), leading to the formation of pyrazolopyrimidines **246** or their mixtures with pyrimidine derivatives **247** and **248**.

The interaction of arylidenepyruvic acids with 5-aminotetrazole and 2-aminobenzimidazole most likely involves the nucleophilic attack of the enone fragment of the unsaturated ketoacid **236** by the endocyclic nitrogen atom.

The three-component reaction of amines **220**–**222** with pyruvic acid **239** and aromatic aldehydes has as the first step the formation of the appropriate azomethine intermediate **D** (Scheme 3.72). This hypothesis was proved by Chebanov et al. [202] through the synthesis of compounds **249**–**251** by the reaction of azomethine **D** with pyruvic acid.

In the case of aminopyrazole **213,** the formation of an azomethine intermediate, according to [202], is not observed, but nucleophilic addition of the

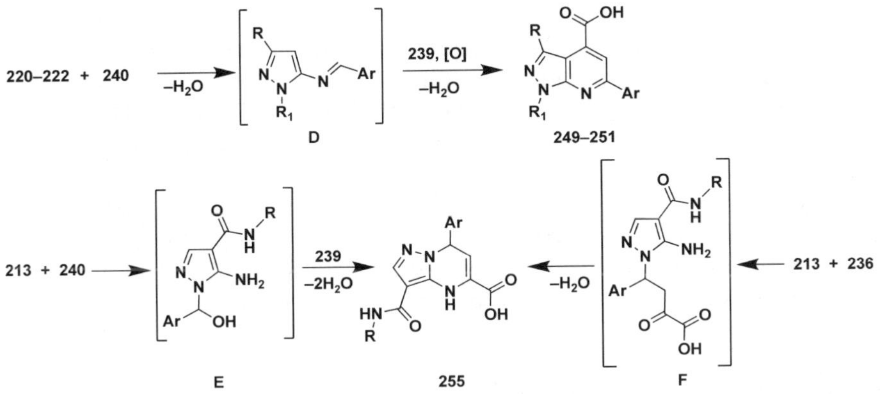


**Scheme 3.72**

endocyclic nitrogen atom at position 1 of pyrazole to the carbonyl group of the aldehyde or enone system of unsaturated ketoacid takes place.

Such behavior of amine **213** in [202] was explained by the presence of an intramolecular hydrogen bond and the electron-withdrawing influence of the carboxamide moiety on the nucleophilicity of the amino group. Owing to the formation of intermediates **E** and **F**, the reaction products of the multicomponent and sequential processes are identical.

The appropriateness of using of microwave-assisted organic synthesis to obtain pyridinecarboxylic acids and pyrimidinecarboxylic acids was also described in [202]. The application of microwave irradiation reduces significantly the reaction time from 2 h to 10 min, increases yields of the target compounds and also allows high-boiling and hard-to-remove solvents (DMF and acetic acid) to be replaced by ethanol.

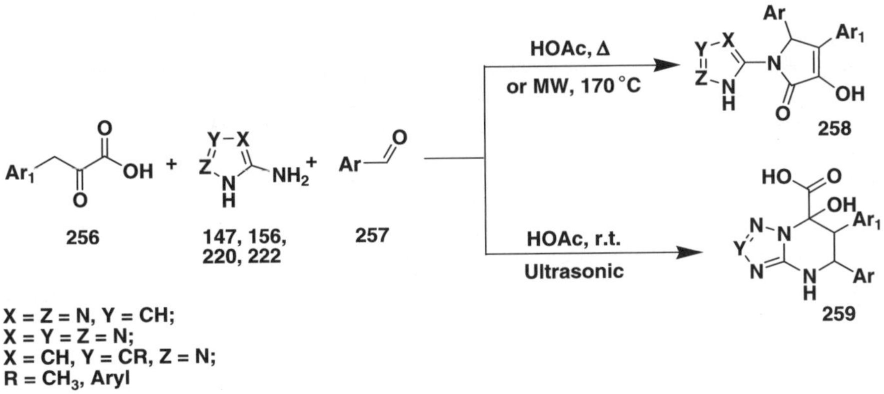


**Scheme 3.73**

The application of arylpyruvic acids **256** in place of pyruvic acid in three-component reactions leads to dramatic changes in the direction of the process. Refluxing of starting compounds for 3 hours of irradiating with microwave at 170°C for 20 minutes in acetic acid yielded 3-hydroxy-4,5-diaryl-1-azolyl-2,5-dihydro-1*H*-2-pyrrolones **258** [203] (Scheme 3.73). Under ultrasonic irradiation in ethanol with the addition of catalytic amounts of hydrochloric acid or in acetic acid, the reaction proceeds in a different direction with the formation of pyrimidinecarboxylic acids **259**. In the case of pyruvic acid the course of the three-component reaction does not so drastically depend on the activation method or solvent type as well as from temperature mode [202].

## 3.5 Reactions of Anilines and Aminoazines

As mentioned in Sect. 3.1, reactions of α,β-unsaturated ketones with aromatic and heteroaromatic amines (Scheme 3.1, pathways f and g) can be considered as enamine reactions, but we decided to describe them in a separate section of the book.

Indeed, aniline derivatives may be assigned to enamines only formally because the directions of reactions with unsaturated carbonyls in both cases are different. Reactions of enamines, which were described in Sect. 3.2, always proceed with the initial nucleophilic addition of the amino group to the carbonyl of the unsaturated ketone, while in the case of anilines another direction (so-called Scraup direction) is observed.

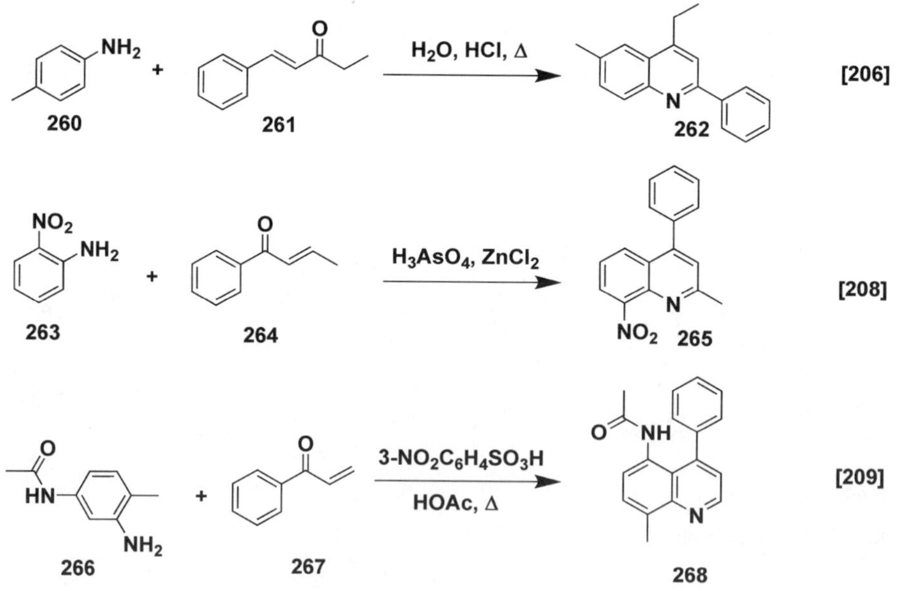


Scheme 3.74

Reactions of α,β-unsaturated ketones with anilines are not very diverse and usually yield the appropriate quinolines [204, 205, 206, 207, 208, 209, 210, 211]. As catalysts in such reactions, hydrochloric acids are often used [204, 205, 206, 207]. In [208] besides hydrochloric acid, the use of arsenic acid and zinc chloride was proposed, which led to increase of the yields. In [209] it was reported that the catalyst used was 3-nitrobenzensulfonic acid (Scheme 3.74).

In [210, 211] the solid-phase reaction of unsaturated ketones and anilines on a surface of silica gel in the presence of indium chloride under microwave irradiation was described. The library consists of 15 quinolines in yields of 55–87% which were generated quickly and characterized. Ranu et al. [211] showed that dihydro derivatives of qiunolines (e.g., **270** and **273**) can also be synthesized, but in this case one component of the reaction must be a 4,4-disubstituted methyl vinyl ketone, for example, mesityl oxide **52** or 1-(2-methylcyclopent-1-enyl)ethanone **272** (Scheme 3.75).

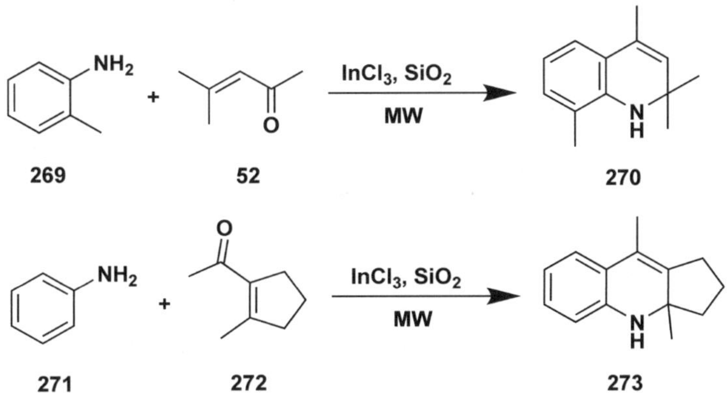


**Scheme 3.75**

The application of 4-oxo-4-arylbut-2-enonic acids **274** as unsaturated carbonyls in reactions with anilines allows the synthesis of quinoline-2-carboxylic acids **275** [212] (Scheme 3.76). Reactions of anilines with arylidenepyruvic acids can also yield quiniline carboxylic acids (or their decarboxylated analogues)

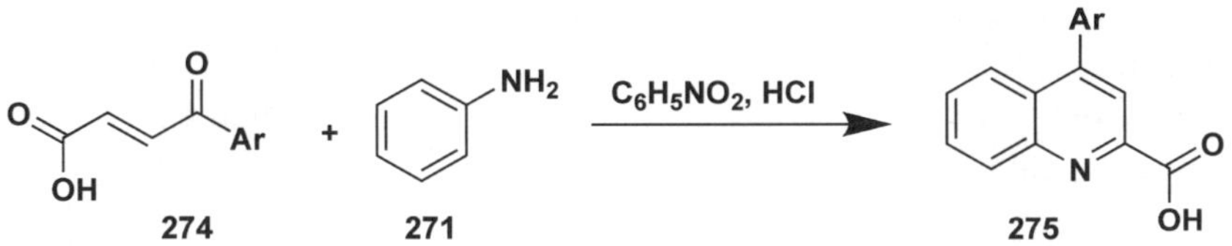


**Scheme 3.76**

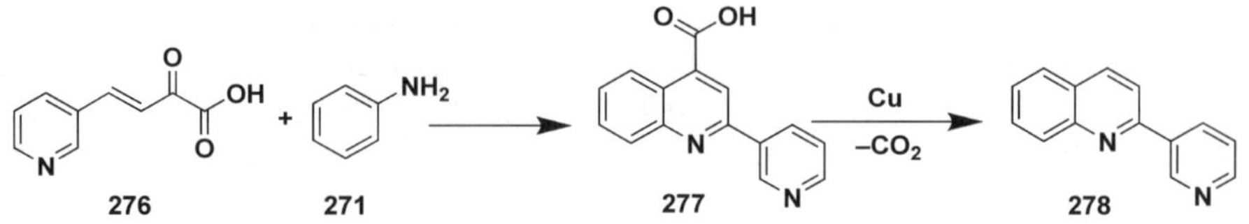


Scheme 3.77

with the reverse location of the carboxylic group and the aryl substituent ("anti-Scraup" direction) [213] (Scheme 3.77).

Besides quinoline, carboxylic acid reactions of anilines with arylidenepyruvic acids can yield other heterocyclic compounds (Scheme 3.78), in particular 5-aryl-3-(2-arylamino)furan-2-ones **280** (by stirring the starting materials in ethanol at room temperature) [214, 215] and 1,5-diarylpyrrolidine-2,3-dione **281** (by refluxing in ethanol) [216].

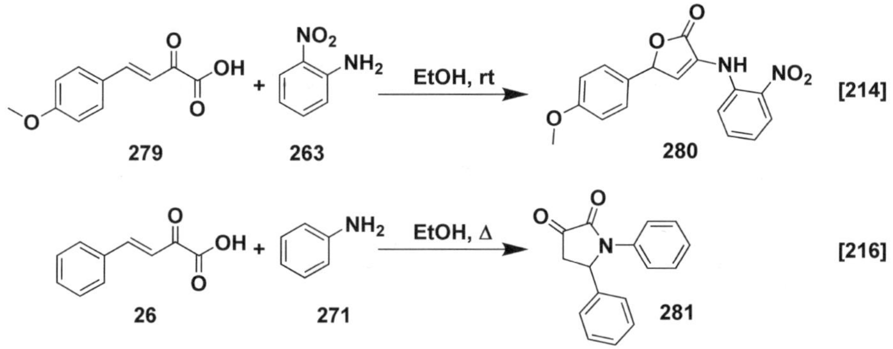


Scheme 3.78

The three-component reaction of pyruvic acid, aldehydes and anilines can also yield pyrrolidine-2,3-dione [198], but in most cases the products of such multicomponent condensations are quinoline carboxylic acids identical to compounds obtained via the reaction of arylidenepyruvic acids [217, 218, 219, 220, 221, 222, 223, 224, 225, 226, 227, 228].

These multicomponent reactions are carried out in boiling ethanol [217, 218, 220, 221, 222, 223, 225, 226, 227, 228], dimethyl sulfoxide (DMSO) [224] or acetic acid with sulfuric acid [219] (Scheme 3.79).

Reactions of α,β-unsaturated ketones with various aminoazines are also described in the literature. Treatments involving 6-aminouracils and chalcones or their analogues are the most investigated among them [229, 230, 231, 232, 233, 234, 235, 236, 237, 238, 239, 240, 241].

These reactions, as observed in the case of aminoazoles, have the "anti-Scraup" direction, i.e., the carbonyl group of an unsaturated ketone reacts

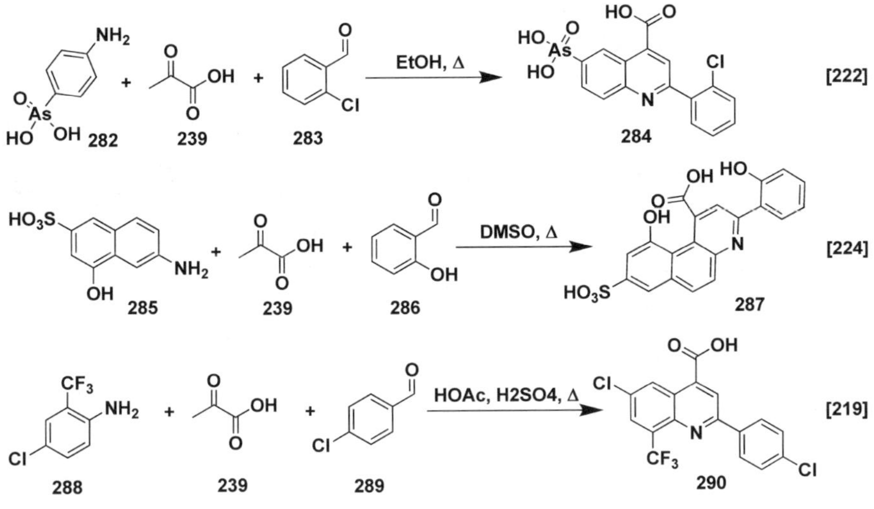


**Scheme 3.79**

with the amino group, while the double bond of the enone is attacked by the CH center at position 5 of uracil, yielding pyrido[2,3-*d*]pyrimidines. Even if the uracil moiety contains several amino groups (for example, 2,6-diaminouracil **291**, Scheme 3.80) only the amino group at position 6 of the heterocycle takes part in the reaction [229, 232]. But reactions of 5,6-diaminouracils with

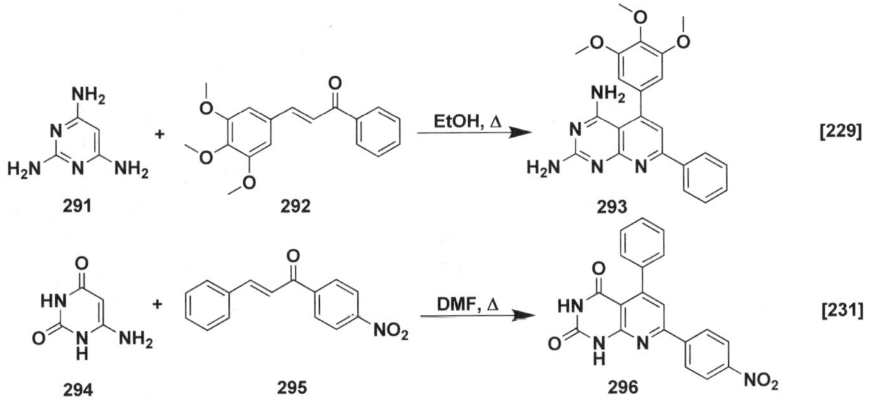


**Scheme 3.80**

α,β-unsaturated ketone are very specific and proceed in a different direction. They will be described in Chap. 4.

According to [231, 232], the reactions of 6-aminouracil **294** with chalcones in air lead to pyrido[2,3-*d*]pyrimidines **296**, while their dihydro analogues can be

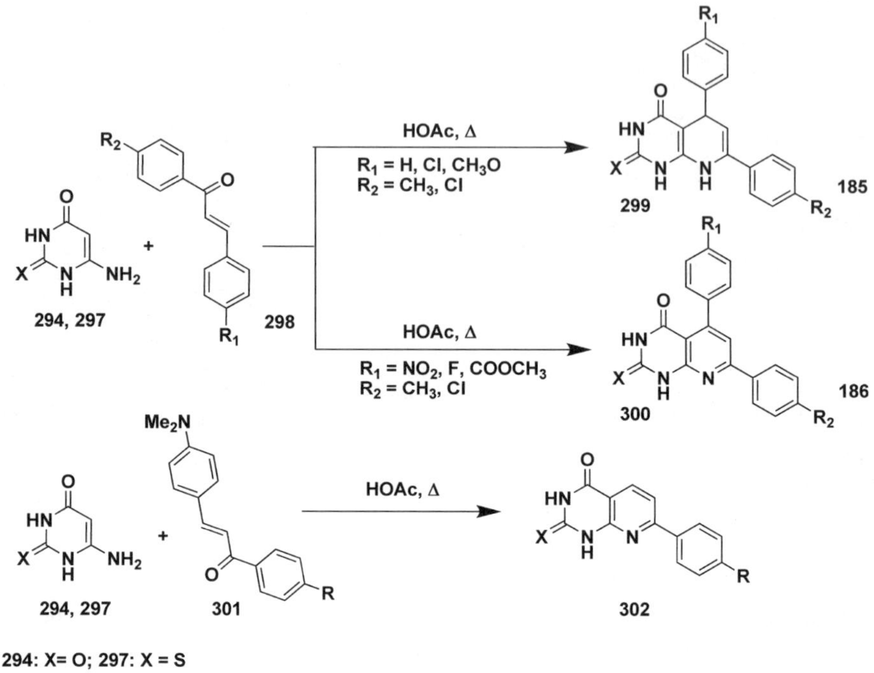


**Scheme 3.81**

synthesized only in an inert atmosphere (Scheme 3.81). But more detailed investigations showed that the oxidation process is influenced by solvent type and electronic character of substituents in the unsaturated ketone [241]. In [241] it was described that dihydro derivatives of pyrido[2,3-*d*]pyrimidines can be obtained without the application of inert gases. It was found that the presence of a basic catalyst in the reaction mixture ($Et_3N$, NaOH) leads to formation of heteroaromatized reaction products, while acidic conditions increase the content of dihydro derivatives in the reaction products. Pure 5,7-diphenyl-5,8-dihydropyrido[2,3-*d*]pyrimidine-2,4(1*H*,3*H*)diones **299** were synthesized by refluxing the starting materials in acetic acid. It is interesting that the presence of strong electron-withdrawing substituents in the aldehyde part of the unsaturated ketones **298** leads to formation of aromatized compounds **300**. But in the case of the presence of electron-donating substituents (compound **301,** for example) in the same part of the chalcone, the reaction proceeded with the elimination of one aryl group and yielded compounds **302** (Scheme 3.81).

The influence of the electronic character of substituents in the acetophenone part of the unsaturated ketones was not studied in [241], but results reported in [240], devoted to the reactions of 6-aminouracil and 2-thio-6-aminouracil with benzimidazole analogues of chalcones **303**, showed that increasing the electron-withdrawing

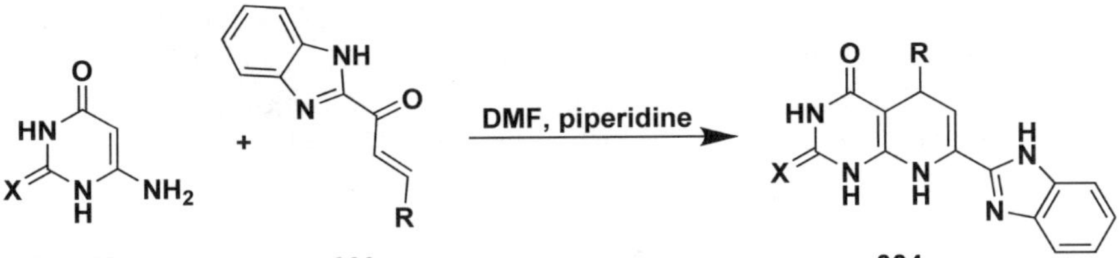

294, 297 303 304

R = 4-$CH_3O$-$C_6H_4$, 4-$NO_2$-$C_6H_4$, 3-pyridyl

**Scheme 3.82**

properties of the substituent on the carbonyl group of the ketone stabilized the dihydroheterocycles **304** (Scheme 3.82).

The reaction of 1,3-dimethyl-6-aminouracil **305** with trifluorobenzylideneacetone, containing a strong σ-acceptor group, in boiling DMSO, yielded a mixture of dihydro (**307**) and heteroaromatized (**306**) heterocycles, with a predominance of the latter [239] (Scheme 3.83).

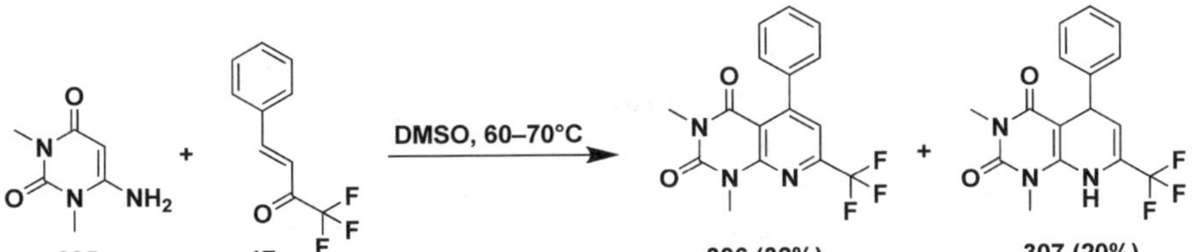

305 17 306 (32%) 307 (20%)

**Scheme 3.83**

The reaction of 4-(2-hydroxyphenyl)-but-3-en-2-one **21** with 1,3-dimethyl-6-aminouracil **305** in trifluoroacetic acid proceeds with the intramolecular addition of the hydroxyl group to the ethylene bond of dihydropyrido[2,3-*d*]pyrimidines and yielded the appropriate tricyclic compounds **308** [230] (Scheme 3.84).

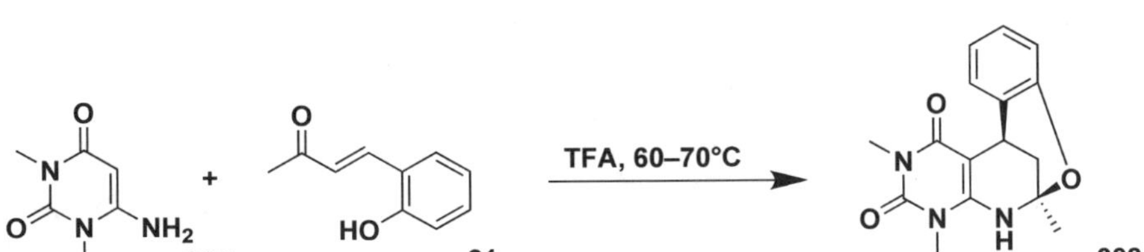

305 21 308

**Scheme 3.84**

Data exists for reactions of benzylidenepyruvic acid **26** with 6-chloropyridazin-3-amine **309** and with 2-amino-6-(4-methoxyphenyl)-4-oxo-1,2,3,4-tetrahydropyrimidine-5-carbonitrile **311** leading to imidazo[1,2-*b*]pyridazine **310** [242] and pyrimido[1,2-*a*]pyrimidine-2-carboxylic acid **312** [243] (Scheme 3.85).

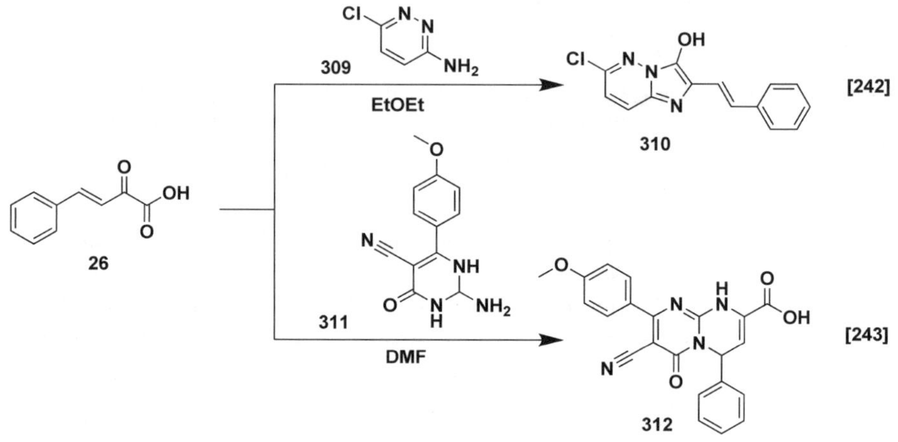


**Scheme 3.85**

Multicomponent reactions of aminoazines with pyruvic acid have not been described in literature at the present time.

Treatment of aminoazines, in particular, 6-aminouracil **305**, with cyclic α,β-unsaturated ketones was described in [235]. The authors showed that 1,3-dimethyl-6-aminouracil reacts easily with cyclooct-2-enone **313**, cyclonon-2-enone **316** and other cyclic unsaturated carbonyls (cyclododec-2-enone, cycloundec-2-enon, etc.), forming the appropriate tricyclic compounds **314**, **317** or analogues (Scheme 3.86). The interaction of 2,3-dichloro-5,6-dicyano-1,4-benzoquinone (DDQ) with compound **314** leads to a rearrangement of the carbon skeleton and formation of another tricyclic structure **315**, while in the

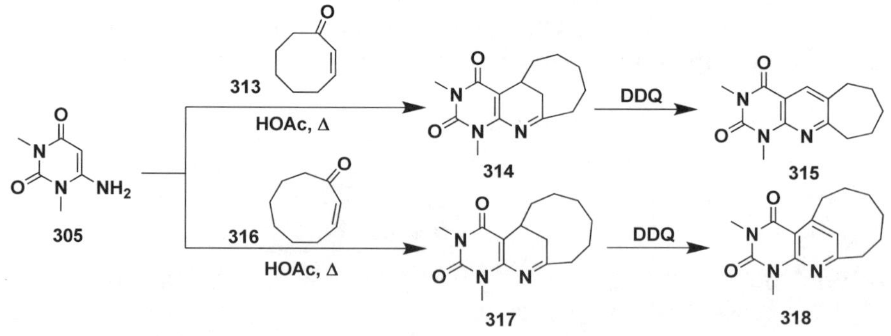


**Scheme 3.86**

case of greater cycles the oxidation occurs without isomerization (e.g., compound **318**).

Reaction of arylidenecycloalkanones with derivatives of 6-aminouracil is not described in the literature but there are examples of multicomponent synthesis involving cyclic 1,3-diketones and aldehydes. In [244] the treatment of 1,3-dimethyl-6-aminouracil **305** with 1,3-diketones **320** and aliphatic aldehydes in boiling acetic acid was reported to lead to the formation of 1,3-dimethyl-8,9-dihydropyrimido[4,5-*b*]quinoline-2,4,6-trione **321** in yields of 50–60% (Scheme 3.87).

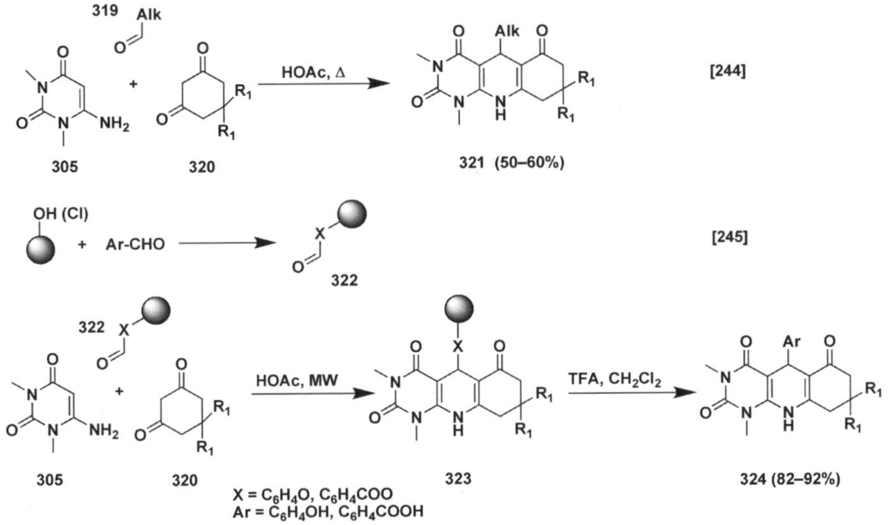


**Scheme 3.87**

Another approach to the synthesis of pyrimido[4,5-*b*]quinoline derivatives was proposed in [245]. To improve yields and the purity of target compounds, Agawal and Chauhan [245] used a combination of solid-supported synthesis (aldehyde component of the reaction) and microwave irradiation. This approach allowed the achievement of yields near 90%. But the significant disadvantages of the methods described in [245] are the limited set of starting aldehydes (for linking to the surface of the polymeric support, i.e., they must contain hydroxyl or carboxyl groups) and the multistage character of the procedure.

These disadvantages were avoided in the procedure proposed in [187]. The authors synthesized target pyrimido[4,5-*b*]quinolines in 90–98% yields and high purity (without any crystallization) by the multicomponent reaction of 6-aminouracil derivatives, aromatic aldehydes and cyclic 1,3-diketones under microwave irradiation in ethanol with $Et_3N$ as the catalyst.

## 3.6 Chemical Properties of Dihydroazines

The properties of substituted dihydropyridines and dihydropyrimidines have been thoroughly described in the literature. There are several reviews devoted to their reactions and modifications [246, 247, 248, 249, 250, 251, 252, 253, 254, 255]. In this section of the book we have systematized the data mentioned in these reviews and supplemented them with results of investigations of dihydroazines fused with azoles.

The most investigated field of the chemistry of dihydroazines is reactions of their heteroaromatization (Scheme 3.88). Interest in these processes is due, first of all, to the very important role of hydrogenation–dehydrogenation reactions in biochemistry, particularly in cell energy interchange.

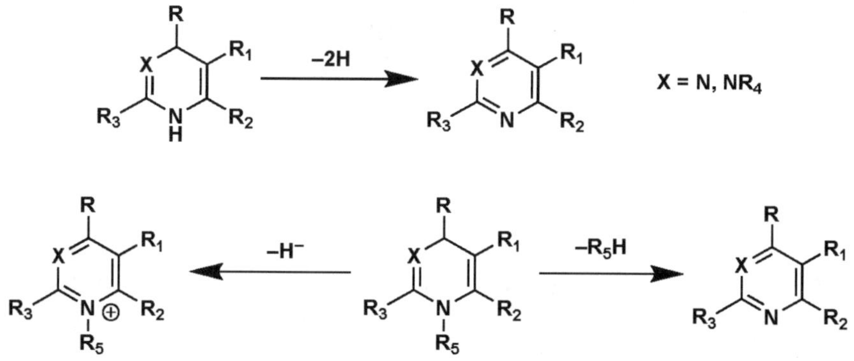


**Scheme 3.88**

The preparative heteroaromatization of dihydroazines (including N-substituted ones) is usually carried out by the action of various oxidative agents, such as air oxygen [81, 256, 257, 258, 259], $HNO_2$ and nitrites [260, 261, 262, 263, 264, 265, 266], $HNO_3$ and nitrates [246, 247, 249, 264, 267, 268, 269, 270, 271], NO [272], $Br_2$ [259], $CrO_3$ [273, 274, 275, 276], sulfur [277], $KMnO_4$ [278, 279], chloranil [264, 270, 280], Pd/C [278, 280, 281, 282], hydroxylamine hydrochloride [283], $Cu(AcO)_2$ [284], $H_2O_2$ / $HClO_4$ [285], nitroaromatic compounds [278, 279], aldehydes and ketones [249], carbocations and some other more specific reagents [246, 247]. Dihydro derivatives of azolopyrimidines are heteroaromatized very easily by air oxygen [286], *N*-bromosuccinimide [158, 159, 162, 171, 172], $Br_2$/HOAc [174], KOH in methanol [287], $SeO_2$ [160], and $MnO_2$ [175].

It should be noted that for N-substituted dihydropyridines and pyrimidines, an alternative aromatization is observed. It includes the elimination of a substituent from the nitrogen atom [246, 247, 288, 289] (Schemes 3.88, 3.89). Heteroaromatized derivatives of pyrimidine, in addition, can be obtained as a result of disproportionation reactions of their dihydro derivatives [246, 247, 290, 291].

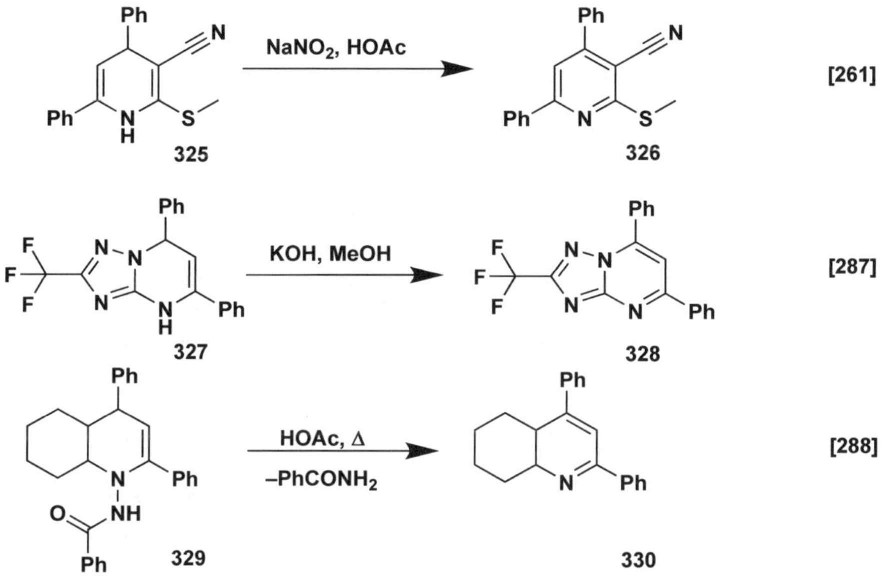


Scheme 3.89

An important feature of the synthesis of heteroaromatized derivatives of azoloazines via their dihydro analogues is high regioselectivity when R is not the same as $R_2$ (Scheme 3.88) in comparison with reactions of aminoazoles with β-diketones with the same R and $R_2$ substituents.

The mechanisms of heteroaromatization of azine dihydro derivatives were studied and described in several publications [250, 271, 292]. At first it was considered that these reactions proceeded via the transfer of a hydride ion (Scheme 3.88). But recently these are regarded as multistage processes starting from single-electron oxidation of the substrate (Scheme 3.90) [250, 271].

The main direction of decomposition of the cation radical formed at the first stage is a deprotonation leading to a neutral free-radical particle which later oxidizes into a heteroaromatized cation (Scheme 3.90, pathway a). The opposite pathway is rarely observed, i.e., oxidation of the cation radical into a dication foregoing the deprotonation stages (Scheme 3.90, pathway b) [250]. When single-electron transfer occurs along with a cation-radical particle, aromatization is also observed (Scheme 3.90, pathway c) [292] at the expense of the elimination of a hydrogen atom in the solvent "cell", i.e., where repeated collisions of two particles take place. A fourth variation is possible and involves the decomposition of the cation radical and the formation of molecular

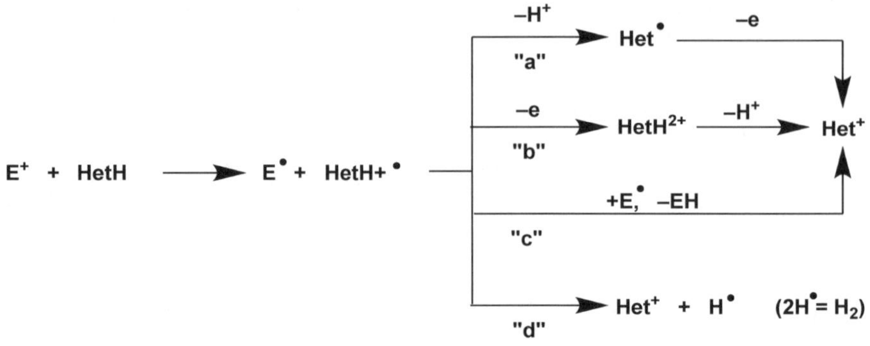


Scheme 3.90

hydrogen (Scheme 3.90, pathway d). This type of process, in particular, was observed during oxidation of 2-phenyl-4,6-dimethyl-1,2-dihydropyrimidine with $AgNO_2$ [271].

Dihydroazolopyrimidine heterocycles possess enhanced stability towards heteroaromatization by air oxygen in comparison with nonfused dihydroazines [158, 162, 172, 174, 175]. Oxidation of azolopyrimidines **331**–**333** becomes easier owing to rapid ionization in alkaline–alcoholic solutions (Scheme 3.91).

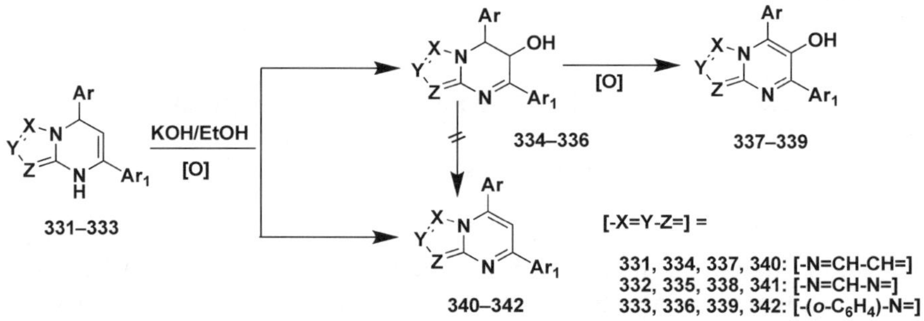


Scheme 3.91

Besides heteroaromatization and isolation of compounds **340**–**342,** the formation of hydroxy derivatives **334**–**336** and **337**–**339** was established [174].

The ability to oxidize to a corresponding hydroxyl-containing heterocycle is typical for dihydro derivatives of pyrazolo[1,5-*a*]pyrimidines **331** and pyrimido[1,2-*a*]benzimidazoles **333**. Simply dissolving these compounds in $CHCl_3$, DMF and alcohols at ambient conditions produces compounds **334** and **336** as major products together with heterocycles **340** and **342** (in some cases the latter products were not observed) [168, 293]. It is interesting to note that compounds **334**–**336** have a very high resistance to the action of dehydration agents. Such stability was explained in [168, 293] by the unfavorability of the dehydration process of the *cis* orientation of the CH–COH fragment of the dihydropyrimidine cycle.

Reduction of dihydroazines **343** can lead both to tetrahydro derivatives **344** and to perhydro heterocyles **345** [246, 247] (Scheme 3.92). It was shown [294] that the reaction of catalytic hydrogenation of dihydroazines **343** can be

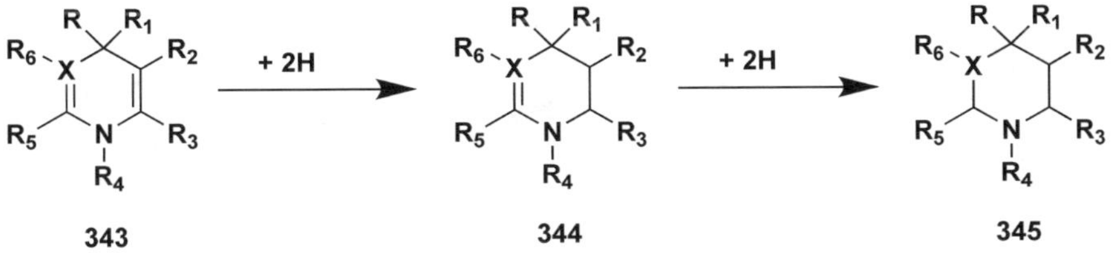

**Scheme 3.92**

stopped at the formation stage of tetrahydro derivatives **344** by deactivation of one of the double bonds with electron-withdrawing substituents ($R_2$ or $R_6$ is COOEt, CN, CONHR).

Stereoselective reduction to the appropriate tetrahydro derivatives is observed in the case of the reaction of heteroannelated dihydroazines ($R^5$and $R^6$ are Het, X is N, for example, compounds **346**, **348** and **350**) with sodium borohydride [174, 295, 296, 297] or with hydrogen in the presence of Pd on $Al_2O_3$ under 3,000 Torr [298] (Scheme 3.93). Reduction of the enamines **346** and

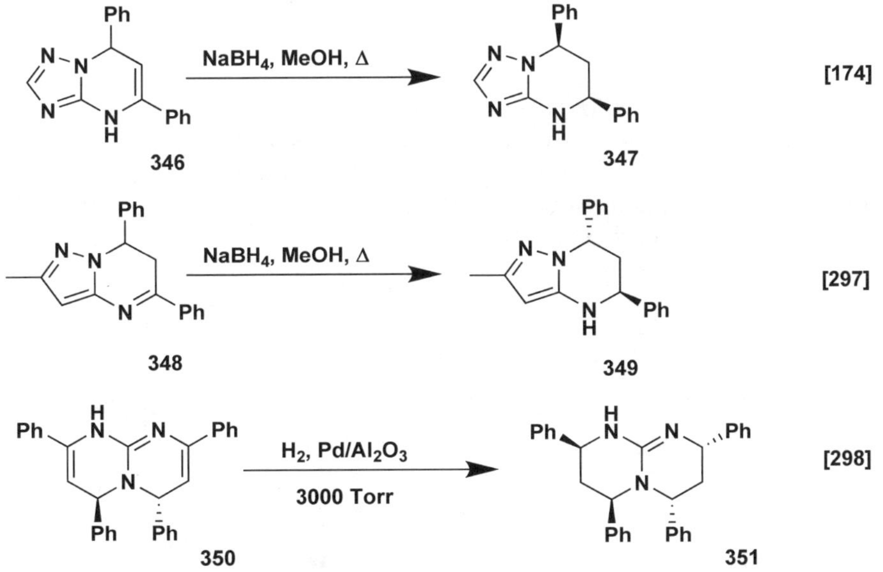

**Scheme 3.93**

**350** leads to a heterocycle with *cis* orientation of the phenyl substituent (compounds **347** and **351**), while imine **348** converts to the *trans* derivative **349**.

When stabilizing substituents are absent or in the case of symmetrically substituted dihydroazines ($R_2$ is the same as $R_6$ and/or $R_3$ is the same as $R_5$), both catalytic hydrogenation and the use of $LiAlH_4$ ($NaBH_4$) lead to a structure like **345** [299, 300, 301, 302, 303, 304, 305, 306]. Sometimes a reduction of the functional groups is also observed [306] (Scheme 3.94).

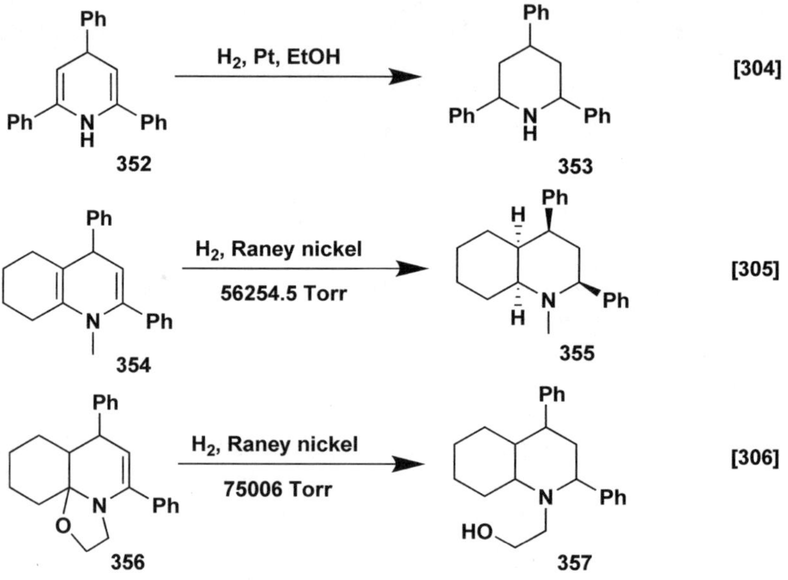


**Scheme 3.94**

A reduction of symmetrically substituted dihydropyridines **358** to the corresponding hexahydro derivatives **359** can be executed at 50 °C under the action of a mild reducing agent —triethylsilane [307]. At room temperature, on the other hand, the reaction stops at the formation of tetrahydro derivatives **360** (Scheme 3.95). Both processes are stereoselective.

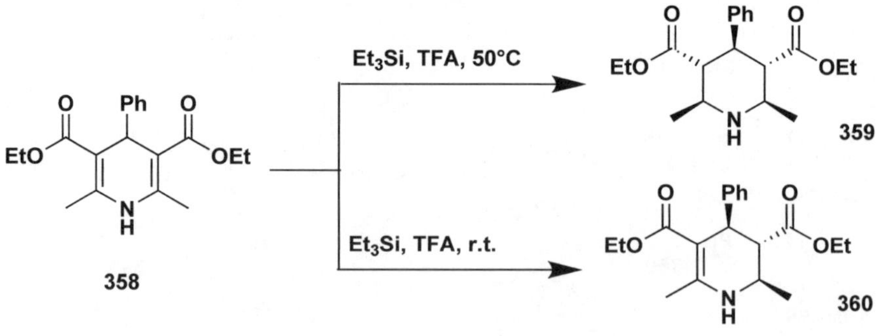


**Scheme 3.95**

Reactions of alkylation and acylation are characteristic for dihydroazines and there are a lot of examples of such processes in the literature. In this book we would like to discuss the general similarities of these reactions with some illustrations without citing dozens of publications.

N-Unsubstituted 1,4-dihydroazines very easily undergo N-alkylation and N-acylation in basic media—KOH [308], NaOH [309], NaOMe [310], LiN(-$SiMe_3)_2$ [311], pyridine [312], DMAP [313], etc. [246, 247].

Only the N-substituted derivatives like **362** and **363** are isolated when heterocycles **361** (4,6-diphenyl-1,4-dihydropyrimidine (R and $R_3$ are Ph; $R_1$, $R_2$, and $R_5$ are H) [248] and dihydrotriazolopyrimidines ($R_5$ and X are –N–CH = N–N; R and $R_3$ are Ar; $R_1$ and $R_2$ are H) [174]) are acylated and alkylated (Scheme 3.96).

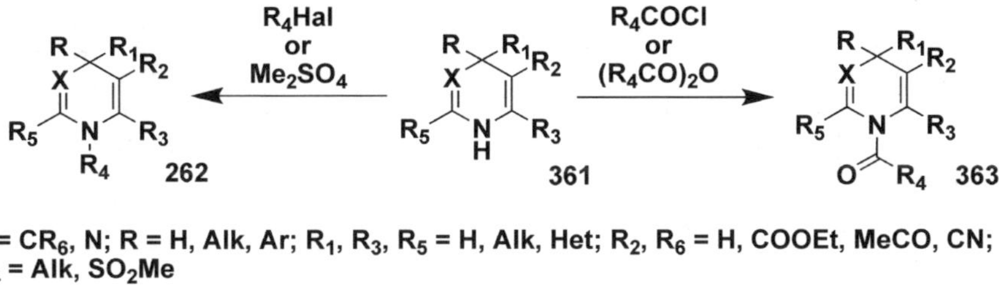


**Scheme 3.96**

The alkylation reaction of dihydroindenopyrimidines **364** can yield both N- and C-alkylated derivatives **365** and **366**, respectively [314]. The primary direction of the alkylation is determined by the nature of the alkylation agent [314]. The presence of two reaction products is related to the ambidentate character of the anion formed in basic medium (Scheme 3.97). When a hard reagent, $(CH_3)_2SO_4$, reacts with this anion, a harder reaction center—nitrogen atom– occurs in the process (charge control). In the reaction with a soft reagent, $CH_3I$, a softer carbanion center is involved in the treatment (orbital control). The absence of C-alkylation in the case of the reaction between

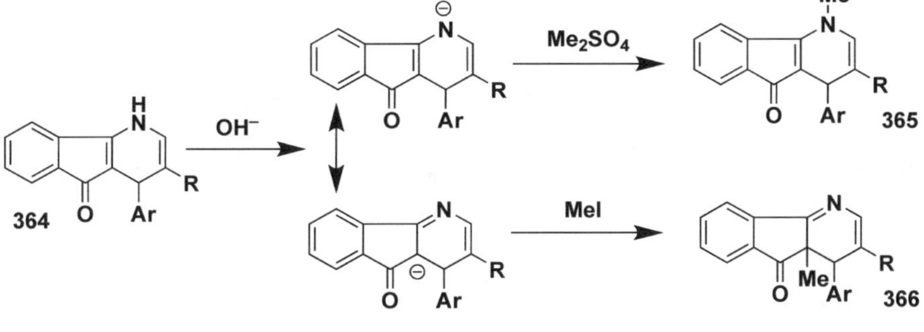


**Scheme 3.97**

compounds **361** and alkyl halogenide most probably can be explained by the deactivation of the carbon reaction centers by the electron-withdrawing substituents $R_2$ and $R_6$.

Alkylation or acylation reactions of dihydroazines can proceed as intramolecular processes and follow with heterocyclizations [308, 310] (Scheme 3.98).

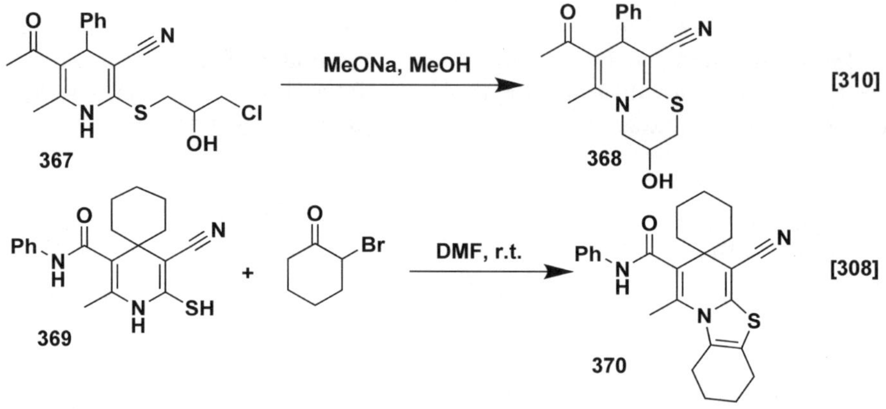


Scheme 3.98

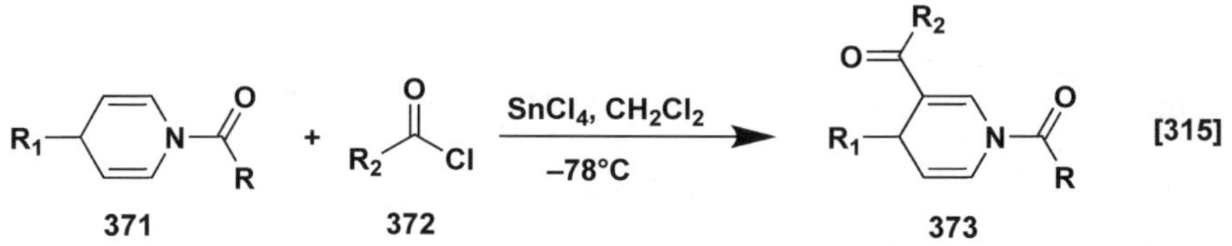


R = NEt$_2$, OEt; R$_1$, R$_2$ = Alkyl, Aryl; R$_3$ = H, Alk

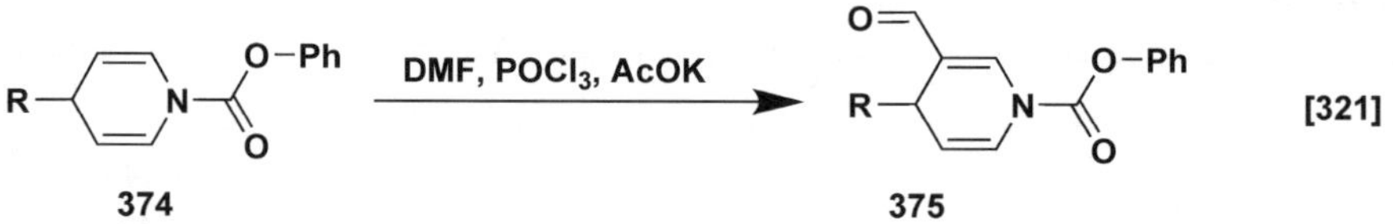


R = Alkyl, Aryl

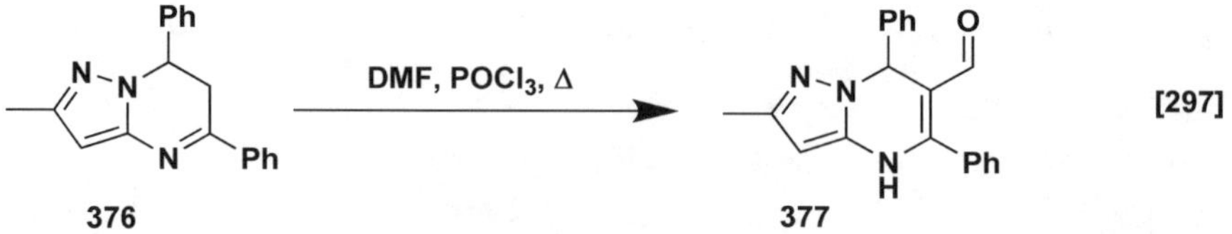


Scheme 3.99

A high density of electrons associated with atoms $C_{(3)}$ and $C_{(5)}$ of 1,4-dihydropyridines and 1,4-dihydropyrimidines is also observed when these heterocycles undergo electrophilic substitutions such as Friedel–Crafts [315, 316, 317, 318, 319, 320] and Vilsmeier [297, 321] reactions (Scheme 3.99). In [315] it was shown that treatment of dihydropyridines **371** with aroyl or acyl chlorides **372** in the presence of $SnCl_4$ leads to acylation of the heterocycle at position 3 (compounds **373**). Dihydropyridines **374** and dihydroazolopyrimidines **376** undergo Vilsmeier reaction with the formation of the corresponding derivatives **375** and **377**. It is interesting that imine heterocycle **376** after Vilsmeier reaction exists in the enamine tautomeric form. The tautomerism of dihydroazines and factors influencing it will be discussed in detail in Sect. 3.8.

Reaction of dihydropyridine **378** with aldehydes like **379** involves $C_{(3)}$ atom [322], but the treatment is more complicated and can lead to further transformations resulting in the formation of 3-alkylpyridines **380** (Scheme 3.100).

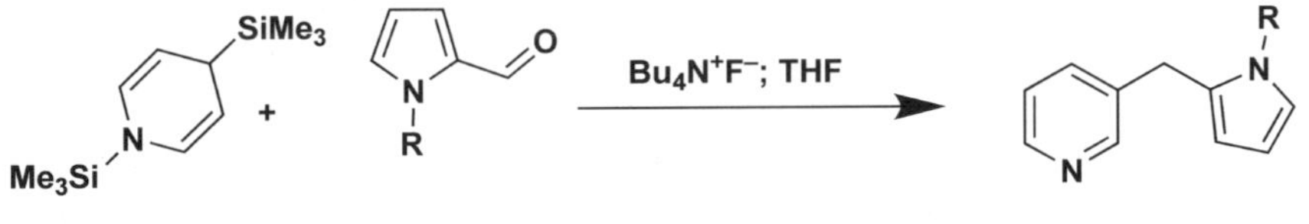


**Scheme 3.100**

As mentioned above, the very stable dihydro systems of triazolopyrimidines and pyrazolopyrimidines proceed in an unusual direction upon treatment with $HNO_2$. Under the conditions of the well-known heteroaromatization method, i.e., the action of $NaNO_2$ in the presence of an acid, compounds **381** and **382** are transformed into their corresponding nitroso derivatives **383** and **384** [174, 297, 323]. Heating compound **383** in the presence of polyphosphoric acid causes cyclization involving aromatic rings at positions 2 and 4 with the formation of a pyrrole ring (compounds **385** and **386**) [323] (Scheme 3.101).

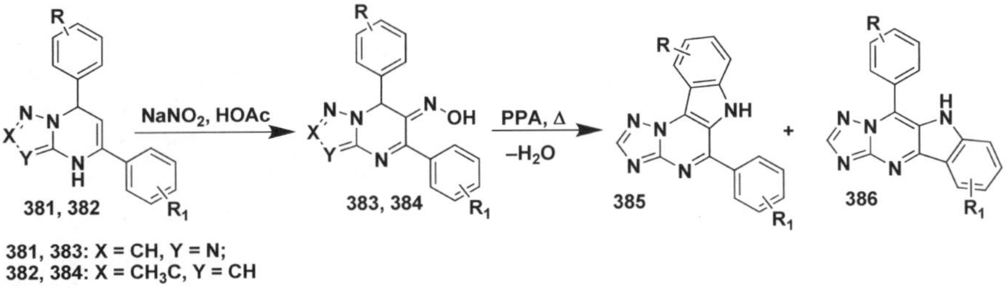


**Scheme 3.101**

Refluxing of dihydroazoloazines and α,β-unsaturated ketones in a methanol solution of sodium methoxide proceeds in a Michael-type addition. For example, reaction of 2-methyl-5,7-diaryl-6,7-dihydropyrazolo[1,5-*a*]pyrimidine **387** and chalcone **5** under these conditions yields adduct **388** [297] (Scheme 3.102). Treatment of 2-methyl-substututed triazolopyrimidine **389** with ketone **5** leads to cyclization with formation of the triazolo[5,1-*b*]quinazoline moiety [324].

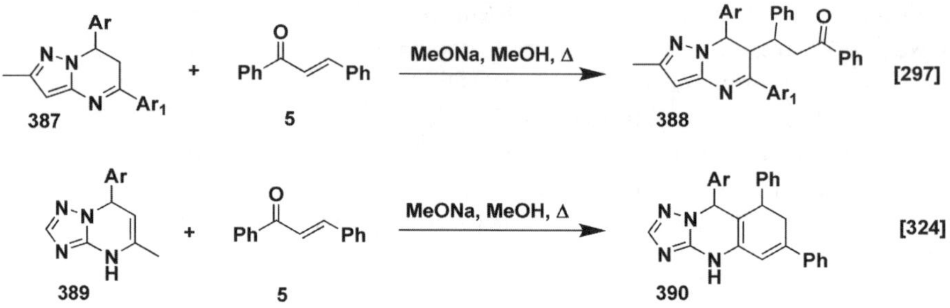


**Scheme 3.102**

Electrophiles can also be introduced into position 2 of the dihydropyridine cycle. The procedure for obtaining compounds **393** involves preliminary metallation of substrate **391** with *tert*-butyl lithium and the subsequent reaction of lithium-substituted heterocycle **392** with the appropriate electrophile [325] (Scheme 3.103).

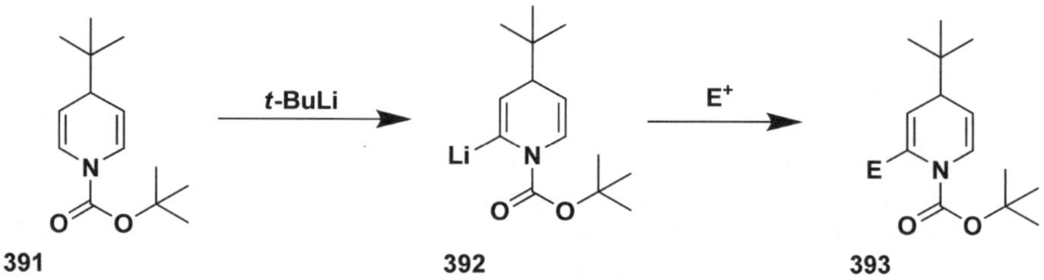


**Scheme 3.103**

Dihydro derivatives of pyridine and pyrimidine are able to add nucleophiles through their double bonds. For example, heating compound **294** in the presence of potassium hydroxide leads not only to complete hydrolysis of the azomethine moiety but, according to Meyers [326], also yields the product **295** formed by the addition of a water molecule (Scheme 3.104).

Dihydropyrimidines **296** react very easily with water, forming 6-oxy derivatives **297** even during storage of **296** in solvents containing water and special catalysts [248]. There are examples of other addition reactions with water [326, 327, 328, 329, 330, 331], alcohols [82, 332], hydrazine and hydroxylamine [333].

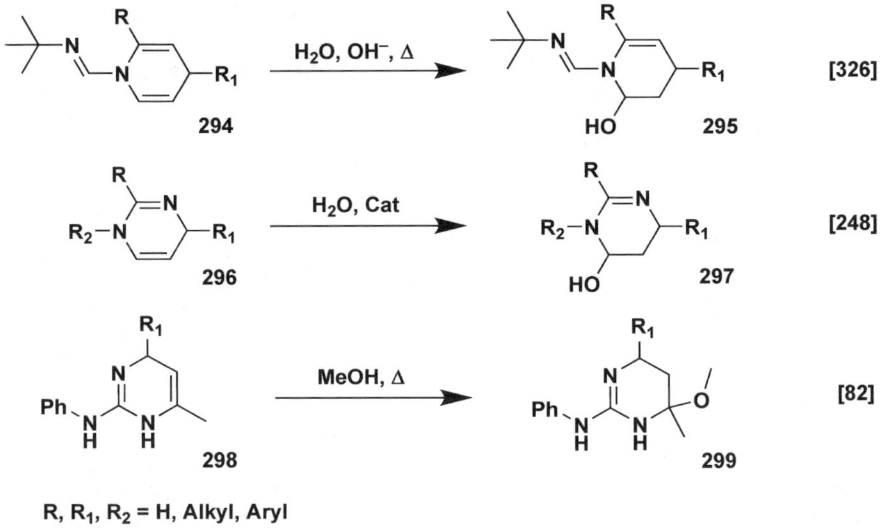

Scheme 3.104

Nucleophilic addition of water to hydrochloric derivatives of pyrimidobenzoimidazole **300** with the formation of compound **301** was observed by El Ella et al. [167]. The hydrolytic cleavage of dihydrotriazolopyrimidine **302** into aminotriazole **147** and the appropriate unsaturated carbonyl **303** most likely occurs via initial addition of water [174, 175] (Scheme 3.105).

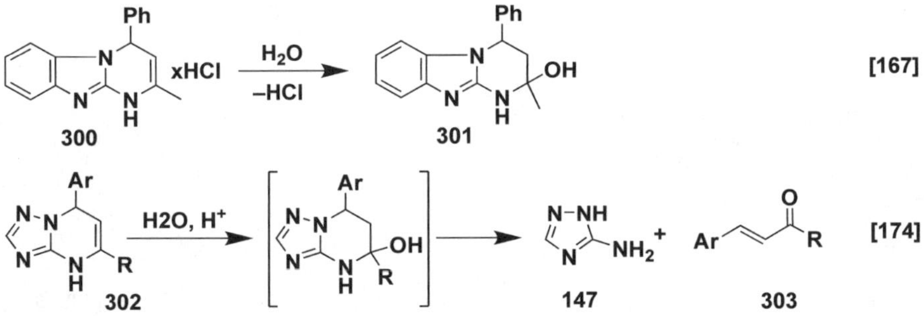

Scheme 3.105

Dihydropyridine **304** under basic conditions reacts with alkylphenols **305** and alkylthiopenols **306** in a different manner [246, 247, 334, 335] (Scheme 3.106): in the first case, the reaction does not involve the hydroxyl group of the phenol, while in the second the product that is formed is due to the interaction with the thiol substituent (compounds **307** and **308**, respectively).

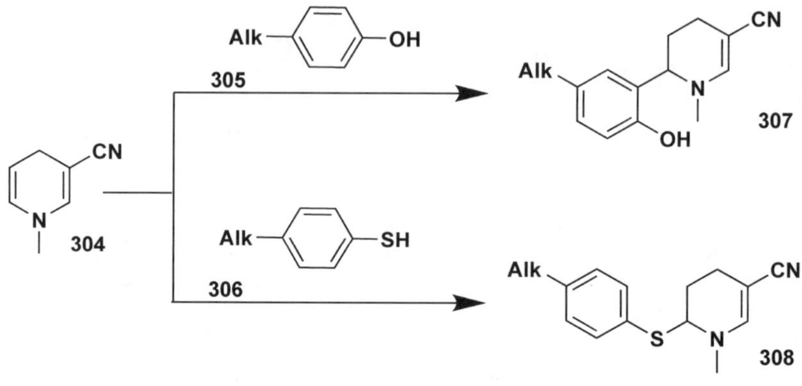


**Scheme 3.106**

The treatment of dihydropyridines with diaryldisulfane in boiling acetonitrile also yields products of an addition reaction like **308** in a mixture with arylsulfides [336].

There are several publications devoted to oxidative additions to dihydropyridines [337, 338, 339, 340, 341]. For instance, the addition of stochiometric amounts of iodine in a methanol solution of dihydropyridine **309** gives iodinated tetrahydropyrimidine **310** in a stereoselective manner [337]. The same result is obtained when the reaction is performed with *N*-iodosuccinimide (NIS) (Scheme 3.107). Interestingly, when the process is carried out in tetrahydrofuran the incorporation of the succinimide moiety at position 2 yields 3-iodo-2-succinimidotetrahydropyridine **311**. Using *N*-bromosuccenimide, *N*-chlorosuccenimide and *N*-fluoropyridinium trifluoromethanesulfonate produces 3-bromo-, 3-chloro- and 3 flouro-substituted pyridines [337].

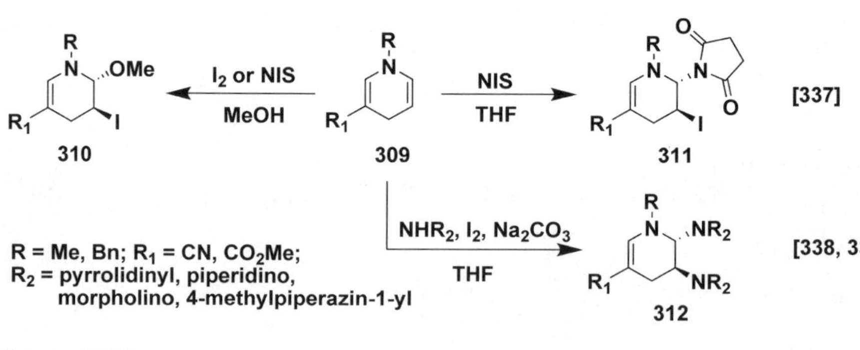


**Scheme 3.107**

Stereoselective vicinal diamination of dihydropyridines **309** by electrophilic interaction with iodine in the presence of secondary amines leading to tetrahydropyrimidines **312** is described in [338, 339] (Scheme 3.107). The formation of **312** can be rationalized by considering the initial formation of a *trans*-2-amino-3-iodotetrahydropyrimidine which undergoes an internal nucleophilic substitution reaction followed by a stereoselective ring opening of the resulting aziridinium ion that is promoted by a second equivalent of the amine [338].

The reaction of the iodine-assisted addition can be carried out as an intramolecular process. The iodocyclizations of N-substituted dihydropyridines **313**, **315** and **317** into corresponding polycyclic compounds **314**, **316** and **318** are described in [340] (Scheme 3.108). The heterocycles **318** can be further modified with

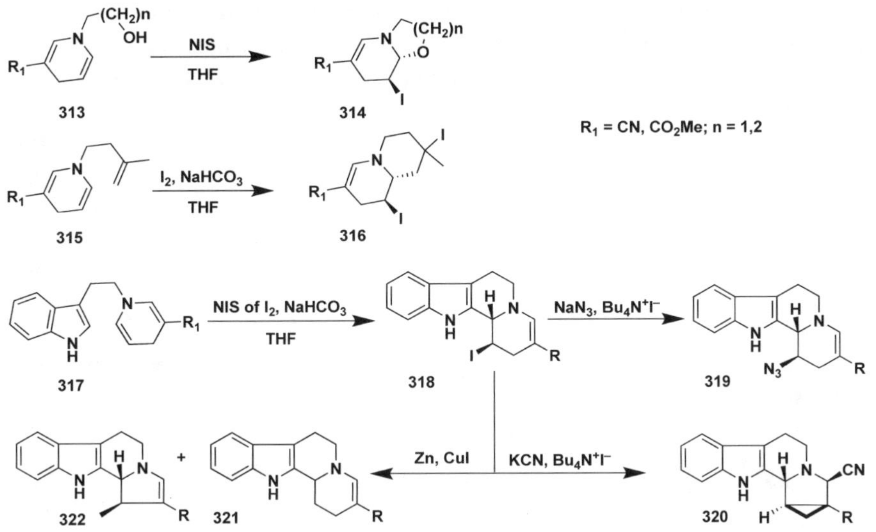


**Scheme 3.108**

sodium azide to give **319**, with potassium cyanide to form cyclopropane derivatives **320** or with zinc and copper(I) iodide to yield a mixture of **321** and **322** [340].

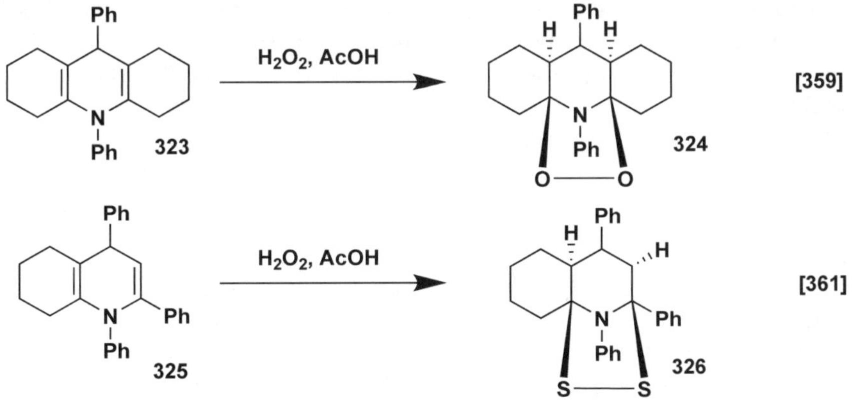


**Scheme 3.109**

The intramolecular cyclizations forming compounds like **318** and **321** were also described in other publications [317, 342, 343, 344, 345, 346, 347, 348, 349, 350, 351, 352, 353, 354, 355, 356, 357, 358].

Reactions of dihydropyridines with $H_2O_2$ [359, 360] and $Na_2S_2$ [361] in acetic acid lead to the formation of compounds like **324** and **326** (Scheme 3.109).

The high reactivity of dihydroazines enhances their ability to cause chemical modifications and to form new heterocyclic systems. The presence in 1,4-dihydropyridines of an enamine double bond increases their ability to cause cycloaddition reactions.

One of the examples of this type of reaction is the addition of 3,4-dichlorothiophene-1,1-dioxide **228** to dihydropyrimidines **327** following a regioselective pathway with the elimination of $SO_2$, yielding tetrahydroquinoline **329** [362] (Scheme 3.110).

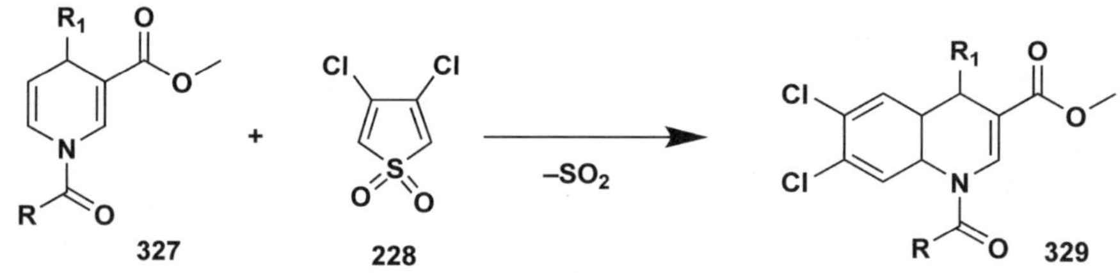

R = H, Me; $R_1$ = Me, CN, COOEt, $SO_2Ph$

**Scheme 3.110**

1,4-Dihydropyridines participate in reactions of 1,3-dipolar cycloaddition with some nitrile oxides [363, 364]. The monoester derivative **330** reacts with several nitrile oxides **331** to produce the corresponding isoxazolo[5,4-*b*]pyridine **332** (Scheme 3.111) in moderate to good yields. The regiochemistry of the cycloaddition was predicted in [363] on the basis of the complementary dipoles of the **331** and enamine double bond and was proven by conversion of **332** (R is Me, $R_1$ is COOH) to the 5-cyano-1,4-dihydropyridine-3-carboxylic acid ester **333**.

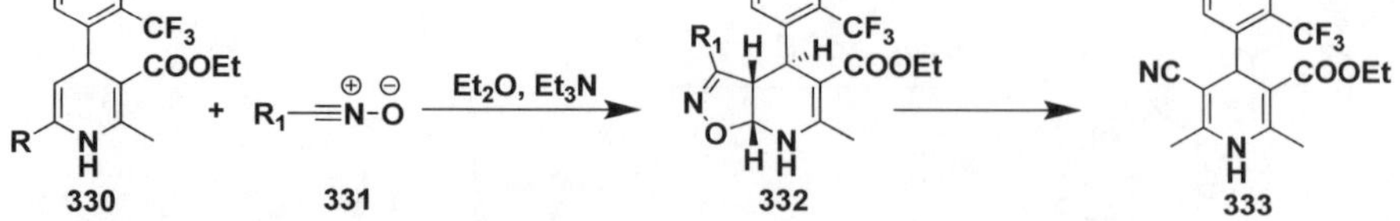

R = H, Me; $R_1$ = COOH, COOEt, COOPr, COOMe, $SO_2Ph$, Oet, CN, COPh, Me

**Scheme 3.111**

Taylor et al. [363] also noted the influence of steric hindrance on the reactivity of 1,4-dihydropyridines. For instance, no reaction is observed between the 3,5-diester derivatives of dihydropyridine and carbethoxyformonitrile oxide.

Regiospecific 1,3-dipolar cycloaddition reactions of dihydropyridines and some organic azides lead to high yields of fused aziridines—2,7-diazabicycli[4.1.0]hept-3-enes **337** [365, 366, 367] (Scheme 3.112). The reaction proceeds via the preliminary formation of an intermediate **336** and the elimination of nitrogen. Reaction of pyrimidine **334** with less reactive methoxycarbonyl and benzoyl azides does not occur [367]. Compounds **337** exhibit significant analgesic, antibacterial and antifungal activities [367].

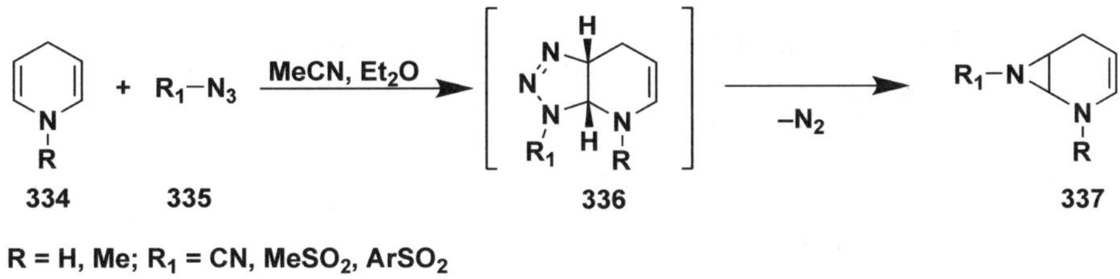


**334** **335** **336** **337**

Scheme 3.112

Cyclocondensations of NADH analogues with *para*-benzoquinone are described in [368, 369]. For example, upon addition of dihydropyridine **338** to an acetonitrile solution of **339** in the presence of scandium trifluoromethanesulfonate , the cycloaddition reaction occurs efficiently at room temperature, yielding cycloadduct **340** [368] (Scheme 3.113). This reaction passes via formation of a complex between azine and scandium trifluoromethanesulfonate.

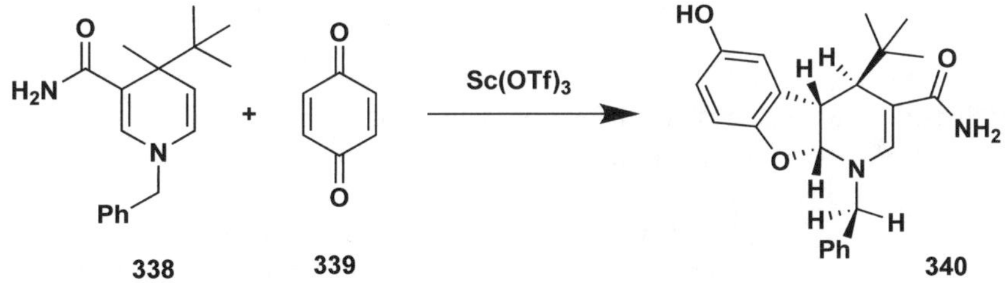


**338** **339** **340**

Scheme 3.113

Several examples of [2 + 2] cycloadditions of 1,4-dihydropyridines are described in the literature [370, 371, 372]. Derivatives of *N*-benzyl-1,4-dihydronicotinic acid **341** can photochemically react ($\lambda = 365$ nm) with acrylonitrile **342**, yielding isomeric 2-benzyl-2-azabicyclo[4.2.0]oct-3-enes **343** and **344** [370], or undergo dimerization, forming tricyclic compounds **345** [371]

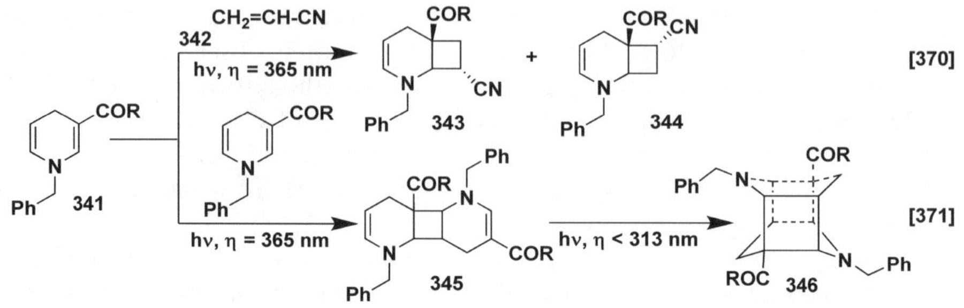


Scheme 3.114

(Scheme 3.114). The use of another wavelength (λ < 313 nm) in the case of the second reaction leads to a similar [2 + 2] cycloaddition yielding the corresponding polycyclic compounds **346** [371].

A similar type of dimerization of dihydropyridines obtained by an electroreduction of the appropriate salts **347** leading to the formation of polyheterocyles **348** is described in [373] (Scheme 3.115).

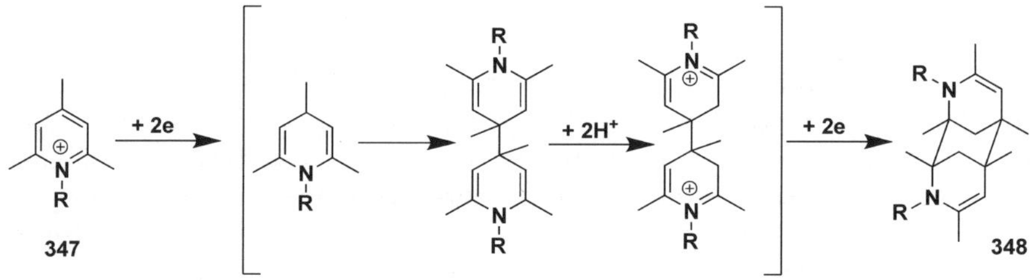


Scheme 3.115

Several publications are devoted to the reactions of olefin cycloaddition to dihydroazines in the presence of Lewis acid and to an intramolecular addition in 1,4-dihydropyridines containing *ortho*-alkenylaryl substituents at position 4 [374, 375, 376, 377, 378, 379].

Reaction between dimethyl 2,6-dimethyl-4-phenyl-1,4-dihydropyridine-3,5-dicarboxylate **349** and alkenes **350** yields tricyclic compounds **351** [374, 375, 376, 377, 378] (Scheme 3.116).

4-Styryl substituted 1,4-dihydropyrimidine **352** treated with hydrogen chloride gas in chloroform leads to a complicated polyheterocyle **354** [379]. The mechanism of the reaction most likely involves formation of a benzyl cation **353** followed by a nucleophilic attack of the aminocrotonate moiety.

Treatment of dihydropyridine **355** with gaseous hydrogen chloride or trifluoromethanesulfonate, i.e., reagents which successfully cyclized **352**, in a variety of solvents produces either no reaction or oxidation to the pyridine.

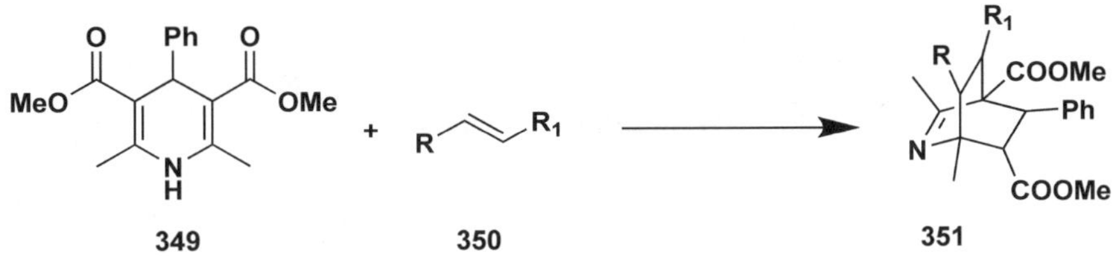


Scheme 3.116

However, treatment of **355** with aluminum chloride in chloroform results in the formation of an ester **357** (Scheme 3.117). No reaction occurs when methylene chloride, rather than chloroform, is used as the solvent.

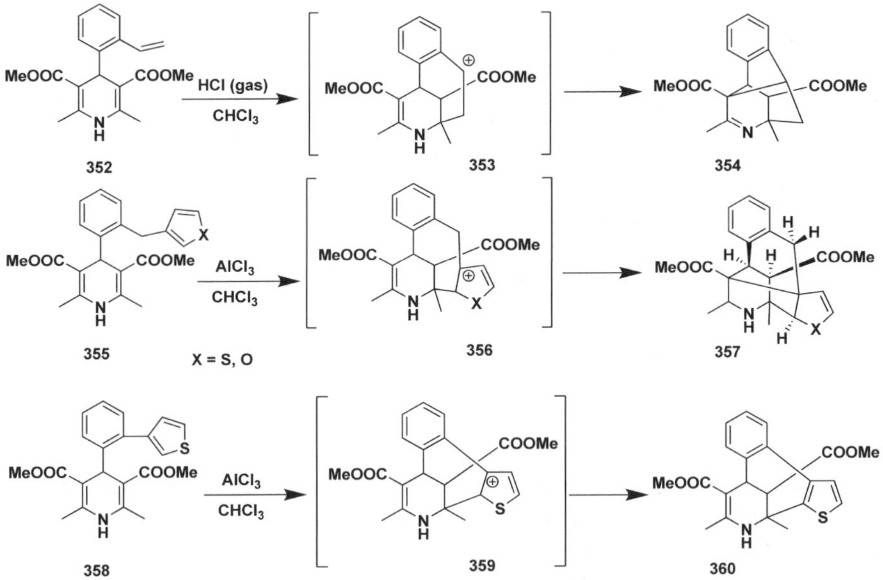


Scheme 3.117

Surprisingly for Hartman et al. [379], reaction of dihydropyrimidine **358** under the same conditions ($AlCl_3$, $CHCl_3$) gives compounds **360**, and the product is produced by the elimination of a proton from cation **359**.

The method of formation of complicated polycyclic compounds by an intramolecular cycloaddition in 1,4-dihydropyridines containing *ortho*-alkenylaryl substituents is very common and is also observed in many other cases. For instance, treatment of dihydropyridines **361** with titanium tetrachloride in

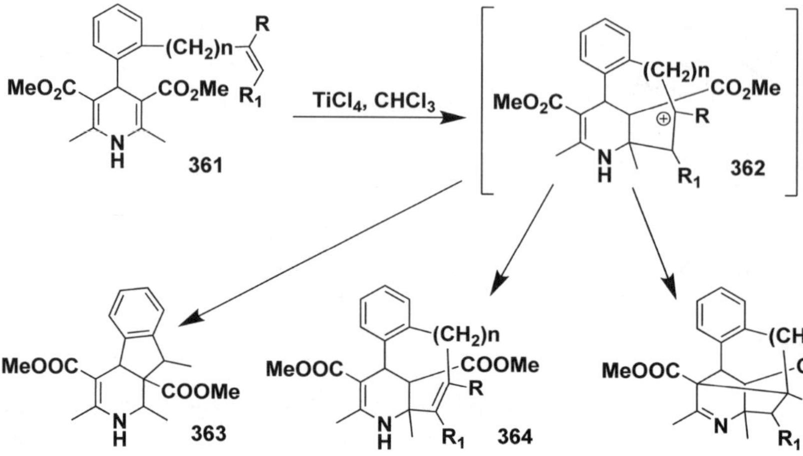


Scheme 3.118

chloroform yields cycloaddition products **363**–**365**, formed via the appropriate carbocation **362** [379] (Scheme 3.118). Cyclization of *ortho*-formyl derivatives **366** under the same conditions leads to compounds **367** [378] (Scheme 3.119).

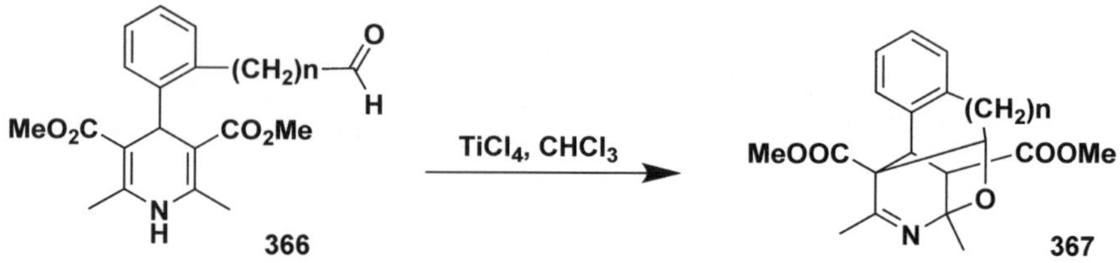


Scheme 3.119

When tozylhydrazone **368** is stirred in benzene at room temperature, within 2 h pyrazoline **369** is formed, while its subsequent refluxing in the same solvent converts the five-membered ring to cyclopropane (compound **370**) [380] (Scheme 3.120). Heterocycle **370** can be obtained directly from hydrazone **368** by heating in benzene for 2 h.

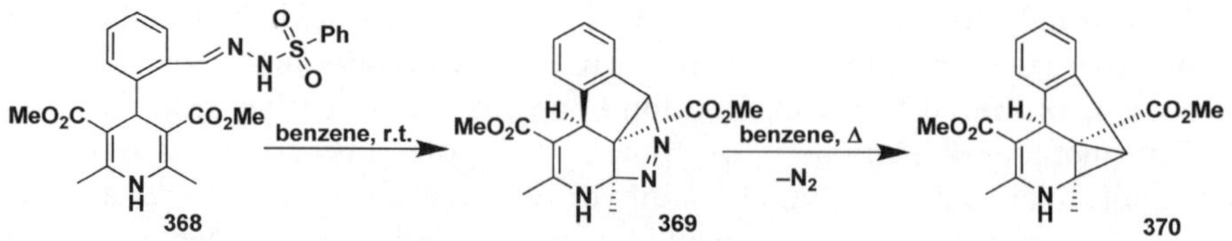


Scheme 3.120

*N*-Benzoyl esters **372** and **374** undergo dihalocarbene addition with phenyl (tribromomethyl)mercury **371** to yield adducts **373** and **375**, respectively [381] (Scheme 3.121). But *N*-methyl- and *N*-benzyl-substituted analogues of **372** and

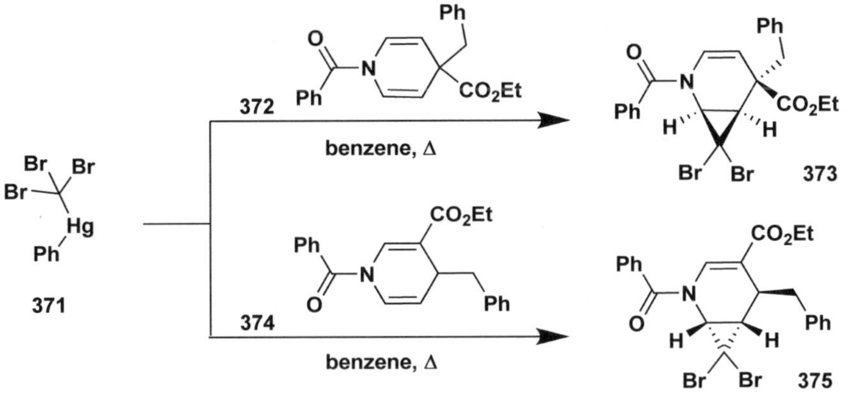


**Scheme 3.121**

**374** fail to give recognizable products when treated with **371**. Adducts **373** and **375** do not undergo a ring-expansion reaction to form corresponding azipine derivatives. In the opinion of McCullough et al. [381] it has to be conceded that assistance from the unpaired electrons on nitrogen would be minimal in the *N*-benzoyl structure.

1,4-Dihydropyridines containing at position 3 a carbonyl group similar to α,β-unsaturated ketones can be involved in cyclocondensation reactions with 1,2-binucleophiles. Dihydropyridine **376** treated with hydroxylamine yields isoxazoline derivatives **377** [363, 382, 383, 384, 385, 386] (Scheme 3.122). Dihydro-1,2,4-triazolo[1,5-*a*]pyrimidine **378** reacts with hydrazine and hydroxylamine in the same manner, giving the condensation products **379** [387].

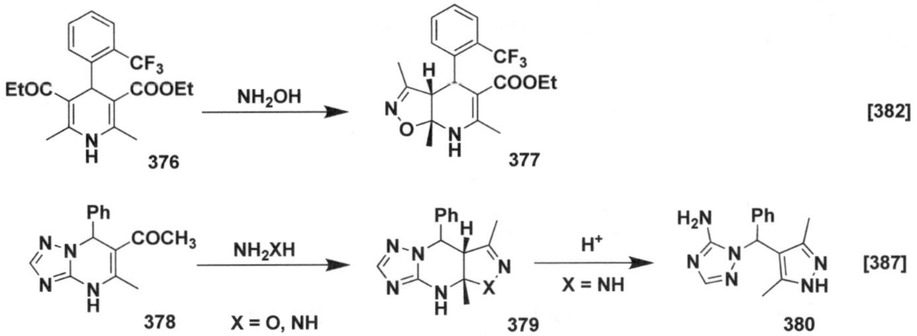


**Scheme 3.122**

It should be noted that compounds **379** are very unstable under acidic conditions: opening of the tetrahydropyrimidine ring with formation of dihetaryarylmethane derivatives **380** occurs even in chloroform by impurities of HCl, which is usually present in chloroform.

Cyclization of 3-carbmethoxy derivatives of 1,4-dihydropiridine **381** after bromination leads to tetrahydrofuro[3,4-*b*]pyridines **383** via the initial formation of bromo derivative **383** [388, 389, 390] (Scheme 3.123). The product upon

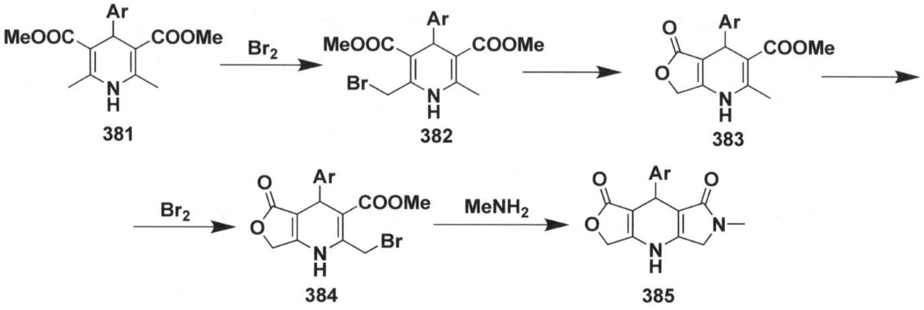


Scheme 3.123

further bromination of **384** treated with primary amines yields derivatives of pyrrolo[3,4-*e*]pyridine **385** [391].

Sulfonation of 4-(2-difluoromethoxyphenyl)-1,4-dihydropyridine **386** is followed by intramolecular cyclization with participation of a hydroxyl group and a carbon atom at position 2 of dihydropyridine [392].

The formation of analogous heterocyclic systems **389** is observed during treatment of the appropriate oxyaryl derivatives of dihydropyridine and pyrimidine **388** under mild conditions [393, 394, 395, 396] (Scheme 3.124). In [393]

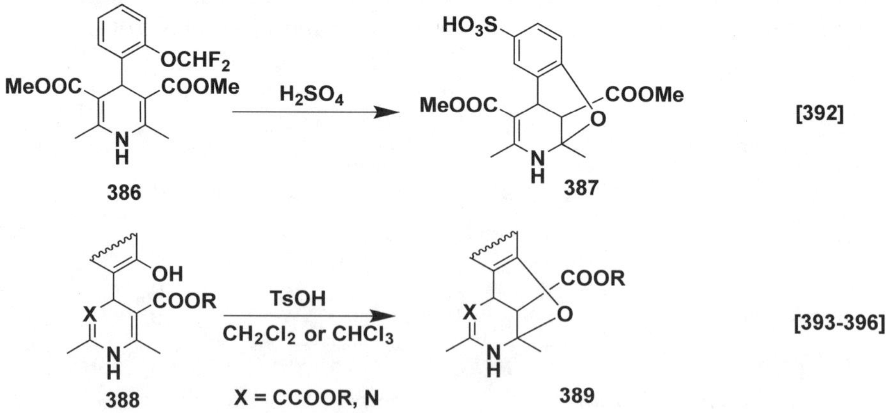


Scheme 3.124

**Scheme 3.125**

formation of both possible stereoisomers is described. Equilibrium between two isomers **A** and **B** (Scheme 3.125) in alcoholic solutions catalyzed with acids and heat was established.

Cyclization of *ortho*-aminophenyl derivatives of 1,4-dihyropyridine **390** leads to analogous heterocycles **391** [397] (Scheme 3.126). But this reaction is

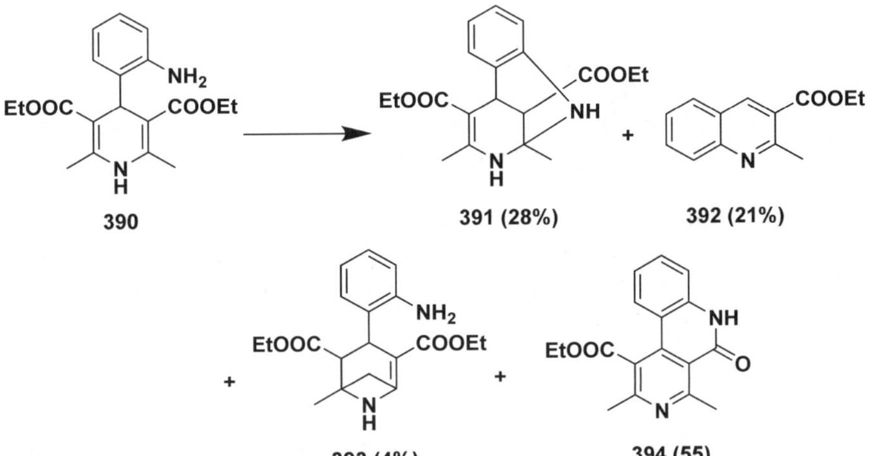


**Scheme 3.126**

followed by several rearrangements and oxidation, forming a lot of side products (**392**–**394**). The structures of several compounds (e.g., heterocycle **393**) described in [397] in our opinion require additional evidence and are doubtful.

Polycyclic lactones based on dihydroazines can be obtained in several ways. For example, trimethylsilyl ester **395** after passing through silica gel is converted to the appropriate lactone **396** [398] (Scheme 3.127). This method is a fast and facile procedure providing lactones **396** in high yields.

Treatment of acids **397** with *meta*-chloroperbenzoic acid in a dichloromethane solution after 2 h at room temperature and silica gel purification

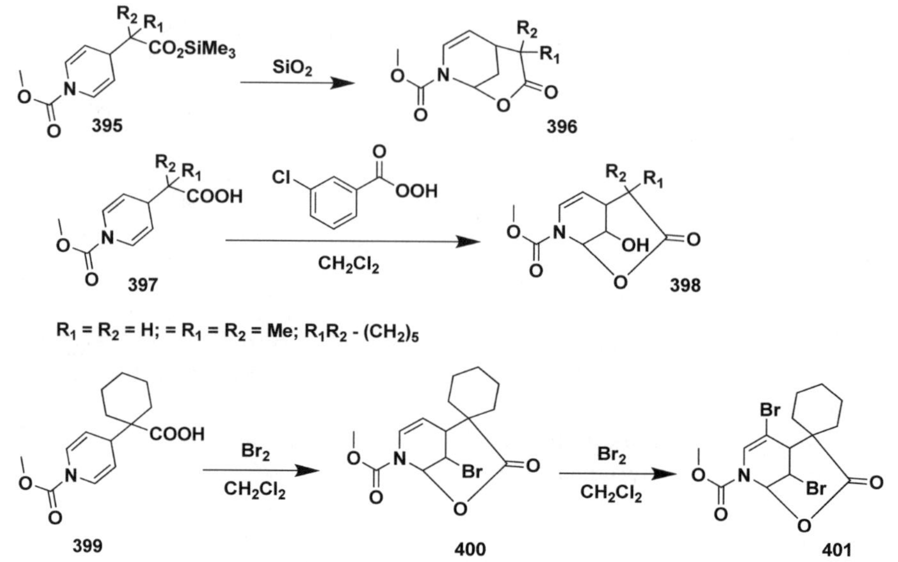


**Scheme 3.127**

gives hydroxylactones **398** in high yields (up to 75%). Bromination of similar acids, for instance compound **399**, leads to bromolactone **400**, while the action of excess bromine gives dibromo derivative **401** [398].

Alcohols containing a dihydropyridine moiety (compounds **402**) very easily undergo intramolecular cyclization to oxazolidines **403**, forming usually a mixture of diastereomers (Scheme 3.128). The reaction occurs during a simple bulb-to-bulb distillation of **402** [399], its filtration over alumina [400] or extraction by ether [401].

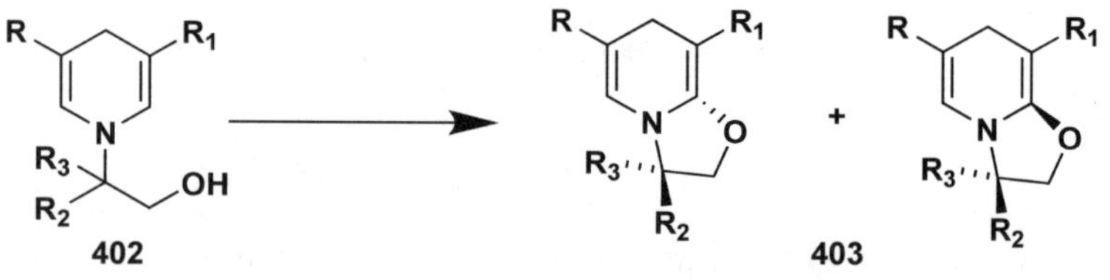


**Scheme 3.128**

Cyclization reactions can involve two functional groups of dihydroazines. For example, treatment of 2-methyl-3,5-biscarbethoxy derivatives of 1,4-dihydropyridine **404** and **405** sequentially with sodium hydride and 1,3,5-triazine yields either 1,4,5,6-tetrahydro-1,6-naphthyridine-3-carboxylate **406** [402, 403, 404, 405] or pyrido[2,3-*d*]pyrimidines **407** [406] (Scheme 3.129).

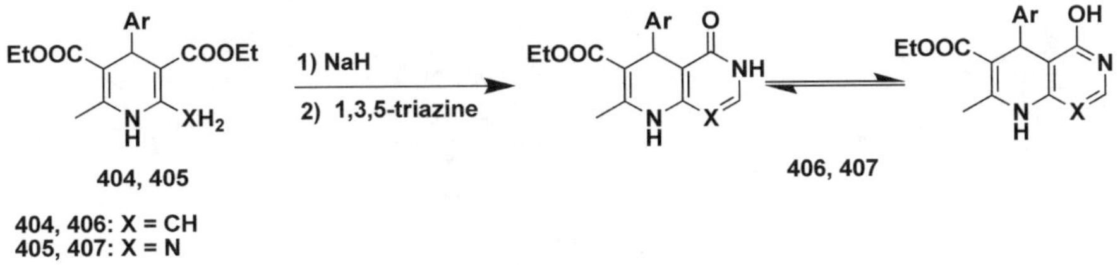


**Scheme 3.129**

Another example of heterocyclizations with the participation of substituents in positions 2 and 3 of the dihydroazine cycle is a treatment of compounds **408** with sodium ethoxide, leading to thienopyridine derivatives **409** [407] (Scheme 3.130).

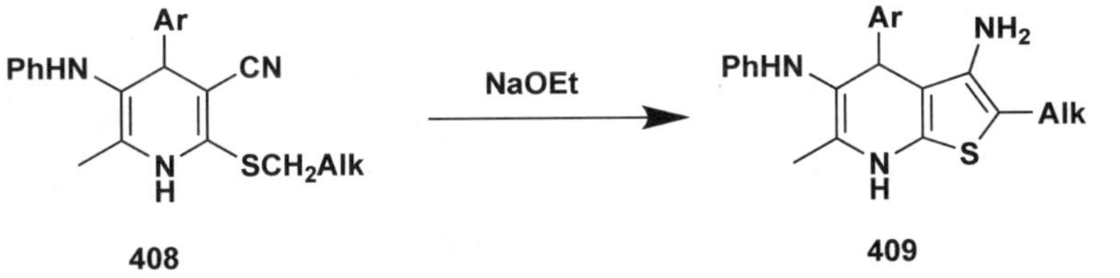


**Scheme 3.130**

The principal stage of the synthesis of alkaloid sesbasine [408] consists of the reaction of starting compounds **410** and **411** by treatment with lithium diisopropilamide (LDA) at –78°C with the formation of an intermediate **412** and subsequently its cyclization into heterocycle **413** (Scheme 3.131). In this case the condensation involves substituents at positions 3 and 4 of the dihydropyridine **412**.

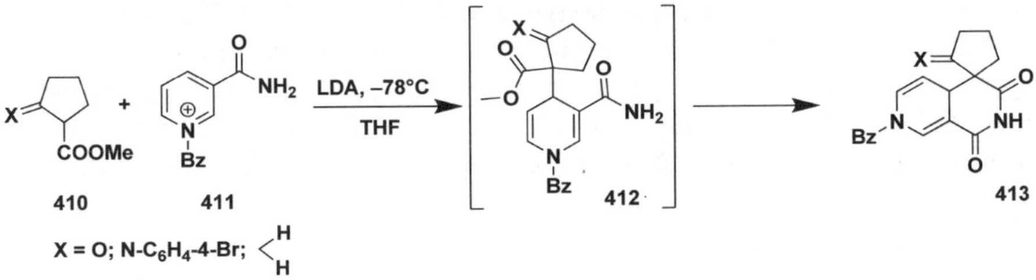


**Scheme 3.131**

Dihydropyridines fused with benzopyrane (**415**) or with quiniline (**416**) moieties are obtained from the corresponding 4-(*ortho*-hyroxyaryl)-substituted

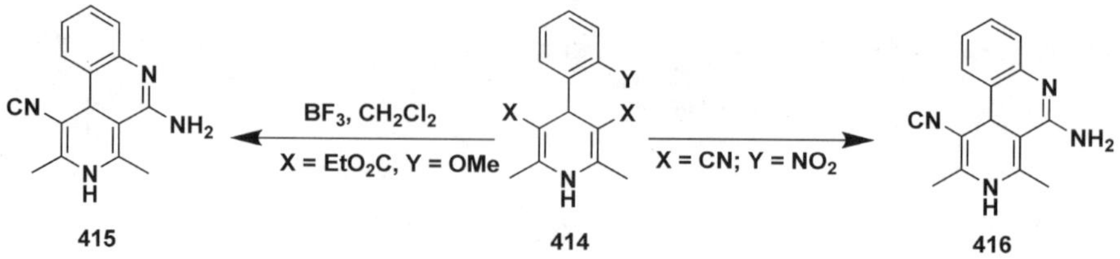


**Scheme 3.132**

1,4-dihydropyrimines **414** [393] with participation of the substituents at positions 3 and 4, similarly to the previous case (Scheme 3.132).

Triazolopyrimidines fused with the benzopyrane ring **419** (usually as mixtures of diastereomers) can be obtained by cyclization of 5-(2-hydroxyaryl)-substituted dihydroazolopyrimidines **417,** with aldehydes **418,** or by direct condensation of 3-amino-1,2,4-triazole **147** with *ortho*-hydroxyacetophenone **421** and 2 mole of the corresponding aldehyde [409, 410] (Scheme 3.133). In the latter case, the

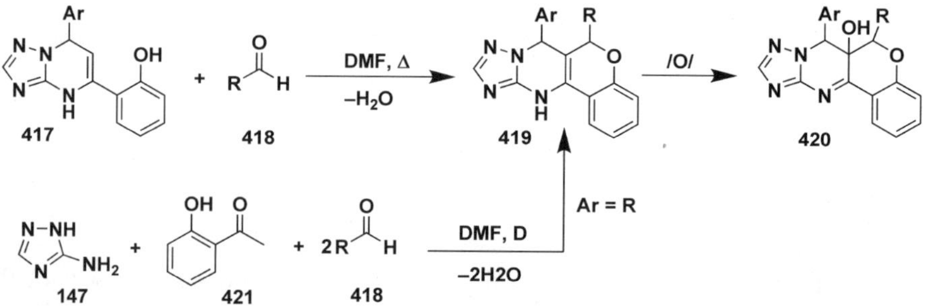


**Scheme 3.133**

formation of the dihydroazine occurs in situ. However, a significant disadvantage of this pathway is the impossibility of introducing two different substituents, Ar and R. Compounds **419** can be very easily oxidized without heteroaromatization by air oxygen into the appropriate hydroxyl derivatives **420** [409, 410].

Thus, at the present time, many effective and facile approaches to the synthesis of diverse heterocycles by modification of dihydroazine derivatives are described. These methods provide strong possibilities using dihydroazines as building-blocks for the construction of new heterocyclic compounds.

## 3.7 Tautomerism of Pyridine and Pyrimidine Dihydro Derivatives

Dihydro derivatives of pyridine and pyrimidine can exist as five isomeric structures **A–F** (Scheme 3.134): for dihydropyridenes **B** is the same as **C**; for dihydropyrimidines **B** is the same as **F**. It is clear that dihydroforms **A** and **B**, **D**

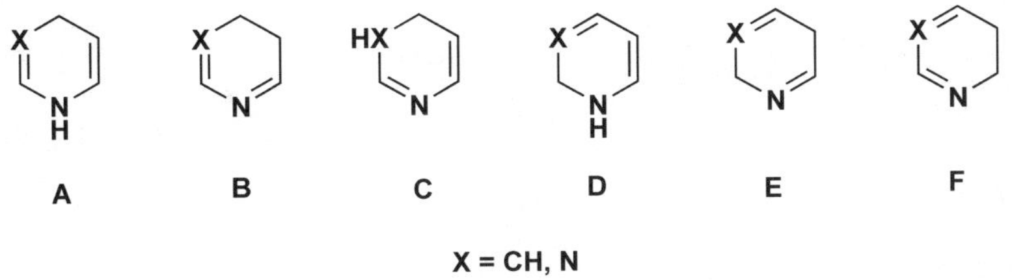


**Scheme 3.134**

and **E** are coupled with an imine–enamine tautomeric equilibrium, while **A** and **C** (X is N) are coupled with an amidine-type tautomerism.

It is a known general rule and also reasonable in most cases that for noncyclic organic compounds, the imine–enamine balance is completely shifted to the imine tautomeric form. It is based on the energetic nonequivalence of specific bonds, that is, the sum of the bond energies of C=N, C–C and C–H calculated according to the additive scheme [411]. The bond energy is 15–20 kcal more than for the additive value of the C–N, C=C and N–H bonds. But in the case of dihydroazines, steric tensions make a sufficient contribution. There is a greater energy stability of the 1,4-dihydroforms than the 1,2-isomers that even exceeds the bond energies (for example, in the case of structures **A** and **B**). It is also very important to note that along with the effects known for noncyclic compounds (conjugation, hydrogen bonds, etc.), for dihydroazines such electronic phenomena as the interaction of "isolated" $\pi$-systems via the bond or the field significantly influence the tautomeric equilibrium [248].

According to *ab initio* quantum-chemical calculations [412] the relative stability of unsubstituted dihydropyridines (X is CH) decreases in the series **A** > **D** > **B**(**C**) > **F** > **E**. Analogous results were obtained with the semiempirical method MINDO/3 [413]. The stability of dihydropyrimidine heterocycles calculated with the *ab initio* methods [248] is different: **C** > **A** > **D** > **E** > **B**(**F**). But for dihydroazines possessing similar energies (dihydropyrimidines **A** and **C**, for instance) substituents can play an important role and change the relative stability. Sausins and Dubur [248] found the following general rules for tautomerism of dihydropyrimidines. The stability of the NH forms is increased by electron-acceptor substituents at the β-position to the NH group. Contrariwise, electron-donor substituents at the α- and γ-positions stabilize the imine tautomeric forms.

In practice, the equilibrium of tautomers in dihydroazines can be observed by experimental methods only in the case of those with similar energies, but this situation occurs rarely. Analysis of the data in the literature shows that for dihydropyridine derivatives **A**(**B**), as a rule, dihydroform **A** is more stable and only in a few cases [414, 415] tautomers **B** are described. For instance, compound **423** was obtained by dehydration of tetrahydropyridine **422** under mild conditions [414] (Scheme 3.135). In the presence of acids, dihydro derivative **423** is converted to a more thermodynamically stable 1,4-dihydro form **424**.

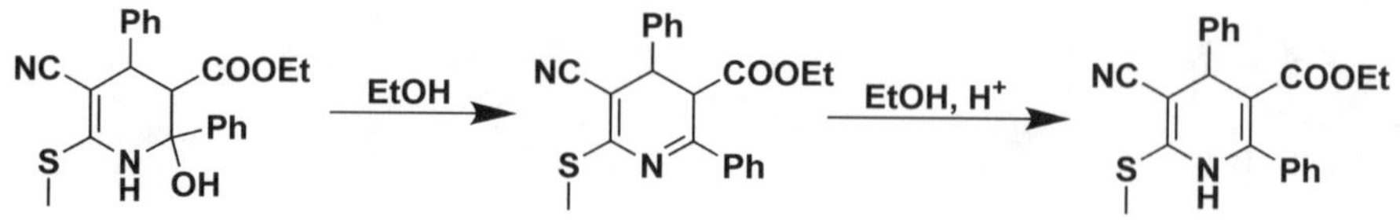


422 423 424

Scheme 3.135

It should be noted that the formation of equilibrium mixtures of **423** and **424** is not observed [414].

But Schwarz et al. [416] showed with NMR spectroscopy that tetraaryl-substituted dihydropyridines **425** in $CDCl_3$ solutions form mixtures of tautomers **A** and **B** in comparable amounts (Scheme 3.136). This detailed study of the

A 425 B

Scheme 3.136

relationship between temperature and composition of the mixture established the thermodynamic characteristics of the tautomeric equilibriums. Replacement of chloroform with DMSO-$d_6$ leads to the formation of compound **425** in the tautomeric form **A**. This phenomenon is associated with formation of hydrogen bonds N–H...O = S, which stabilizes the 1,4-dihydro derivatives [416]. Unfortunately an analysis of the influence of substituents on the tautomeric equilibrium was not carried out.

The most investigated type of tautomerism of dihydropyrimidines is the amidine equilibrium (Scheme 3.137). The energies of substituted dihydropyrimidines with these structures are usually similar and the existence of mixtures of tautomeric forms is typical [248].

Interconversions of dihydroforms **A** and **C** are connected with proton migration from one heteroatom to another and are characterized by a high-speed

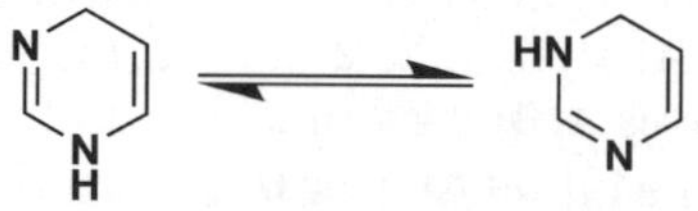


A C

Scheme 3.137

tautomeric transition. For example, Van der Stoel and Van der Plas [417, 418] even at −88 °C using $^{1}$H NMR spectra of dihydropyrimidines observed averaged signals of both tautomeric forms. Amidine tautomerism of dihydropyrimidine was intensively investigated by Weis et al. [419, 420, 421], Kashima et al. [86] and Cho et al. [422, 423]. It was established that tautomeric transformations are effectively catalyzed by even trace amounts of acids or water and their speed increased as the temperature increased. Kinetic investigations showed [421] that amidine-type tautomerism can be explained by two competitive mechanisms in accordance with first- and second-order kinetics equations. The first-order reaction undoubtedly needs the participation of the solvent in the proton transfer process. The second-order reaction is an intermolecular proton exchange between two molecules of dihydropyrimidine and causes acceleration of the tautomeric transition with increasing concentration [248, 419]. For the dihydropyrimidine derivative, two types on imine–enamine tautomeric equilibria are also possible (Scheme 3.138)

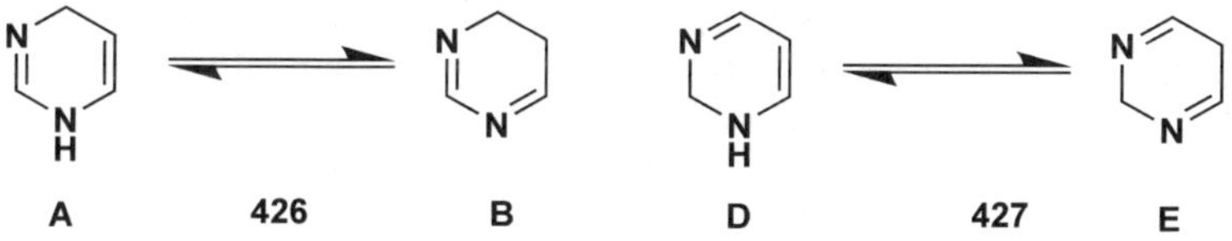


**Scheme 3.138**

It has been shown [248] that unsubstituted dihydropyrimidine **427** existed in tautomeric form **D**. But introduction of phenyl substituents at positions 2 and 4 leads to a convergence of energies of the tautomeric forms and 4,6-diphenyl-1,2(2,5)-dihydropyrimidine is observed in solutions of $CDCl_3$ as a mixture of **D** and **E** in a ratio of 2:1 [248]. Pyrimidines **427** containing electron-donors at positions 4 and 6 exist in the dihydro form **E** [424].

Data concerning the tautomeric equilibrium **A** ⇔ **B** in the case of dihydropyrimidines **426** and their substituted derivatives are absent from the literature, while for heteroannelated pyrimidines (for instance, compounds **428**) the formation of mixtures of tautomers **A** and **B** is a common situation [425] (Scheme 3.139). And as a consequence, even slight structural changes can significantly influence

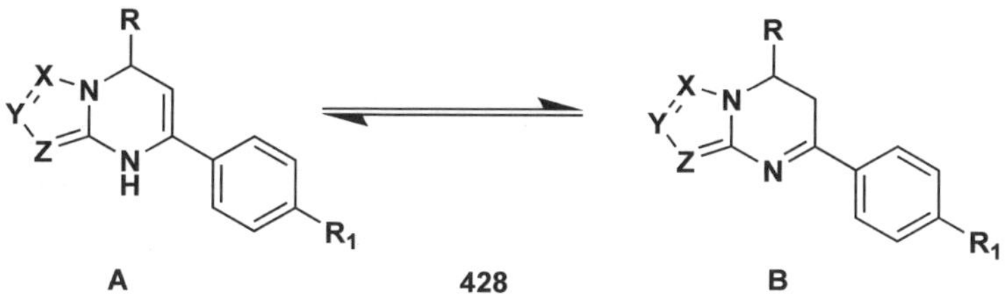


**Scheme 3.139**

the tautomerism of such heterocycles. This fact together with ease of experimental determination of relative amounts of tautomeric forms by NMR spectral methods makes dihydroazolopyrimidines **428** very convenient models to study specific tautomeric equilibria [156, 158, 159, 160, 162, 168, 169, 171, 172, 293, 426, 427, 428, 429, 430, 431].

The key factor affecting tautomeric equilibrium of heterocycles **428** is the nature of the fused azole ring. Decreasing the electron-withdrawing properties of the azole moiety shifts the balance to the **B** form. For example, most dihydro derivatives of pyrazolo[1,5-*a*]pyrimidines [X is N, Y, Z are C(R] and imidazo[1,2-*a*]pyrimidines [X, Z are C(R), Y is N] both in the solid state and in solution exist in the imine tautomeric form **B**, while for dihydropyrimidines fused with the 1,2,4-triazole ring [X, Z are N; Y is C(R)] the tautomers **A** are dominant [156].

Increasing the electron-donor properties of the $R_1$ substituent in the case of dihydro derivatives of 1,2,4-triazolo[1,5-*a*]pyrimidines [X, Z are N, Y is C(R)], tetrazolo[1,5-*a*]pyrimidines (X, Y, Z are N), and pyrimido[1,2-*a*]benzimidazoles (X, Y are *o*-$C_6H_4$, Z is N) tends to stabilize the **B** form. This can be explained by the conjugation of $R_1$ with the electron-withdrawing azomethine group and the azole moiety. Reduction of the electron-accepting properties of the azole fragment (imidazopyrimidines and pyrazolopyrimidines) decreases the influence of the $R_1$ substituent on tautomeric composition.

Knowledge of these rules was used to isolate individual tautomers of compound **429** [431] (Scheme 3.140). This heterocycle is crystallized from alcohols and chloroform in the imine tautomeric form **B**, while in trifluoroacetic acid solution the dimethylamino group undergoes a protonation leading to

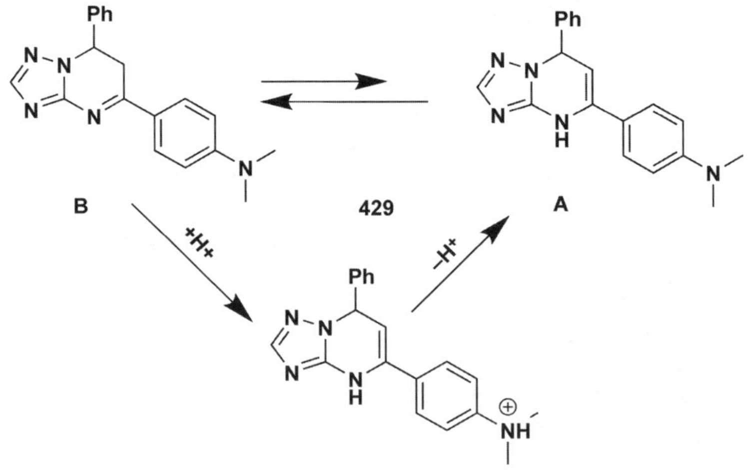


**Scheme 3.140**

formation of electron-withdrawing properties. This stimulates almost complete transformation of **429** into the dihydroform **A**, which can be isolated after rapid neutralization of the solution.

Tautomeric equilibrium of dihydro-1,2,4-triazolopyrimidines is practically insensitive to the electronic character of substituent R [156, 428, 432] (Scheme 3.139), but in some cases the tendency to shift the equilibrium to the dihydroform **B** is observed with increasing donor properties. The influcnce of the size of the substituent R on tautomerism is described in [427, 433]. The most probable reason for the increase in the concentration of form **A** with increasing size of the substituent at position 7 is the flexibility of the enamine tautomer allowing the R group to occupy a more favorable conformation that in the case of form **B**.

The influence of steric factors is more significant for dihydroazolopyrimidines **430** fused with a hydrogenated carbocycle [156] (Scheme 3.141). Regardless of the nature of the azole and carbocycle moieties, these compounds usually exist in the enamine tautomeric form.

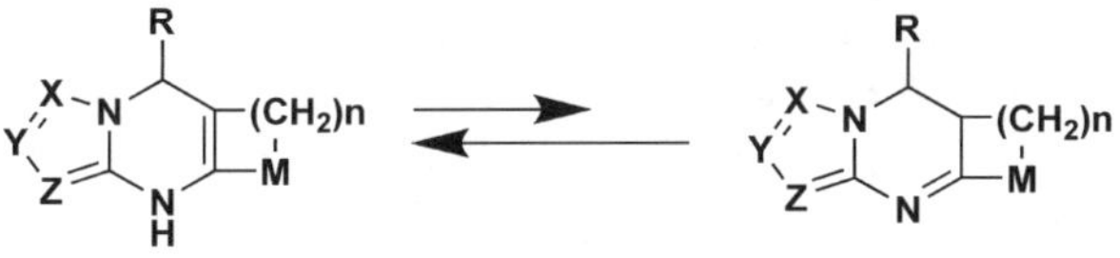


A B

**430**: M = –($CH_2$)–; *o*-$C_6H_4$; n = 1–3

**Scheme 3.141**

In derivatives of dihydroazolopyrimidines containing the *ortho*-oxyaryl substituent (for instance, compounds **431**) a sufficient factor stabilizing the imine tautomeric form is an intamolecular hydrogen bond [428, 429, 430] (Scheme 3.142).

A **431** B

**Scheme 3.142**

The influence of the nature of the solvent on tautomeric balance is an additional demonstration of the significance of the hydrogen bond. The main reason for shifting of the equilibrium to the enamine form **A** when changing the solvent from $CHCl_3$ to *i*-PrOH and DMSO is the competition of intermolecular hydrogen bonds and specific solvation. On the other hand, the absence of well-defined symbiosis between the changing tautomeric composition and the proton-acceptor properties of the solvent (DMSO, DMF, pyridine) points to a significant role of nonspecific solvation [429]. Effects of solvation were used to isolate both forms of heterocycles **431** by crystallization from alcohols or chloroform (**B**) and from DMSO (**A**) [434]. Dissolving of individual crystals of compounds **A** or **B** in different solvents (MeOH, *i*-PrOH, DMSO) leads to the formation of equilibrium mixtures of both forms (Scheme 3.142).

## References

1. Sausins, A.; Duburs, G. *Khim. Geterotsikl. Soedin.*, 1992, *4*, 435.
2. Sausins, A.; Duburs, G. *Heterocycles*, 1988, *27*, 269.
3. Miyabara, H.; Takayasu, T.; Nitta, M. *J. Chem. Soc. Perkin Trans.1*, 1999, *21*, 3199.
4. Miyabara, H.; Takayasu, T.; Nitta, M. *Heterocycles*, 1999, *51*, 983.
5. Zhang, J.; Wang, M.; Huang, Z. *J. Chem. Soc. Perkin Trans. 1*, 1999, *15*, 2087.
6. Zhang, J.; Wang, M.; Huang, Z. *Tetrahedron Lett.*, 1998, *39*, 9237.
7. Troschuetz, R.; Lueckel, A. *Arch. Pharm. (Weinheim)*, 1992, *325*, 785.
8. Tokumitsu, T. *Bull. Chem. Soc .Jpn.*, 1990, *63*, 1921.
9. Katsuyama, I.; Ogawa, S.; Yamaguchi, Y.; Funabiki, K.; Matsui, M. *Synthesis*, 1997, *11*, 1321.
10. Case, F. H.; Sasin, R. *J. Org. Chem.*, 1955, *20*, 1330.
11. Shibata, K.; Urano, K.; Matsui, M. *Bull. Chem. Soc. Jpn.*, 1988, *61*, 2199.
12. Chatterjea, S.; Prasad, Y. *J. Indian Chem. Soc.*, 1957, *34*, 375.
13. Palit, S. R. *J. Indian Chem. Soc.*, 1949, *26*, 501.
14. Kastron, V. V.; Vitolinya, R. O.; Skrastin'sh, I. P.; Dubur, G. Y. *Khim. Farm. Zh.*, 1993, *27*, 20.
15. Caballero, E.; Puebla, P.; Medarde, M.; Feliciano, A. S. *Tetrahedron*, 1993, *49*, 10079.
16. Soliman, F. M. A.; El-Hashash, M. A.; Souka, L.; Salman, A. S. *Rev. Roum. Chim.*, 1996, *41*, 109.
17. Willems, A.; Vandenberghe, N. *Congr. Int. Chim. Ind. Luettich*, 1958, *2*, 476.
18. Sammour, A. *J. Prakt. Chem.*, 1975, *317*, 387.
19. El-Maghraby, M. A.; Abbady, M. A. *Indian J. Chem. Sect. B*, 1978, *16*, 57.
20. Khalil, Z. H.; Yanni, A. S.; Abdel-Hafez, A. A.; Khalaf, A. A. *J. Indian Chem. Soc.*, 1990, *67*, 821.
21. Naik, S. K.; Behera, R. K.; Nayak, A. *Indian J. Chem. Sect. B*, 1982, *21*, 1124.
22. Fahmy, A. M.; Hassan, K. M.; Khalaf, A. A.; Ahmed, R. A. *Indian J. Chem. Sect. B*, 1987, *26*, 884.
23. El-Hashash, M. A.; El-Nagdy, S.; Amine, M. S. *Phosphorus Sulfur Silicon Relat. Elem.*, 1991, *55*, 279.
24. Bayoumy, B. E.; El-Bahie, S.; El-Mobayed, M.; El-Latif, G. A. *Rev. Roum. Chim.*, 1993, *38*, 701.
25. Bayoumy, B. E.; El-Bahaie, S.; El-Mobayed, M.; El-Latif, G. A. *Pol. J. Chem.*, 1991, *65*, 1409.
26. Kulkarni, S. E.; Mane, R. A.; Ingle, D. B. *Indian J. Chem. Sect. B*, 1986, *25*, 452.
27. Mohamed, T. A.; Kandeel, M. M.; Awad, I. M. A.; Youssef, M. S. K. *Collect. Czech. Chem. Commun.*, 1991, *56*, 2999.
28. Kamel, M.; Nabih, I.; Gadalla, K. Z.; Ashour, M. M. *Pharmazie* 1986, *41*, 55.
29. Hammouda, H. A.; Hussain, S. M. *Indian J. Chem. Sect. B*, 1985, *24*, 632.
30. Kamel, M. M.; Gadalla, K. Z.; El-Hamouly, W. S.; El-Wassimi, M. T. *Z. Naturforsch. B Anorg. Chem. Org. Chem.*, 1985, *40*, 661.

31. Etmann, H. A.; Sofan, M. A.; Metwally, M. A. *Boll. Chim. Farm.*, 1995, *134*. 249.
32. Ahmed, G. A. *J. Indian Chem. Soc.*, 1995, *72*, 181.
33. Yagodinets, P. I.; Skripskaya, O. V.; Prodanchuk, N. G.; Chernyuk, I. N.; Sinchenko, V. G. *Khim. Farm. Zh.*, 1995, *29*, 49.
34. Sammour, A. *J. Prakt. Chem.*, 1972, *314*, 941.
35. Hammam, A. E. G.; Hussain, S. M.; Kotob, I. R. *Phosphorus Sulfur Silicon Relat. Elem.*, 1990, *47*, 47.
36. El-Rayyes, N. R.; Ramadan, H. M. *J. Heterocycl. Chem.*, 1987, *24*, 589.
37. Ali, M. I.; Hammam, A. E. G.; Youssef, N. M. *J. Chem. Eng. Data*, 1981, *26*, 214.
38. Hammam, A. E. G.; Ali, M. I. *J. Chem. Eng. Data*, 1981, *26*, 101.
39. Ali, M. I.; Hammam, A. E. G. *J. Chem. Eng. Data*, 1981, *26*, 342.
40. Albar, H. A.; Makki, M. S. I.; Faidallah, H. M. *J. Chem. Res. Miniprint*, 1997, *2*, 336.
41. Peesapati, V.; Rupavani, G. *Indian J. Chem. Sect. B*, 1998, *37*, 468.
42. Sayed, H. H. *Indian J. Chem. Sect. B*, 1998, *37*, 1054.
43. Zigeuner, G.; Frank, A.; Dujmovits, H.; Adam, W. *Monatsh. Chem.*, 1970, *101*, 1415.
44. Mahmoud, M. R.; Madkour, H. M. F. *Synth. Commun.*, 1996, *26*, 3799.
45. Amine, M. S.; Nassar, S. A.; El-Hashash, M. A.; Essawy, S. A.; Hanishish, A. A. *Indian J. Chem. Sect. B*, 1996, *35*, 388
46. El-Hashash, M. A.; Abdel-Rahman, T. M.; Soliman, F. M.; *Boll. Chim. Farm.*, 1998, *137*, 87.
47. Gorbyleva, O. I.; Filipenko, T. Y.; Mikhlina, E. E.; Turchin, K. F.; Sheinker, Y. N.; Yakhontov, L. N. *Khim. Geterotsikl. Soedin.*, 1982, *18*, 793.
48. El-Ablak, F. Z.; Etman, H. A.; Metwally, M. A. *Boll. Chim. Farm.*, 1995, *134*, 77
49. Gautier, J.-A. *Bull. Soc. Chim. Fr.*; 1972, 682
50. Simon, D.; Lafont, O.; Farnoux, C. C.; Miocque, M. *J. Heterocycl. Chem.*, 1985, *22*, 1551.
51. Khalil, M. A.; Soliman, R.; Farghaly, A. M.; Bekhit, A. A. *Arch. Pharm. (Weinheim)*, 1994, *327*, 27
52. Latif, F. M.; Maghraby, A. S.; Barsy, M. A.; Badr, M. Z. A.; Dopp, D. *Collect. Czech. Chem. Commun.*, 1994, *59*, 247.
53. El-Maghraby, M. A.; Khalafalla, A. K.; Hassan, M. E.; Soleiman, H. A. *J. Indian Chem. Soc.*, 1986, *63*, 910.
54. El-Latif, F. M. A; Barsi, M. A.; Maghraby, A. S.; Badr, M. Z. A.; Doepp, D. *J. Indian Chem. Soc.*, 1995, *72*, 641.
55. Geies, A. A.; Bakhite, E. A.; El-Kashef, H. S. *Pharmazie*, 1998, *53*, 686.
56. Zandersons, A. Z.; Lusis, V. K.; Liepin'sh, E. E.; Mutsenietse, D. K.; Khanina, E. L.; Dubur, G. Y. *Khim. Geterotsikl. Soedin.*, 1998, *24*, 1136.
57. Salama, M. A.; El-Essa, S. A. *Indian J. Chem. Sect. B*, 1999, *38*, 739.
58. Ali, M. I.; El-Fotooh, A.; Hamman, G.; Mohamed, S. F. *Phosphorus Sulfur Silicon Relat. Elem.*, 1988, *39*, 211.
59. Kidwai, M.; Misra, P. *Synth. Commun.*, 1999, *29*, 3237.
60. Abdel-Ghani, E. *J. Chem. Res. Miniprint*, 1999, *3*, 1135.
61. El-Hashash, M. A.; Mahmoud, M. R.; Madboli, S. A. *Indian J. Chem. Sect. B,* 1993, *32*, 449.
62. Osske, G.; Szabo, D.; Lorand, T. *Pharmazie*, 1988, *43*, 164.
63. Perjesi, P.; Batta, G.; Foeldesi, A. *Monatsh. Chem.*, 1994, *125*, 433.
64. Lorand, T.; Szabo, D.; Foeldesi, A.; Osske, G. *Pharmazie*, 1984, *39*, 535.
65. Augustin, M.; Jeschke, P. *J. Prakt. Chem.*, 1987, *329*, 626.
66. Pandeya, S. N.; Nair,G. V. *Indian J. Chem.*, 1965, *3*, 165.
67. Singh, H.; Kumar, S.; Singh, P. *J. Chem. Res. Miniprint*, 1984, *5*, 1641.
68. Aziz, G. *Indian J. Chem. Sect. B*, 1976, *14*, 286.
69. Nenajdenko, V. G.; Sanin, A. V.; Kuz'min, V .S.; Balenkova, E. S. *Zh. Org. Khim.*, 1996, *32*, 1579.
70. Perjesi, P.; Foeldesi, A.; Szabo, D.; Zschunke, A.; Mak, M., *Chem. Ber.*, 1987, *120*, 1449.

71. Rovnyak, G.; Shu, V.; Schwartz, J. *J. Heterocycl. Chem.*, 1981, *18*, 327.
72. Hatt K. *Aust. J. Chem.*, 1970, *23*, 561.
73. Sabri, S. S.; Hussein, A. Q.; Al-Hajjar, F. H. *J. Chem. Eng. Data*, 1985, *30*, 512.
74. El-Maati, T. M. A. *Boll. Chim. Farm.*, 1999, *138*, 272.
75. Rehman, A. U.; Qureshi, M. A. *Pak. J. Sci. Ind. Res.*, 1994, *37*, 503.
76. Hasaneen, H.; Kamel, M. M.; El-Zahar, M. I.; Emam, M. *Egypt. J. Chem.*, 1995, *38*, 241.
77. Essawy, S. A.; Wasfy, A. A. F. *Egypt. J. Chem.*, 1994, *37*, 283.
78. Kim, Y. H.; Yoon, C. M.; Lee, N. J. *Heterocycles*, 1981, *16*, 49.
79. El-Rayyes, N. R. *J. Heterocycl. Chem.*, 1982, *19*, 415.
80. Usifoh, C. O.; Olugbade, T. A.; Onawumi, G. O.; Oluwadiya, J.O.; Reisch, J. *J. Heterocycl. Chem.*, 1989, *26*, 1069.
81. Wendelin, W.; Kerbl, H. *Monatsh. Chem.*, 1984, *115*, 309.
82. Wendelin, W.; Schermanz, K. *J. Heterocycl. Chem.*, 1984, *21*, 65.
83. Wendelin, W.; Schramm, H.-W.; Blasi-Rabassa, A. *Monatsh. Chem.*, 1985, *116*, 385.
84. Wendelin, W.; Harler, A. *Monash. Chem.*, 1976, *107*, 133.
85. Wendelin, W.; Schermanz, K.; Fuchsgruber, A.; Harler, A. *Monatsh. Chem.*, 1980, *111*, 1399.
86. Kashima, C.; Shimizu, M.; Omote, Y. *Tetrahedron Lett.*, 1985, *26*, 5057.
87. Weis, A.; Frolow, F. J. *Chem. Soc. Chem. Commun.*, 1982, 89.
88. Schmidt, R. R.; Dimmler, M. *Chem. Ber.*, 1975, *108*, 6.
89. Barrett, G. C. *J. Chem. Soc.*, 1964, 788.
90. Giordano, C. *Gazz. Chim. Ital.*, 1975, *105*, 1265.
91. Perjesi, P.; Foeldesi, A.; Tamas, J. *Monatsh. Chem.*, 1993, *124*, 167.
92. El-Rayyes, N. R., Al-Saleh, B.; Al-Omran, F. *J. Chem. Eng. Data*, 1987, *32*, 280.
93. Weis, A. L.; Frolow, F. *J. Org. Chem.*, 1984, *49*, 3635.
94. Asinger, F.; Thiel, M.; Baltz H. *Monatsh. Chem.* 1957, *88*, 464.
95. Chebanov V. A., Muravyova E. A., Sadchikova Y. V., Desenko S. M., Chernenko V. N., Kobzar K. M., Musatov V. I. *Collect. Czech. Chem. Commun.* 2004, *69*, 897.
96. Muravyova, E. A.; Desenko, S. M.; Musatov, V. I.; Knyazeva, I. V.; Shishkina, S. V.; Shishkin, O. V.; Chebanov, V. A. *J. Comb. Chem.*, 2007, *9*, 798.
97. Al-Hajjar, F. H.; Al-Farkh, Y. A.; Hamoud, H. S. *Can. J. Chem.* 1979, *57*, 2734.
98. Mamaev, V. P.; Weis, A. L. *Khim. Geterotsikl. Soedin.*, 1975, 1555.
99. Weis, A. L.; Mamaev, V. P. *Khim. Geterotsikl. Soedin.*, 1977, 674.
100. Butler, A. R.; Hussain, J.; Leitch E. J.; *Chem. Soc. Perkin Trans. II.*, 1980, 106.
101. Wendelin, W.; Zmolnig, I.; Schram, H.W. *Monatch. Chem.*, 1980, *111*, 1189.
102. Sanin, A. V.; Nenaidenko, V. G.; Krasovskii, A. L.; Churakov, A. V.; Howard, J. A. K.; Balenkova, E. S. *Zh. Org. Khim.*, 1997, *33*, 236.
103. Grant, N.; Mishriky, N.; Asaad, F. M.; Fawzy, N. G. *Pharmazie*, 1998, *53*, 543.
104. Singh, G.; Swati; Mishra, A. K.; Prakash, L. *Indian J. Chem. Sect. B*, 1998, *37*, 517.
105. Attia, A.; Abdel-Salam, O. I.; Abo-Ghalia, M. H.; Amr, A. E. *Egypt. J. Chem.*, 1995, *38*, 543.
106. Manna, F.; Chimenti, F.; Bolasco, A.; Bizzarri, B.; Filippelli, W. *Eur. J. Med. Chem. Chim. Ther.*, 1999, *34*, 245.
107. Singh, G.; Sharma, A. K.; Mishra, A. K.; Prakash, L. *Phosphorus Sulfur Silicon Relat. Elem.*, 1998, *133*, 83.
108. Gadaginamath, G. S., Shyadligeri, A. S.; Kavali, R. R. *Indian J. Chem. Sect. B*, 1998, *37*, 1137.
109. Patel, V. B.; Sorathiya, S. D.; Parikh, A. R.; *Indian J. Chem. Sect. B*, 1997, *36,* 822.
110. Doshi, R.; Kagthara, P.; Parekh, H. *Indian J. Chem. Sect. B*, 1999, *38*, 348.
111. Manna, F; Chimenti, F; Bolasco, A; Filippelli, A; Palla, A; Filippelli, W; Lampa, E; Mercantini, R. *Eur. J. Med. Chem. Chim. Ther.*, 1992, *27*, 627.
112. Zayed, S. M. A. D.; Attia, A. *J. Heterocycl. Chem.*, 1983, *20*, 129.

113. Ankhiwala, M. D. J. *Indian Chem. Soc.*, 1992, *69*, 166.
114. Yassin, F. A. *Egypt. J. Chem.*, 1991, *34*, 539.
115. Shaihla, S. M.; Verma, S. S.; Malik, S.; Mital, R. L.; Prakash, L. *J. Indian Chem. Soc.*, 1990, *67*, 687.
116. Sharma, A. K.; Yadav, A. K.; Prakash, L. *Indian J. Chem. Sect. B*, 1995, *34*, 740.
117. Patel, P.; Koregaokar, S.; Shah, M.; Parekh, H. *Farmaco*, 1996, *51*, 59.
118. Murugan, P.; Raghukumar, V.; Ramakrishnan, V. T. *Synth. Commun.*, 1999, *29*, 3881.
119. Sankaranarayanan, R.; Raj, S. S. S.; Velmurugan, D.; Fun, H.-K.; Raghukumar, V.; Ramakrishnan, V. T. *Acta Crystallogr. Sect. C Cryst. Struct. Commun.*, 1999, *55*, 1670.
120. Wang, D.-L.; Imafuku, K. *Heterocycles*, 1999, *51*, 2093.
121. Nakib, T.; Tyndall, D. V.; Meegan, M. J. *J. Chem. Res. Miniprint*, 1988, 301.
122. Tyndall, D. V.; Al Nakib, T.; Meegan, M. J. *Tetrahedron Lett.*, 1988, *29*, 2703.
123. Nakib, T. A.; Bezjak, V.; Rashid, S.; Fullam, B.; Meegan, M. J. *Eur. J. Med. Chem. Chim. Ther.*, 1991, *26*, 221.
124. Nakib, T.; Tyndall, D. V.; Meegan, M. J. *J. Chem. Res. Miniprint*, 1987, 2743.
125. Al-Arab, M. M. *J. Heterocycl. Chem.*, 1989, *26*, 1665.
126. Latif, N.; Mishriky, N.; Basyouni, W. *Heterocycles*, 1984, *21*, 459.
127. Latif, N.; Mishriky, N.; Haggag, B.; Basyouni, W. Indian *J. Chem. Sect. B*, 1985, *24*, 1230.
128. El-Sadany, S. K.; Sharaf, S. M.; Darwish, A. I.; Youssef, A. A.; *Indian J. Chem. Sect. B*, 1991, 576.
129. Al-Arab, M. M. *J. Heterocycl. Chem.*, 1990, *27*, 523.
130. Kalme, Z. A.; Liepin'sh, E. E.; Pelcher, Y. E.; Dubur, G. Y. *Khim. Geterotsikl. Soedin.*, 1989, 620.
131. Krauze, A. A.; Liepin'sh, E. E.; Dubur, G. Y. *Khim. Geterotsikl. Soedin.*, 1987, 994.
132. Krauze, A. A.; Liepin'sh, E. E.; Dubur, G. Y. *Khim. Geterotsikl. Soedin.*, 1989, 787.
133. Barat J. *J. Indian Chem. Soc.*, 1930, 321.
134. Al-Hajjar, F. H.; Jarrar, A. A. *J. Heterocycl. Chem.* 1980, *17*, 1521.
135. Elgemeie, G. E. H.; Zohdi, H. F.; Sherif, S. M. *Phosphorus Sulfur Silicon Relat. Elem.*, 1990, *54*, 215.
136. Hassan, M. A.; Elkafrawy, A. F.; Elborai, M. *Pol. J. Chem.*, 1982, *56*, 573.
137. Vieweg, H.; Leistner, S.; Wagner, G. *Pharmazie*, 1988, *43*, 358.
138. Wagner, G.; Vieweg, H.; Leistner, S. *Pharmazie*, 1993, *48*, 63.
139. Shestopalov, A. M.; Sharanin, Y. A. *Zh. Org. Khim.*, 1986, *22*, 1291.
140. Geies, A. A.; Abdel-Hafez, A. A.; Lancelot, J. C.; El-Kashef, H. S. *Bull. Chem. Soc. Jpn.*, 1993, *66*, 3716.
141. Geies, A. A.; El-Dean, A. M. K. *Bull. Pol. Acad. Sci. Chem.*, 1997, *45*, 381.
142. Geies, A. A. *Phosphorus Sulfur Silicon Relat. Elem.*, 1999, *148*, 201.
143. Moustafa, A. H. *J. Prakt. Chem.*, 1978, *320*, 97.
144. Moustafa, A. H. *J. Prakt. Chem.*, 1975, *317*, 545
145. Krauze, A.; Germane, S.; Eberlins, O.; Sturms, I.; Klusa, V.; Duburs, G. *Eur. J. Med. Chem. Chim. Ther.*, 1999, *34*, 301.
146. O'Callaghan, C. N.; McMurry, T. B. H.; Cardin, C. J.; Wilcock, D. J. J. *Chem. Soc. Perkin Trans.1*, 1993, 2479.
147. Shestopalov, A. M.; Mortikov, V. Y.; Sharanin, Y. A.; Turov, A. V.; Litvinov, V. P. *Zh. Org. Khim.*, 1989, *25*, 1980.
148. Bakhite, E. A.; Abdel-Monem, M. I. *Phosphorus Sulfur Silicon Relat. Elem.*, 1993, *85*, 129.
149. Assy, M. G.; Youssif, Sh. A.; Ouf, N. H. *Pol. J. Chem.*, 1995, *69*, 896.
150. Moustafa, A. H.; Kaddah, A. M.; El-Abbady, S. A.; Gado, S. H. *J. Prakt. Chem.*, 1982, *324*, 1045.
151. Eweiss, N. F. *J. Heterocycl. Chem.*, 1982, *19*, 273.
152. Krauze, A. A.; Liepin'sh, E. E.; Dubur, G. Y. *Khim. Geterotsikl. Soedin.*, 1990, 115.

153. Otto H. H. *Monatsh. Chem.*, 1978, *109*, 681.
154. Bomika, Z. A.; Pelcher, Yu. E.; Dubur. G. Y.; Krauze, A. A.; Liepin'sh, E. E. *Khim. Geterotsikl. Soedin.*, 1979, 1377.
155. Desenko, S. M.; Orlov, V. D. *Azaheterocycles Based on Aromatic Unsaturated Ketones*, Folio: Kharkov, 1998.
156. Desenko, S. M. *Khim. Geterotsikl. Soedin.*. 1995, 147.
157. Belec, C.; Lhommet, G. *J. Heterocycl. Chem.*, 1995, *32*, 1793.
158. Orlov, V. D.; Desenko, S. M.; Potekhin, K. A.; Struchkov, Y. T. *Khim. Geterotsikl. Soedin.*, 1988, 229.
159. Orlov, V. D.; Desenko, S. M. *Khim. Geterotsikl. Soedin.*, 1989, 1071.
160. Orlov, V. D.; Kolos, N. N.; Kiroga, K.; Desenko, S. M. *Khim. Geterotsikl. Soedin.*, 1988, 962.
161. Lipson, V. V.; Desenko, S. M.; Shirobokova, M. G.; Borodina, V. V.; Karnozhitskaya, T. M.; Krupchitsky, D. A. *Vestnik Khark. Univ. Ser. Khim.*, 2000, 477, 71.
162. Orlov, V. D.; Desenko, S. M.; Pivnenko, N. S. *Khim. Geterotsikl. Soedin.*, 1988, 1489.
163. Desenko, S. M.; Lipson, V. V.; Kaganovsky, A. S.; Borovskoj, V. A.; Ryndina, E. N.; Shirobokova, M. G.; Orlov, V. D. *Fiziol. Akt. Rechov.*, 1999, *2*, 23.
164. Kost, A. N. *Khim. Geterotsikl. Soedin.*, 1980, 1200.
165. Sagitullin, R. S.; Kost, A. N.; Simonova, N. A. *Khim. Geterotsikl. Soedin.* 1971, 62.
166. Sagitullin, R. S.; Kost, A. N.; Mel'nikova, T. V.; Sharbaten, P. A. *Khim. Geterotsikl. Soedin.*, 1977, 945.
167. El Ella, D. A.; Goessnitzer, E.; Werdelin, W. *J. Heterocycl. Chem.* 33. 373.
168. Desenko, S. M.; Orlov, V. D.; Lipson V. V.; Shishkin, O. V.; Potekhin, K. A., Struchkov, Y. T. *Khim. Geterotsikl. Soedin.*, 1999, 109.
169. Desenko, S. M.; Orlov, V. D.; Lipson, V. V.; Kaganovskii, A. A.; Van-Tue, Z.; Ivkov, S. M. *Dokl. Akad. Nauk USSR*, 1990, 44.
170. Desenko, S. M.; Orlov, V. D.; Lipson, V. V.; Estrada, K. *Khim. Geterotsikl. Soedin.*, 1991, 1251.
171. Desenko, S. M.; Kolos, N. N.; Tueni, M.; Orlov, V. D. *Khim. Geterotsikl. Soedin.*, 1990, 938.
172. Desenko, S. M.; Orlov, V. D.; Estrsda, K.; Ponomarev, O. A. *Khim. Geterotsikl. Soedin.*, 1991, 105.
173. Kurihara, T.; Nasu, T.; Tani, T. *J. Heterocycl. Chem.* 1982, *19*, 519 .
174. Desenko, S. M.; Orlov, V. D.; Lipson, V. V. *Khim. Geterotsikl. Soedin.*, 1990, 1638.
175. Lipson, V. V. *PhD thesis,* Kharkov State University, Kharkov, 1991.
176. Desenko, S. M.; Orlov, V. D.; Getmanskii, N. V.; Shishkin, O. V.; Lindeman, S. V.; Struchkov, Y. T. *Khim. Geterotsikl. Soedin.*, 1993, 481.
177. Getmanskii, N. V. *PhD thesis,* Kharkov State University, Kharkov, 1986
178. Wermann, K.; Hartman, M. *Synthesis*, 1991, *3*, 189.
179. Desenko, S. M.; Orlov, V. D.; Getmanskii, N. V.; Shishkin, O. V.; Lindeman, S. V.; Struchkov, Y. T. *Dokl. Akad. Nauk*, 1992, *324*, 801.
180. Desenko, S. M.; Orlov, V. D.; Estrada, Kh. *Khim. Geterotsikl. Soedin.*, 1990, 999.
181. Desenko, S. M.; Orlov, V. D.; Estrada, K., Ivkov, S. M. *Khim. Geterotsikl. Soedin.*, 1991, 694.
182. Lipson, V. V.; Desenko, S. M.; Shirobokova, M. G.; Borodina, V. V. *Khim. Geterotsikl. Soedin.*, 2003, 1383.
183. Lipson, V. V.; Borodina, V. V.; Shirobokova, M. G.; Karnozhitskaya, T. M. *Fiziol. Akt. Rechov.*, 2002, *2*, 10.
184. Lipson, V. V.; Desenko, S. M.; Borodina, V. V.; Shirobokova, M. G.; Musatov, V. I. *Zh. Org. Khim.*, 2005, 115.
185. Lipson, V. V.; Desenko, S. M.; Shishkina, S. V.; Shirobokova, M. G.; Shishkin, O. V.; Orlov, V. D. *Khim. Geterotsikl. Soedin.*, 2003, 1194.
186. Borodina, V. V. *PhD thesis,* Kharkov State University, Kharkov, 2006.

187. Chebanov, V. A.; Desenko, S. M.; Sakhno, Y. I.; Muravyova, E. A.; Saraev, V. E.; Chernenko, V. N.; Kappe, C. O. In *Abstracts of the Third International Conference on Multi-component Reactions and Related Chemistry*, Amsterdam, Netherlands, 2006, p. 52.
188. Chebanov, V. A.; Saraev, V. E.; Desenko, S. M.; Chernenko, V. N.,; Shishkin O. V.; Lukinova, E. V.; Kappe, C. O. In *Abstracts of the 2nd Symposium on Microwave Accelerated Synthesis*, Dusseldorf, Germany, 2006, p. 32
189. Quiroga, J; Mejia, D.; Insuasty, B.; Abonia, R.; Nogueras, M.; Sanchez, A.; Cobo, J.; Low, J. N. *Tetrahedron*, 2001, 6949.
190. Quiroga, J.; Insuasty, B.; Hormaza, A.; Saitz, C.; Jullian, C. *J. Heterocycl. Chem.*, 1998, *35*, 575.
191. Shirobokova, M. G. *PhD thesis,* Kharkov State University, Kharkov, 2001.
192. Drizin, I; Holladay, M. W.; Yi, L.; Zhang, G. Q.; Gopalakrishnan, S.; Gopalakrishnan, M.; Whiteaker, K. L.; Buckner, S. A.; Sullivan, J. P.; Carroll, W. A. *Bioorg. Med. Chem. Lett.*, 2002, *12*, 1481.
193. Chebanov, V. A.; Saraev, V. E.; Desenko, S. M.; Chernenko, V. N.; Shishkina, S. V.; Shishkin, O. V.; Kobzar, K. M.; Kappe, C. O. *Org. Lett.*, 2007, *9*, 1691.
194. Chebanov, V. A.; Saraev, V. E.; Desenko, S. M.; Chernenko, V. N.; Knyazeva, I. V.; Groth, U.; Glasnov, T. N.; Kappe, C. O. *J. Org. Chem.*, 2008, 73, 5110.
195. Drizin, I.; Altenbach, R. J.; Buckner, S. A.; Whiteaker, K. L.; Scott, V. E.; Darbyshire, J. F.; Jayanti, V.; Henry, R. F., Coghlan, M. J.; Gopalakrishnan, M.; Carroll, W. A. *Bioorg. Med. Chem.*, 2004, *12*, 1895.
196. Gladkov, E. S.; Chebanov, V. A.; Desenko, S. M.; Shishkin, O. V.; Shishkina, S. V.; Dallinger, D. Kappe, C. O. *Heterocycles*, 2007, *73*, 469.
197. Chebanov, V. A.; Desenko, S. M.; Kuzmenko, S. A.; Borovskoy, V. A.; Musatov, V. I.; Sadchikova, Yu.V. *Russ. Chem. Bull.*, 2004, *53*, 2845.
198. Borsche, W. *Ber. Dtsch. Chem. Ges.* 1908, *41*, 3886.
199. Chebanov, V. A.; Desenko, S. M.; Sakhno, Y. I.; Panchenko, E. S.; Saraev, V. E.; Musatov, V. I.; Konev, V. F. *Fiziol. Akt. Rechov.*, 2002, 10.
200. Annan, N.; Paris, R.; Jordan, F. *J. Am. Chem. Soc.*, 1999, *111*, 8895.
201. Chebanov, V. A.; Sakhno, Y. I.; Desenko, S. M.; Shishkina, S. V.; Musatov, V. I.; Shishkin, O. V.; Knyazeva, I. V. *Synthesis*, 2005, 2597.
202. Chebanov, V. A.; Sakhno, Y. I.; Desenko, S. M.; Chernenko, V. N.; Musatov, V. I.; Shishkina, S. V.; Shishkin, O. V.; Kappe, C. O. *Tetrahedron*, 2007, *63*, 1229.
203. Sakhno, Y. I.; Chebanov, V. A.; Desenko, S. M.; Chernenko, V. N.; Shishkina, S. V.; Shishkin, O. V.; Lukinova, E. V. In *Abstracts of International Conference "Chemistry of Nitrogen Containing Heterocycles"*, Kharkiv, Ukraine, 2006, p. 113.
204. John, H.; Noziczka, Fr. *J. Prakt. Chem.*, 1925, *111*, 77.
205. John, H.; Weber, G. *J. Prakt. Chem.*, 1925, *111*, 87.
206. John, H. *J. Prakt. Chem.*, 1929, *121*, 179.
207. John, H. *J. Prakt. Chem.*, 1929, *121*, 185.
208. Case, F. H.; Brennan, J. A. *J. Org. Chem.*, 1954, *19*, 919.
209. Gellerman, G.; Rudi, A.; Kashman, Y. *Tetrahedron Lett.*, 1992, *33*, 5577.
210. Ranu, B. C.; Hajra, A.; Jana, U *Tetrahedron Lett.*, 2000, *41*, 531.
211. Ranu, B. C.; Hajra, A.; Dey, S. S.; Jana, U. *Tetrahedron*, 2003, *59*, 813.
212. Koenigs, W.; Jaegle, G. *Chem. Ber.*, 1895, *28*, 1046
213. Strell, M.; Kopp, E. *Chem. Ber.*, 1958, *91*, 2854.
214. Vaughan, W. R.; Tripp, R. C. *J. Am. Chem. Soc.*, 1960, *82*, 4370.
215. Meyer, W. L.; Vaughn, W. R. *J. Org. Chem.*, 1957, *22*, 1560.
216. Vaughan, W. R.; Peters, L. R. *J. Org. Chem.*, 1953, *18*, 382.
217. Garzarolli-Thurnlackh, K. *Monatsh. Chem.* 1899, *20*, 487.
218. Jablonski, W. *Ber. Dtsch. Ges.* 1900, *33*, 2925
219. Saggiomo, A. J.; Kato, K.; Kaiya, T. *J. Med. Chem.*, 1968, *11*, 277.

220. Belen'kaya R. S.; Boreko, E. I.; Zemtsova, M. N.; Kalinina, M. I.; Timofeeva, M. M.; Trakhtenberg, P. L.; Chelnov, V. M.; Lipkin, A. E.; Votyakov, V. I. *Khim. Farm. Zh.*, 1981, *15*, 29.
221. Bhatt, D. J.; Kamdar, G. C.; Parikh, A. R. *J. Indian Chem. Soc.*, 1984, *61*, 816.
222. Bhatt, D. J.; Kamdar, G. C.; Parikh, A. R. *J. Inst. Chem.*, 1984, *56*, 233.
223. Epling, G. A.; Lin, K. Y. *J. Heterocycl. Chem.*, 1987, *24*, 853.
224. Letunov, V. I. *Izv. Vyssh. Uchebn. Zaved., Khim. Khim. Tekhnol.*, 1987, *30*, 27.
225. Atwell, G. J.; Baguley, B. C.; Denny, W. A. *J. Med. Chem.*, 1989, *32*, 396.
226. Raulfs, F. W.; Mayer, U. Eur. Patent Appl. 384313, 1990.
227. Epling, G. A.; Provatas, A. A. *Chem. Commun.*, 2002, *10*, 1036.
228. Um, S.-J.; Park, S.-H.; Park, C.-H.; Chang, B.-H.; Yoon, J.-H.; Sin, H.-S. *Bull. Korean Chem. Soc.*, 2003, *24*, 677.
229. Troschuetz, R. *Arch. Pharm. (Weinheim)*, 1989, *322*, 285.
230. Svetlik, J.; Hanus, V.; Bella, J. *J. Chem. Res. Miniprint*, 1991, *1*, 101.
231. Quiroga, J.; Insuasty, B.; Sanchez A.; Nogueras, M.; Meier H. *J. Heterocycl. Chem.*, 1992, *29*, 1045.
232. Quiroga, J. P.; Garcia, J. O.; Insuasty, B. O.; Mendoza, N. L.; Pungo, M.; Meier, H. *An. Quim.*, 1994, *90*, 300.
233. Rodrigues, R.; Suarez, M.; Ochoa, E.; Morales, A.; Gonzalez, L.; Martin, N.; Quinteiro, M.; Seoane, C.; Soto, J. L. *J. Heterocycl. Chem.*, 1996, *33*, 45.
234. Powers, D. G.; Casebier, D. S.; Fokas, D.; Ryan, W. J.; Troth, J. R.; Coffen, D. L.; *Tetrahedron*, 1998, *54*, 4085.
235. Yamamoto, H.; Takeda, H.; Nitta, M. *Heterocycles*, 2000, *53*, 1891.
236. Ram, V. J.; Saxena, A. S.; Srivastava, S.; Chandra, S. *Bioorg. Med. Chem. Lett.*, 2000, *10*, 2159.
237. Shaker, R. M.; Mahmoud, A. F.; Abdel-Latif, F. F. *Phosphorus Sulfur Silicon Relat. Elem.*, 2000, *160*, 207.
238. El-Gazzar, A. B. A.; Gaafar, A. M.; Aly, A. S. *Phosphorus Sulfur Silicon Relat. Elem.*, 2002, *177*, 45.
239. Takahashi, M.; Nagaoka, H.; Inoue, K. *J. Heterocycl. Chem.*, 2004, *41*, 525.
240. Hassanien, A. Z. A.; Ibrahim, E.-S. I.; Afifi, M. E. *Croat. Chem. Acta*, 2005, *78*, 63.
241. Chebanov, V. A.; Saraev, V. E.; Gura, E. A.; Desenko, S. M.; Musatov, V. I. *Collect. Czech. Chem. Comm.*, 2005, *70*, 350.
242. Barlin, G. B.; Davies, L. P.; Harrison, P. W.; Jacobsen, N. W.; Willis, A. C. *Aust. J. Chem.*, 1994, *47*, 1989.
243. Abdel-Megid, M. *Pharmazie*, 2000, *55*, 263.
244. Grinstein, E. E.; Stankevich, E. I.; Dubur, G. Y. *Khim. Geterotsikl. Soedin. Collect. Vol. 1*, 1967, 395.
245. Agawal, A.; Chauhan, P. M. S. *Tetrahedron Lett.*, 2005, *46*, 1345.
246. Sausins, A.; Duburs, G. *Heterocycles*, 1988, *27*, 291.
247. Sausins, A. E.; Dubur, G. Y. *Khim. Geterotsikl. Soedin.*, 1993, 579.
248. Sausins, A. E.; Dubur, G. Y. *Khim. Geterotsikl. Soedin.*, 1992, 781.
249. Weis, A. L.; Van der Plas, H. C. *Heterocycles*, 1986, *24*, 1433.
250. Berberova, N. T., Okhlobystin O. Y. *Khim. Geterotsikl. Soedin.*, 1984, 1011.
251. Eisner, U.; Kuthan, J. *Chem. Rev.*, 1972, *72*, 1.
252. Stout, D. M.; Meyers, A. I. *Chem. Rev.*, 1982, *82*, 223.
253. Alekseeva, N. V.; Yakhontov, L. N. *Russ. Chem. Rev.*, 1990, *59*, 514.
254. Charushin, V. N.; Chupakhin, O. N.; Van der Plas, H. C., *Heterocycles*, 1988, *43*, 304.
255. Chupakhin, O. N.; Beresnev, D. G. *Russ. Chem. Rev.*, 2002, *71*, 707.
256. Hauck, A. E.; Giam, C. S. *J. Chem. Soc. Perkin Trans. 1*, 1984, 2227.
257. Marchalin, S.; Kuthan, J. *Coll. Chech. Chem. Commun.*, 1985, *50*, 1962.
258. Nomura, Y.; Takeuchi, Y.; Tomoda, S.; Ito, M. M. *Bull. Chem. Soc. Japan*, 1981, *54*, 2779.

259. Snyder, C. A.; Thorn, M. A.; Klijanowicz, J. E.; Southwick, P. L. *J. Heterocycl. Chem.*, 1982, *19*, 603.
260. Lusis, V. K.; Zandersons, A. Z.; Mutsenietse, D. K.; Dubur, G. Y. *Khim. Geterotsikl. Soedin.*, 1983, 508.
261. Kastron, V. V.; Dubur, G. Y.; Vitolins, R.; Kimenis, A.; Selga, M.; Kondratenko, N. V.; Yagupol'skii, L. M.; Tirzite, D.; Fialkov, Y. A.; Shelezhenko, S. V. *Khim. Farm. Zh.*, 1982, *16*, 1322.
262. Krauze, A. A.; Pelcher, Y. E.; Kalme, Z. A.; Dubur, G. Y. *Khim. Geterotsikl. Soedin.*, 1984, 1140.
263. Krauze, A. A.; Liepin'sh, E. E.; Pelcher, Y. E.; Kalme, Z. A.; Dubur, G. Y. *Khim. Geterotsikl. Soedin.*, 1986, 630.
264. Vigante, B. A.; Ozols, Y. Y.; Dubur, G. Y.; Belash, E. M.; Beilis, Y. I. *Khim. Geterotsikl. Soedin.*, 1984, 210.
265. Sagitullina, G. P.; Glizdinskaya, L. V.; Sitnikov, G. V.; Sagitullin, R. S. *Khim.Geterotsikl. Soedin.*, 2002, 1518.
266. Marchalin, S.; Faehnrich, J.; Popl, M.; Kuthan, J. *Collect. Czech. Chem. Commun.*, 1986, *51*; 1061.
267. Carabateas, P. M.; Brundage, R. P.; Gelotte, K. O.; Gruett, M. D.; Lorenz, R. R.; Opalka, C. J. Jr.; Singh, B.; Thielking, W. H.; Williams, G. L.; Lesher, G. Y. *J. Heterocycl. Chem.*, 1984, *21*, 1849.
268. Krauze, A. A.; Liepin'sh, E. E.; Pelcher, Y. E.; Kalme, Z. A.; Dipan, I. V.; Dubur, G. Y. *Khim. Geterotsikl. Soedin.*, 1985, 95
269. Balogh, M.; Hermech, I.; Meszaros, Z; Laszlo, P. *Helv. Chim. Acta.*, 1984. *67*, 2270.
270. Sato, Y.; Shimoji, Y.; Fujita, H.; Mizuno, H.; Kumakura, S. *Yakugaku Zasshi.*, 1978. *98*, 448.
271. Morkovnik, A. S.; Ivakhnenko, E. P.; Bogachev, Y. G.; Tertov, B. A.; Berberova, N. T.; Okhlobystin, O. Y. *Khim. Geterotsikl. Soedin.*, 1988, 203.
272. Zhu, X.-Q.; Zhao, B.-J.; Cheng, J.-P. *J. Org. Chem.*, 2000, *65*, 8158.
273. Meyer, H.; Wehinger, E.; Bossert, F.; Scherling, D. *Arzneim. Forsh.*, 1983, *33*, 106.
274. Meyer, H.; Scherling, D.; Karl, W. *Arzneim. Forsh.*, 1983, 33, 1528.
275. Towart, R.; Wehinger, E.; Meyer, H.; Kaszda, S. *Arzneim. Forsh.*, 1982, *32*, 338.
276. Goerlitzer, K.; Duewel, W. *Arch. Pharm. (Weinheim)*, 1992, *325*, 361.
277. Watana, M.; Fushimi, M.; Baba, N.; Inouye, Y. *Agric. Biol. Chem.*, 1985, *49*, 3533.
278. Hauck, A. E.; Giam, C. S. *J. Chem. Soc. Perkin Trans. 1*, 1980, 2070.
279. Van der Plas, H. C.; Charushin, V. N.; Van Veldhuizen, B. *J. Org. Chem.*, 1983, *48*, 1354.
280. Rosenberg, S. H.; Rapoport, H. *J. Org. Chem.*, 1984, *49*, 56.
281. Shono, T.; Matsumura, Y.; Tsubata, K.; Sugihara, Y.; Yamane, S.; Kanazawa, T.; Aoki, T. *J. Am. Chem. Soc.*, 1982, *104*, 6697.
282. Endo, T.; Ageishi, K.; Furusawa, K.; Okawara, M. *J. Mol. Catal.*, 1983, *18*, 351.
283. Kazarinova, T. D.; Markova, L. I.; Kharchenko, V. G. *Khim. Geterotsikl. Soedin.*, 1990, 511.
284. Konakahara, T.; Hojahmat, M.; Tamura, S. *J. Chem. Soc. Perkin Trans. 1*, 1999, 2803.
285. Muceniece, D.; Lusis, V.; Dubur, G. *Khim. Geterotsikl. Soedin.*, 1982, 1225.
286. Quiroga, J.; Cruz, S.; Insuasty, B.; Abonia, R.; Cobo, J.; Sanchez, A.; Nogueras, M.; Low, J. N. *J. Heterocycl. Chem.*, 2001, *38*, 53.
287. Lipson, V. V.; Desenko, S. M.; Orlov, V. D.; Shishkin, O. V.; Shirobokova, M. G.; Chernenko, V. N.; Zinov'eva, L.I. *Khim. Geterotsikl. Soedin.*, 2000, 1542.
288. Moskovkina, T. V. *Zh. Org. Khim.*, 1999, *35*, 1101.
289. Joergensen, L. V.; Lund, H. *Acta Chem. Scand.*, 1994, *48*, 759.
290. Sedova, V. F.; Aseeva, V. N.; Mamaev, V. P. *Izv. Sib. Otd. Akad. Nauk SSSR, Ser. Khim. Nauk,* 1975. 97.
291. Borovik, V. P.; Mamaev, V. P. *Izv. Sib. Otd. Akad. Nauk SSSR, Ser. Khim. Nauk,* 1990, 40.

292. Ballester, M.; Riera-Figueras, J.; Castafier, J.; Rodriquez, A. *Tetrahedron Lett.*, 1971, *12*, 2079.
293. Desenko, S. M.; Orlo, V. D.; Lipson, V.V.; Shishkin, O.V.; Lindeman, S. V.; Struchkov, Y. T. *Khim. Geterotsikl. Soedin.*, 1993, 688.
294. Kuthan, J.; Kurfust, A. *Ind. Eng. Chem., Prod. Res. Dev.*, 1982, *21*, 191.
295. Desenko, S. M.; Orlo, V. D.; Lipson, V.V.; Shishkin, O.V.; Lindeman, S. V.; Struchkov, Y. T. *Khim. Geterotsikl. Soedin.*, 1994, 981.
296. Desenko, S. M.; Gladkov, E. S.; Komykhov, S. A.; Shishkin, O. V.; Orlov, V. D. *Khim. Geterotsikl. Soedin.*, 2001, 811.
297. Lipson, V. V.; Desenko, S. M.; Shirobokova, M. G.; Borodina, V. V.; Musatov, V. I. *Khim. Geterotsikl. Soedin.*, 2005, 577.
298. Wendelin, W.; Schermanz, K.; Kerbl, J. *Monatsh. Chem.*, 1983, *114*, 717.
299. Courtois, G.; Al-Arnaut, A.; Miginiac, L. *Tetrahedron Lett.*, 1985, *26*, 1027.
300. Comins, D. L.; Abdullah, A. H. *J. Org. Chem.*, 1982, *47*, 4315.
301. Weller, D. D.; Stirchak, E. P.; Weller, D. L. *J. Org. Chem.*, 1983, *48*, 4597.
302. Lubig, R.; Kusch, P.; Rober, K.; Zahn, H. *Monatsch. Chem.*, 1981, *112*, 1313.
303. Claremon, D. A.; Young, S. D. *Tetrahedron Lett.*, 1985, *26*, 5417.
304. Merz; R. *Arch. Pharm. (Weinheim)*, 1937, 294.
305. Nikolaeva, T. G.; Petrova, N. V.; Kriven'ko, A. P. *Khim. Geterotsikl. Soedin.*, 1999, 929.
306. Golikov, A. G.; Reshetov, P. V.; Kriven'ko, A. P. *Khim. Geterotsikl. Soedin.*, 2001, 1339.
307. Rosentreter U. *Synthesis*, 1985, 210.
308. Dyachenko, A. D.; Desenko, S. M.; Dyachenko, V. D.; Chernega, A. N. *Khim. Geterotsikl. Soedin.*, 2004, 765.
309. Aksenov, A. V.; Moiseev, D. V.; Borovlev, I. V.; Nadein, O. N. *Khim. Geterotsikl. Soedin.*, 2000, 1084.
310. Krauze, A.; Sile, L.; Danne, R.; Duburs, G. *Khim. Geterotsikl. Soedin.*, 2005, 1394.
311. Ferrer, S.; Naughton, D. P.; Threadgill, M. D. *Tetrahedron*, 2003, *59*, 3437.
312. Roman, S. V.; Dyachenko, V. D.; Litvinov, V. P. *Izv. Akad. Nauk Ser. Khim.* 2001, *50*, 118.
313. Carroll, W. A.; Altenbach, R. J.; Bai, H.; Brioni, J. D.; Brune, M. E.; Buckner, S. A.; Cassidy, C.; Chen, Y.; Coghlan, M. J.; Daza, A. V.; Drizin, I.; Fey, T. A.; Fitzgerald, M.; Gopalakrishnan, M.; Gregg, R. J.; Henry, R. F.; Holladay, M. W.; King, L. L.; Kort, M. E.; Kym, P. R.; Milicic, I.; Tang, R.; Turner, S. C.; Whiteaker, K. L.; Yi, L.; Zhang, H.; Sullivan, J. P. *J. Med. Chem.*, 2004, *47*, 3163.
314. Lusis, V. K.; Muceniece, D. K, Zandesons, A. Z.; Mazhejka, I. B.; Dubur, G. Y. *Khim. Geterotsikl. Soedin.*, 1984, 393.
315. Comins, D. L.; Mantlo, N. B. *Tetrahedron Lett.*, 1983, *24*, 3683.
316. Akiba, K.; Iseki, Y.; Wada, M. *Bull. Chem. Soc. Jpn.*, 1984, *57*, 1994.
317. Bennasar, M. L.; Vidal, B.; Bosch, J. *J. Chem. Soc. Chem. Commun.*, 1995, 125.
318. Bennasar, M. L.; Juan, C.; Bosch, J. *Tetrahedron Lett.*, 1998, *39,* 9275.
319. Bennasar, M. L.; Zulaica, E.; Roca, T.; Alonso, Y.; Monerris, M. *Tetrahedron Lett.*, 2003, *44*, 4711.
320. Klapars, A.; Vedejs, E. *Khim. Geterotsikl. Soedin.*, 2004, *40*, 887.
321. Comins, D. L.; Herrick, J. J. *Heterocycles*, 1097, *26*, 2159.
322. Martinez, G. R.; Hirschfeld, D. R.; Maloney, P. J.; Yang, D. S.; Rosenkranz, R. P.; Walker, K. A. M. *J. Med. Chem.*, 1989, *32*, 890.
323. Lipson, V. V.; Desenko, S. M.; Shishkin, O. V. *Mendeleev Commun.*, 2006, 280.
324. Lipson, V. V.; Ignatenko, I. V.; Desenko, S. M.; Shishkina, S. V.; Shishkin, O. V.; Komykhov, S. A.; Logvinenko, N. V.; Orlov, V. D.; Meier, H. *J. Heterocycl. Chem.*, 2003, *40*, 1081.
325. Comins, D. L. *Tetrahedron Lett.*, 1983, *24*, 2807.
326. Meyers, A. L.; Edwards, P.; Bailey, T. R.; Jagdman, G. E. *J. Org. Chem.*, 1985, *50*, 1019.
327. Anderson, A. G., Jr.; Berkelhammer, G. *J. Am. Chem. Soc.*, 1958, *80*, 992.

328. Gutsche, C. D.; Alam, I. *Tetrahedron*, 1988, *44*, 4689-4694.
329. Skibo, E. B.; Bruice, T. C. *J. Am. Chem. Soc.*, 1983, *105*, 3316.
330. Shinkai, S.; Koreishi, H.; Mori, S.; Sone, T.; Manabe, O. *Chem. Lett.*, 1985, 1033.
331. Hamachi, I.; Kobuke, Y.; Tabushi, I. *Bull. Chem. Soc. Jpn.*, 1988, *61*, 3613.
332. Steele, T. G.; Coburn, C. A.; Patane, M. A.; Bock, M. G. *Tetrahedron Lett.*, 1998, *39*, 9315.
333. Singh, H.; Kumar, S. *Indian J. Chem.*, 1986, *25B*, 688.
334. Butt, G. L.; Deady, L. W. *J. Heterocycl. Chem.*, 1984, *21*, 981.
335. Wallenfels, K.; Hofmann, D.; Schuly, H. *Ann. Chem.*, 1959, 188.
336. Oae, S.; Nagata, T.; Yoshimura, T.; Fujimori, K. *Tetrahedron Lett.*, 1982, *23*, 3189.
337. Lavilla, R.; Coll, O.; Kumar, R.; Bosch, J. *J. Org. Chem.*, 1998, *63*, 2728.
338. Lavilla, R.; Kumar, R.; Coll, O.; Masdeu, C.; Bosch, J. *Chem. Commun.*, 1998, 2715.
339. Lavilla, R.; Kumar, R.; Coll, O.; Masdeu, C.; Spada, A.; Bosch, J.; Espinosa, E.; Molins, E. *Chem. Eur. J.*, 2000, *6*, 1763.
340. Lavilla, R.; Coll, O.; Nicolas, M.; Sufi, B. A.; Torrents, J.; Bosch, J. *Eur. J. Org. Chem.*, 1999, 2997.
341. Lavilla, R.; Baron, X.; Coll, O.; Gullon, F.; Masdeu, C.; Bosch, J. *J. Org. Chem.*, 1998, *63*, 10001.
342. Alvarez, M.; Lavilla, R.; Bosch, J. *Tetrahedron Lett.*, 1987, *28*, 4457.
343. Spitzner, D.; Wenkert, E. *Angew. Chem.*, 1984, *96*, 972.
344. Wanner, M. J.; Koomen, G. J.; Pandit, U. K. *Tetrahedron*, 1983, *39*, 3673.
345. Wenkert, E.; Pyrek, J. St.; Uesato, S.; Vankar, Y. D. *J. Am. Chem. Soc.*, 1982, *104*, 2244.
346. Jimenez, J.-M.; Zulaica, E.; Bennasar, M. L.; Bosch, J. *J. Chem. Soc. Chem. Commun.*, 1993, 732.
347. Mangeney, P.; Gosmini, R.; Raussou, S.; Commercon, M.; Alexakis, A. *J. Org. Chem.*, 1994, *59*, 1877.
348. Bennasar, M. L.; Zulaica, E.; Ramirez, A.; Bosch, J. *J. Org. Chem.*, 1996, *61*, 1239.
349. Miettinen, J.; Jokela, R.; Lounasmaa, M. *Planta Med.*, 1996, *62*, 42.
350. Bennasar, M. L.; Jimenez, J.-M.; Sufi, B. A.; Bosch, J. *Tetrahedron Lett.*, 1996, *37*, 7653.
351. Bennasar, M. L.; Jimenez, J.-M.; Sufi, B. A.; Bosch, J. *Tetrahedron Lett.*, 1996, *37*, 9105.
352. Bennasar, M. L.; Jimenez, J.-M.; Vidal, B.; Sufi, B. A.; Bosch, J. *J. Org. Chem.*, 1999, *64*, 9605.
353. Lavilla, R.; Coll, O.; Nicolas, M.; Bosch, J. *Tetrahedron Lett.*, 1998, *39*, 5089.
354. Bennasar, M. L.; Juan, C.; Roca, T.; Monerris, M.; Bosch, J. *Tetrahedron*, 2001, *57*, 10125.
355. Lavilla, R.; Bernabeu, M. C.; Carranco, I.; Diaz, J. L. *Org. Lett.*, 2003, *5*, 717.
356. Putkonen, T.; Tolvanen, A.; Jokela, R.; Caccamese, S.; Parrinello, N. *Tetrahedron*, 2003, *59*, 8589.
357. Rosentreter, U.; Born, L.; Kurz, J. *J. Org. Chem.*, 1986, *51*, 1165.
358. Hartman, G. D.; Halczenko, W.; Phillips, B. T.; Pitzenberger, S. M. *J. Org. Chem.*, 1986, *51*, 2202.
359. Shumakov, S. A.; Kaminskii, V. A.; Tilichenko, M. N. *Khim. Geterotsikl. Soedin.*, 1985, 89.
360. Shumakov, S. A.; Kaminskii, V. A.; *Khim. Geterotsikl. Soedin.*, 1991, 697.
361. Kaminskii, V. A.; Kostyukevich, T. A.; Shumakov, S. A.; Tilichenko, M. N. *Khim. Geterotsikl. Soedin.*, 1983, 1692.
362. Akiba, K.; Ohtani, A.; Yamamoto, Y. *J. Org. Chem.*, 1986, *51*, 5328.
363. Taylor, M. D.; Himmelsbach, R. J.; Kornberg, B. E.; Quin, J.; Lunney, E.; Michel, A. *J. Org. Chem.*, 1989, *54*, 5585.
364. Soufiaoui, M.; Jazouli, H.; *J. Heterocycl. Chem.*, 1990, *27*, 1931.
365. Ondrus, T. A.; Knaus, E. E.; Giam, C. S. *Can. J. Chem.*, 1978, *56*, 1026.
366. Ondrus, T. A.; Knaus, E. E.; Giam, C. S. *Can. J. Chem.*, 1979, *57*, 2342.
367. Warren, B. K.; Knaus, E. E. *J. Med. Chem.*, 1981, *24*, 462.

368. Fukuzumi, S.; Fujii, Y.; Suenobu, T. *J. Am. Chem. Soc.*, 2001, *123*, 10191.
369. Hilgeroth, A.; Brachwitz, K.; Baumeister, U. *Heterocycles*, 2001, *55*, 661.
370. Adembri, G.; Donati, D.; Fusi, S.; Ponticelli, F. *Heterocycles,* 1987, *26*, 3221.
371. Adembri, G.; Camparini, A.; Donati, D.; Fusi, S.; Ponticelli, F.; Scotton, M. *Tetrahedron Lett.*, 1983, *24*, 5399.
372. Nisole, C.; Uriac, P.; Huet, J.; Toupet, L. *J. Chem. Res. Miniprint*, 1991, 1914.
373. Pragst, F.; Koeppel, H.; Walkhoff, E.; Boche, E. *J. Prakt. Chem.*, 1987, *329*, 665.
374. Hartman, G. D.; Halczenko, W.; Phillips, B. T. US Patent 4,808,718, 1989.
375. Hartman, G. D.; Phillips, B. T.; Halczenko, W.; Springer, J. P.; Hirshfield, J. *J. Org. Chem.*, 1987, *52*, 1136.
376. Hartman, G. D.; Phillips, B. T.; Halczenko, W. *J. Org. Chem.*, 1989, *54*, 4134.
377. Hartman, G. D.; Halczenko, W.; Cochran, D. W *Can. J. Chem.*, 1986, 556.
378. Hartman, G. D.; Halczenko, W.; Phillips, B. T. *J. Org. Chem.*, 1985, *50*, 2427.
379. Hartman, G. D.; Halczenko, W.; Phillips, B. T. *J. Org. Chem.*, 1986, *51*, 142.
380. Remy, D. C.; King, S. W.; Cochran, D.; Springer, J. P.; Hirshfield, J. *J. Org. Chem.*, 1985, *50*, 4120.
381. McCullough, K. J.; MacTavish, J.; Proctor, G. R.; Redpath, J. *J. Chem. Soc. Perkin Trans. 1*, 1996, *20*, 2553.
382. Kim, D.H. US Patent 4,593,100, 1984.
383. Toth, G.; Szollosy, A.; Levai, A.; Kotovich. G. *J. Chem. Soc. Perkin Trans. 2,* 1986, 1895.
384. Toth, G.; Levai, A.; Duddeck, H. *Magn. Reson. Chem.*, 1992, *30*, 235.
385. Toth, G.; Levai, A.; Szollosy. A.; Duddeck, H. *Tetrahedron*, 1993, *49*, 863.
386. Padwa, A.; Eastman, D.; Hamilton, L. *J. Org. Chem.*, 1968, *33*, 1317.
387. Desenko, S. M.; Komykhov, S. A.; Orlov, V. D.; Meier, H. *J. Heterocycl. Chem.*, 1998, *35*, 989.
388. Skrastins, I P.; Kastron, V. V.; Dubur, G. Y.; Mazhejka, I. B.; Kadysh, V. P *Khim. Geterotsikl. Soedin.*, 1987, 1227.
389. Skrastins, I. P.; Kastron, V. V.; Dubur, G. Y.; Mazhejka, I. B.; Liepins, E.E. *Khim. Geterotsikl. Soedin.*, 1989, 948.
390. Young, S. D. *Synthesis*, 1984, 617.
391. Skrastins, I P.; Kastron, V. V.; Dubur, G. Y. *Khim. Geterotsikl. Soedin.*, 1992, 570.
392. Liepins, E. E.; Skrastins, I. P.; Kastron, V. V.; Dubur, G. Y. *Khim. Geterotsikl. Soedin.*, 1989, 1568.
393. Claremon, D. A.; Hirshfield, J.; Lumma, P. K.; McClure, D. E.; Springer, J. P. *Synthesis*, 1986, 144.
394. Claremon, D. A. Eur. Patent 163238, 1985.
395. Claremon, D. A. US Patent 4,579,851, 1984.
396. Baldwin, J. J.; Claremon, D. A.; McClure. D. E. US Patent 4,609,494, 1985.
397. Kim, D. H. *J. Heterocyl. Chem.* 1986, *23*, 1471.
398. Rudler, H.; Denise, B.; Xu, Y.; Parlier, A.; Vaissermann, J. *Eur. J. Org. Chem.*, 2005, 3724.
399. Wong, Y.-S.; Marazano, C.; Gnecco, D.; Genisson, Y.; Chiaroni, A.; Das, B. C. *J. Org. Chem.*, 1997, 62, 729.
400. Gnecco, D.; Marazano, C.; Das, B. C. *J. Chem. Soc. Chem. Commun.*, 1991, 625.
401. Wong, Y.-S.; Marazano, C.; Gnecco, D.; Das, B. C. *Tetrahedron Lett.*, 1994, *35*, 707.
402. Satzinger, G.; Hartenstein, J.; Mannhardt, K.; Kleinschroth, J.; Herrmann, M.; Fritschi, E.; Tauschel, H. D.; Wagner, B.; Wolf, G. Eur. Patent 189898, 1986.
403. Satzinger, G.; Hartenstein, J.; Mannhardt, K.; Kleinschroth, J.; Osswald, H.; Weinheimer, G.; Fritschi, E. Eur. Patent 173933, 1986.
404. Satzinger, G.; Hartenstein, J.; Mannhardt, K.; Kleinschroth, J.; Fritschi, E.; Osswald, H.; Weinheimer, G. Ger. Patent 3602655, 1987.
405. Kleinschroth, J.; Mannhardt, K.; Wagner, B.; Weinheimer, G. Ger. Patent 3609795, 1987.

406. Kleinschroth, J.; Satzinger, G.; Mannhardt, K.; Hartenstein, J.; Osswald, H.; Weinheimer, G.; Fritschi, E. Ger. Patent 3438350, 1986.
407. Docenko, V. V.; Krivokolysko, S. G.; Dyachenko, V. D.; Sharanin, Y. O. In XVII Ukrainian Conference in Organic Chemistry, 1995, Kharkiv, Ukraine, p. 155.
408. Wanner, M. J.; Koomen, G. J.; Pandit, U. K. *Tetrahedron*, 1982, *38*, 2741.
409. Desenko, S. M.; Orlov, V. D.; Getmanskii, N. V.; Komykhov, S. A. *Khim. Geterotsikl. Soedin.*, 1993, 1433.
410. Desenko S. M.; Orlov, V. D.; Getmanskii, N. V.; Komykhov, S. A.; Paponov, B. V.; Kovalevsky, A. Y.; Shishkin, O. V.; Struchkov, Y. T. *Khim. Geterotsikl. Soedin.*, 1996, 240.
411. Carey, F.A.; Sundberg, R.J. *Advanced Organic Chemistry. Part A: Structure and Mechanisms*, Springer: New York, 2007.
412. Bohm, S.; Kuthan, J. *Collect. Czech. Chem. Commun.*, 1981, *46*, 2068.
413. Bodor, N.; Pearlman, R. *J. Am. Chem. Soc.*, 1978, *100*, 4946.
414. Krauze, A. A.; Liepin'sh, E. E.; Pelcher, Y. E.; Kalme, Z. A.; Dubur, G. Y. *Khim. Geterotsikl. Soedin.*, 1987, 75.
415. Krauze, A. A.; Liepin'sh, E. E.; Dubur, G. Y. *Khim. Geterotsikl. Soedin.*, 1987, 563.
416. Schwarz, M.; Trska, P.; Kuthan, J. *Collect. Czech. Chem. Commun.*, 1989, *54*, 1854.
417. Van der Stoel, R. E.; Van der Plas, H. C. *Recl. Trav. Chim. Pays-Bas.*, 1978, *97*, 116.
418. Van der Stoel, R. E.; Van der Plas, H. C. *J. Chem. Soc. Perkin Trans. 1.*, 1979, 1228.
419. Weis, A. L. *Tetrahedron Lett.*, 1982, *23*, 449.
420. Weis, A. L.; Frolow, F. *Heterocycles*, 1982, 493.
421. Weis, A. L.; Porat, Z.; Luz, Z. *J. Am. Chem. Soc.*, 1984. *106*, 8021.
422. Cho, H.; Hamamoto, Y.; Shima, K.; Taira, Z. *Tetrahedron*, 1986, *42*, 6429.
423. Cho, H.; Iwashita, T.; Ueda, M.; Mizuno, A.; Mizukawa, K.; Hamaguchi, M. *J. Am. Chem. Soc.*, 1988, *110*, 4832.
424. Weis, A. L.; Vishkautsan, R. *Heterocycles,* 1982, *23*, 1077.
425. Shishkin, O. V. *PhD thesis,* Kharkov State University, Kharkov, 1986.
426. Desenko, S. M.; Orlov, V. D.; Lipson, V. V.; Estrada, K. *Khim. Geterotsikl. Soedin.*, 1991, 1215.
427. Desenko, S. M.; Orlov, V. D.; Shishkin, O. V.; Barykin, K. E.; Lindeman, S. V.; Struchkov, Y. T. *Khim. Geterotsikl. Soedin.*, 1993, 1357.
428. Desenko, S. M.; Orlov, V. D.; Getmanskii, N. V.; Komykhov, S. A. *Khim. Geterotsikl. Soedin.*, 1993, 1353.
429. Desenko, S. M.; Orlov, V. D.; Beskorovainyi, V. I. Uspenskii, B.V. *Dokl. Akad. Nauk USSR*, 1989, 38.
430. Desenko, S. M.; Orlov, V. D.; Beskorovainyi, V. I. *Khim. Geterotsikl. Soedin.*, 1991, 245.
431. Desenko, S. M.; Orlov, V. D.; Shishkin, O. V.; Lipson, V. V.; Lindeman, S. V.; Struchkov, Y. T. *Khim. Geterotsikl. Soedin.*, 1992, 933.
432. Desenko, S. M.; Orlov, V. D.; Ryndina, N. E.; Popov, V. I.; Ygupolskii, L. M. *Dokl. Akad. Nauk USSR*, 1993, 122.
433. Desenko, S. M.; Orlov, V. D.; Lipson, V. V.; Shishkin, O. V.; Lindeman, S. V.; Struchkov, Y. T. *Khim. Geterotsikl. Soedin.*, 1991, 1539
434. Desenko, S. M.; Orlov, V. D. *Khim. Geterotsikl. Soedin.*, 1989, 1000.

# Chapter 4
# Cyclocondensations of 1,2-Diamines

In the series of heterocyclizations of unsaturated carbonyl compounds, cyclocondensations based on *ortho*-diamines (or more specifically 1,2-diamines) are of particular significance. These reactions are characterized by the formation of different heterocycles with various (sometimes unexpected) structures.

First of all, it should be noted that the products of the "normal" interaction of unsaturated ketones with 1,2-diamines, i.e., dihydrated diazepine and triazepine systems, possess considerable chemical lability and are able to undergo further transformations (multistage, as a rule). Moreover, if there exists an alternative, then the process of seven-membered heterocycle formation is noticeably less advantageous thermodynamically as opposed to that of six- or five-membered structures, especially heteroaromatic ones. Owing to this, the interaction of *ortho*-diamines with chalcones is characterized either by superimposition of numerous secondary chemical transformations during condensation or by alternative reaction pathways. This has given rise to prolonged discussions concerning the structures of the products formed: in some papers three or more structures based on corresponding spectral and chemical data have been accredited to certain products.

In the present chapter, an attempt is made to systemize the literature data and to analyze the rules of the reactions of 1,2-diamines, both the aromatic and the heteroaromatic nature, with chalcones for the synthesis of specific heterocycles.

## 4.1 Synthesis of 1,5-Dihydroazepines

The interaction of *ortho*-phenylenediamine (*o*-PDA) with aliphatic α,β-unsaturated ketones, in particular, mesityl oxide, was reported for the first time as long ago as 1905 [1]. Nevertheless, only in the 1950s was the reaction product convincingly proved to have the structure of dihydrobenzodiazepine [2, 3]. Meanwhile, the condensation reactions of aromatic unsaturated ketones with aromatic and heterocyclic *ortho*-diamines remained unstudied for a long time. One of the reasons for this situation was attributed to the impossibility of the formation of a dihydrobenzodiazepine cycle by the interaction between *o*-PDA **1** and chalcone

V.A. Chebanov et al., *Azaheterocycles Based on α,β-Unsaturated Carbonyls*, 

[2, 4]. According to the data obtained, such a reaction either does not proceed at all [4] or yields the corresponding *β*-amino adduct **2** or 2-phenylbenzimidazole **4** [2], depending on the reaction conditions. These results were treated as a general feature of the interaction between *o*-PDA and chalcones [2]. The impossibility of the formation of aromatic dihydrobenzodiazepine derivatives was explained by the essentially lower reactivity of the carbonyl group in chalcone molecules in comparison with that of the C = C bond.

The first reliable report on the synthesis of 2,4-diphenyl-2,3-dihydro-1*H*-1,5-benzodiazepine **3** for the reaction of *o*-PDA and chalcone was published in 1975 [5].

A more detailed study [6] showed the β-amination of an unsaturated ketone and the formation of adduct **2** (Scheme 4.1) to be the first stage of this reaction. Later the general character of the interaction between *o*-PDA and substituted ketones or their vinylogs was established [7, 8]. Unsuccessful attempts [7] to synthesize aromatic dihydrobenzodiazepine derivatives from 1,2-diaminoanthraquinone, 4-nitro-, 4-nitrile-4,6-dichlor- and *N*-phenyl-substituted *o*-PDAs were also reported. Condensation of *o*-PDA with *ortho*-hydroxychalcone **5**

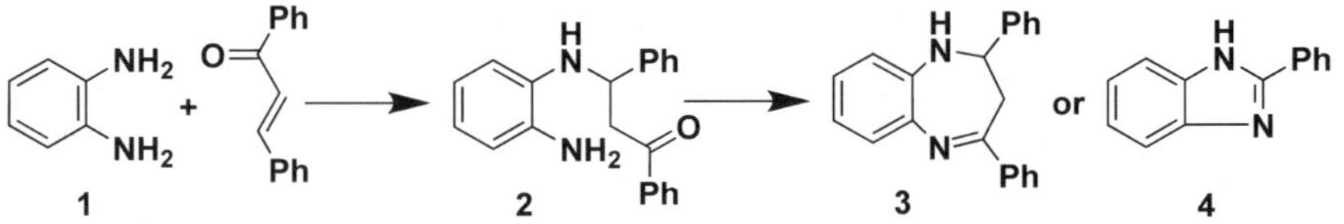


**Scheme 4.1**

(Scheme 4.2) does not produce diazepine derivatives, but 2-(2-oxyphenyl)benzimidazole **6** was found to be the only product isolated. However, there are known data on the interaction between *o*-PDA and *ortho*-oxybenzalacetone **7** (the structure of which is similar to that of *ortho*-oxychalcone) resulting in the formation of benzodiazepine **8** [9]. The reaction involves an intramolecular cyclization in which the hydroxyl group and the azomethine bond participate:

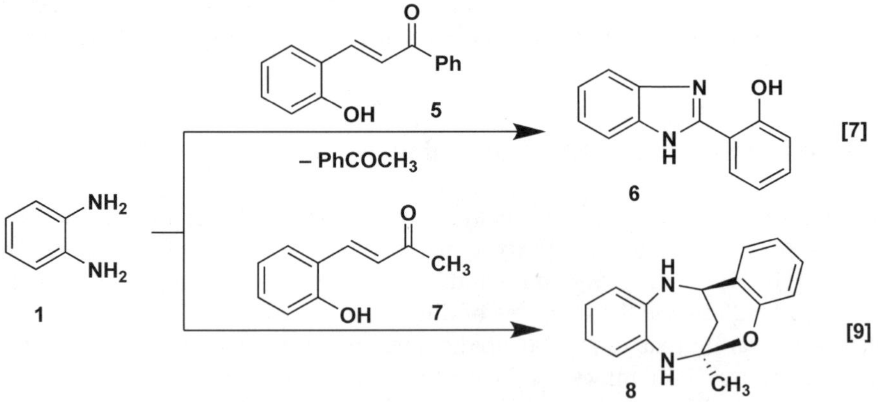


**Scheme 4.2**

Thus, the alternative direction of the reaction of *ortho*-hydroxychalcone **5** is not associated only with a specific influence of hydroxyl group, but most likely is due to several other reasons. Indeed, the presence of a hydroxyl group in other fragments of an unsaturated ketone does not change the reaction direction.

For example, treatment of 1-(2-hydroxyphenyl)-3-arylpropenones **9** with *o*-PDA **1** carried out in dimethyl sulfoxide in the presence of sodium acetate under microwave irradiation produces dihydrodiazepines **10** in good yields [10] (Scheme 4.3). Chalcone analogues of dihydroacetic acid **11**, as well, containing an *ortho*-hydroxyl group, react with *o*-PDA in boiling ethanol to produce dihydrobenzodiazepines **12** [11, 12].

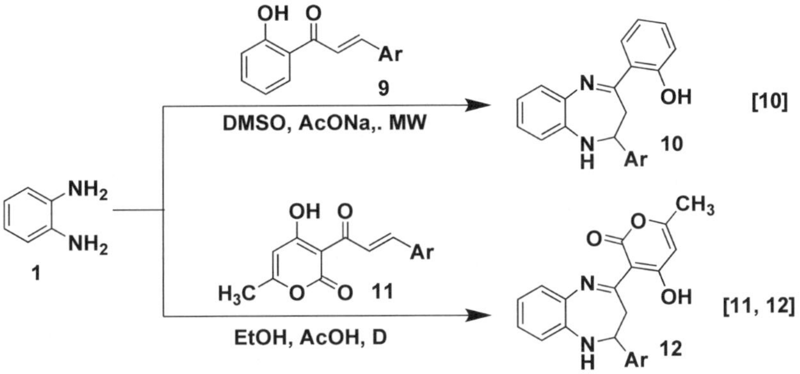


**Scheme 4.3**

The reaction of *o*-PDA with benzalcyclohexane **13** gives phenoxazine **14** but not dihydrodiazepine, which one would expect to obtain [7]. However, under the same conditions benzoannelated cyclic unsaturated ketones **15** readily react with *o*-PDA and yield the seven-membered dihydrocycle **16** [13] (Scheme 4.4).

There are many conditions described for the preparation of 1,5-benzodiazepines in the presence of acid catalysts such as $BF_3$-etherate [14], $NaBH_4$ [15], polyphosphoric acid or $SiO_2$ [16], $MgO/POCl_3$ [17], $Yb(OTf)_3$ [18], Amberlyst-15 [19], $Al_2O_3/P_2O_5$, acetic acid [20, 21, 22], the $TiCl_4$/Sm/tetrahydofuran system [23], silica gel [24] or $CeCl_3$/NaI/silica gel [25]; many of the known procedures suffer from limitations such as requiring harsh conditions, expensive reagents, high catalyst loading, toxicity, low or moderate yields and side reactions. However, recently some new synthetic approaches were elaborated. Microwave-assisted organic synthesis [26, 27, 28, 29, 30] (including solvent-free reactions) and application of ionic liquids [31, 32, 33, 34] are the most promising among them. Microwave irradiation was used to synthesize 1,5-benzodiazepines in several works. Besides the abovementioned example involving 1-(2-hydroxyphenyl)-3-arylpropenones **9** [10, 20, 21, 22] a synthesis of dihydrodiazepines in a microwave field was also described in [35, 36, 37]. For example,

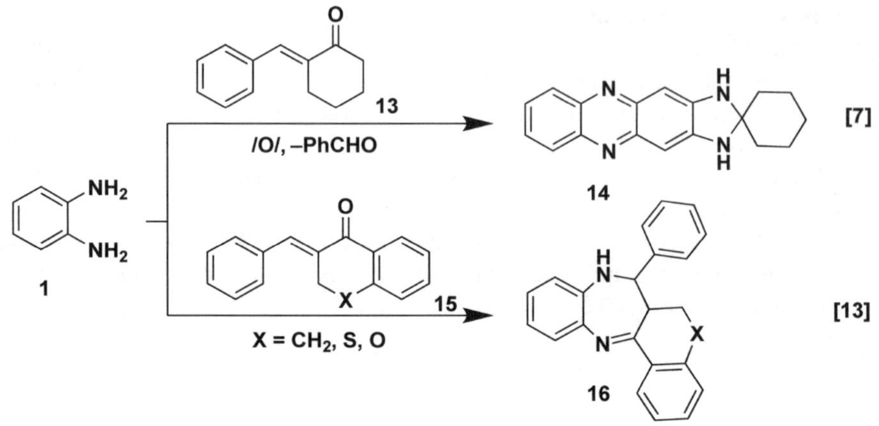


Scheme 4.4

Sucheta and Rao [36] investigated the reaction of *o*-PDA with heterocyclic α,β-unsaturated ketone **17**, based on the *N*-methylquinolin-2-one moiety (Scheme 4.5). The reaction proceeds rapidly (40 s) in solvent-free conditions on acidic alumina under microwave irradiation.

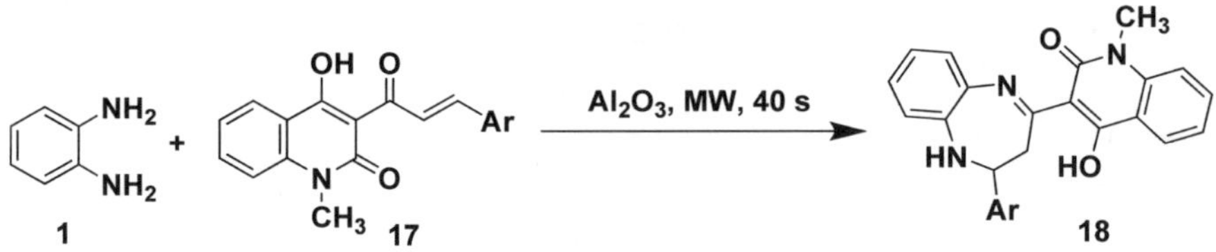


Scheme 4.5

Solvent-free conditions can also be applied without microwave assistance. In [24] reactions of *o*-PDA with several substituted chalcones were reported to have been carried out on neutral $Al_2O_3$ at 80°C with high yields (60–80%) of the target dihydrobenzodiazepines. Besides diazepines, some samples of benzothiazepines **21** were also synthesized by the treatment of *ortho*-aminothiophenol **19** with α,β-unsaturated ketones **20** on a silica gel surface (Scheme 4.6). The application of other solid supports like acidic, basic or neutral alumina,

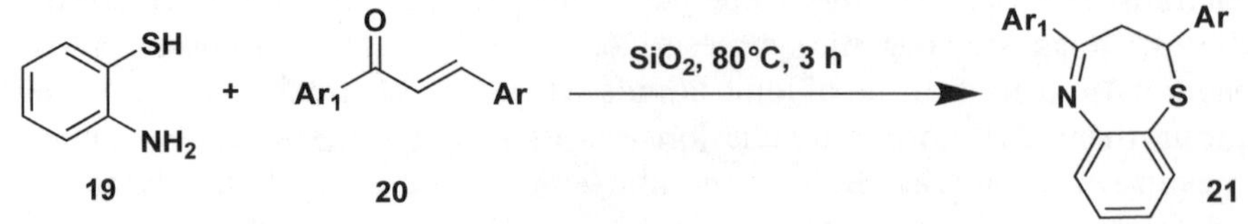


Scheme 4.6

molecular sieves and kieselguhr in the case of thiophenol led to a drastic decrease in the reaction yields (up to 2%).

Ionic liquids and room-temperature ionic liquids were also used as the media to synthesize dihydro-1,5-benzodiazepines [38, 39]. Du et al. [39] described two new acidic ionic liquids based on 1-butylpyridinium hydrogen sulfate and 1-butylisoquinolinium hydrogen sulfate as efficient catalysts to promote the formation of 1,5-benzodiazepines. The reactions were carried out in ethyl acetate at 80°C under an inert atmosphere.

According to [40], cyclocondensation leading to the formation of benzodiazepines **23** can be realized using cyclic unsaturated ketones with a steroid nature **22** (Scheme 4.7). However, in [41] the conclusions drawn on the formation of seven-membered heterocycles in the reaction mentioned [40] were shown to be inconsistent.

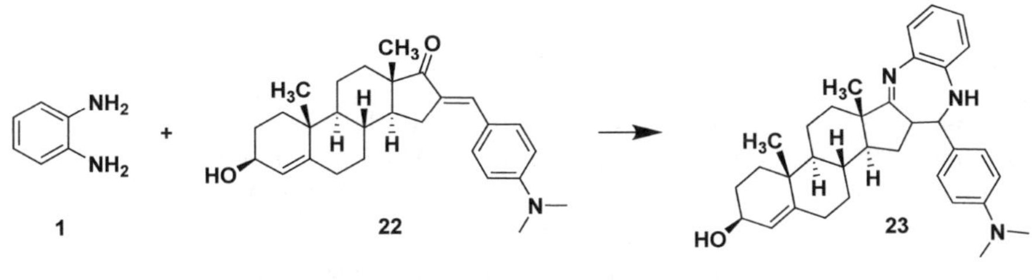


**Scheme 4.7**

As mentioned many times above [11, 12, 13, 36] heterocyclic analogues of chalcones can be used to synthesize dihydrodiazepines. In [42, 43] it was reported that open-chain **24** and cyclic **26** α,β-unsaturated ketones were introduced into the treatment with *o*-PDA to obtain seven-membered heterocycles **25** and **27**, respectively (Scheme 4.8).

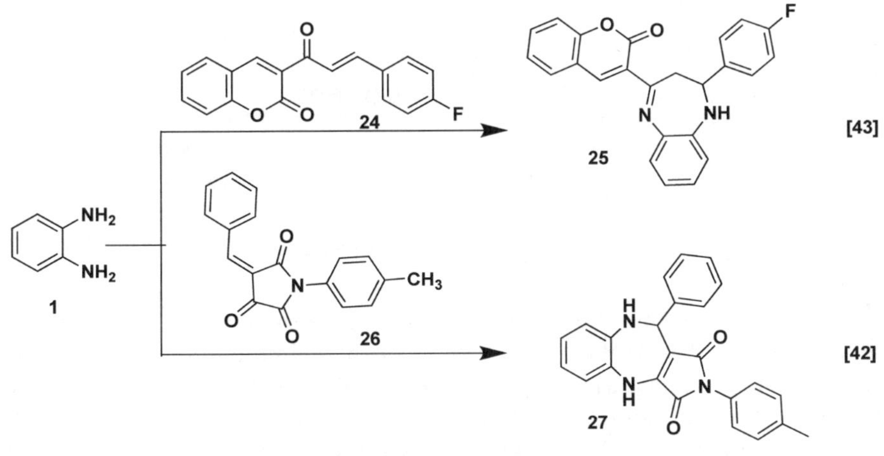


**Scheme 4.8**

The possibility of opening the cyclic unsaturated ketone ring was described for the reaction of (*R*)-(+)-puligone **29** with substituted *o*-PDA **28** and benzodiazepine **30** annelated by the eight-membered ring which has a crown conformation is formed [44] (Scheme 4.9).

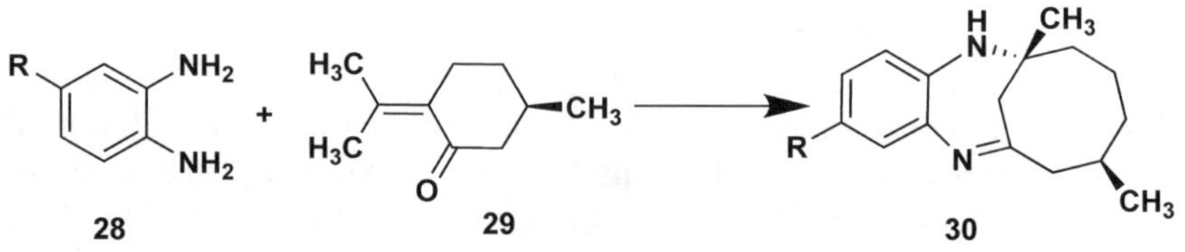

$R = H, CH_3, Cl$

**Scheme 4.9**

Besides aromatic 1,2-diamines and aminothiophenols, aliphatic diamines, in particular, ethylene diamine and its derivatives **31** or 2-aminoethanthiole **34**, were used for the synthesis of 1,2-dihydrodiazepines **33** [45, 46] and tetrahydrothiazepine **36** [24, 47] (Scheme 4.10). In [47] it was reported the target heterocycles **36** were formed along with other noncyclic β-adduct impurities.

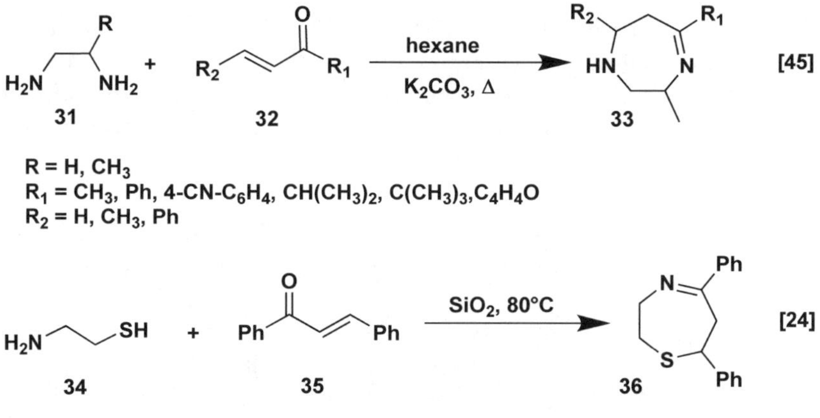

$R = H, CH_3$
$R_1 = CH_3, Ph, 4\text{-}CN\text{-}C_6H_4, CH(CH_3)_2, C(CH_3)_3, C_4H_4O$
$R_2 = H, CH_3, Ph$

**Scheme 4.10**

1,5-Dihydrodiazepines can also be obtained in the reactions with unsaturated ketones if one of the amino groups of the diamine in the heterocycle is endocyclic (e.g., amine **37**, Scheme 4.11) [48]. This reaction is achieved under rather severe conditions (long-duration heating in dimethylformamide). An interesting case using sydnones **40** in the synthesis of 1,5-dihydrodiazepines **41** was described by Kavali and Badami [49] (Scheme 4.12).

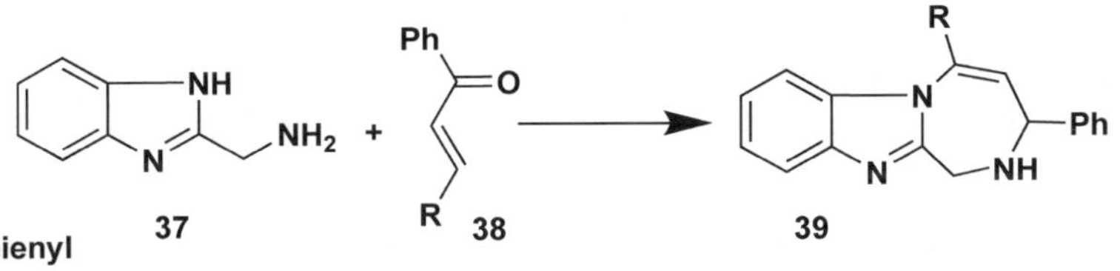

R = Ph, 2-thienyl

**Scheme 4.11**

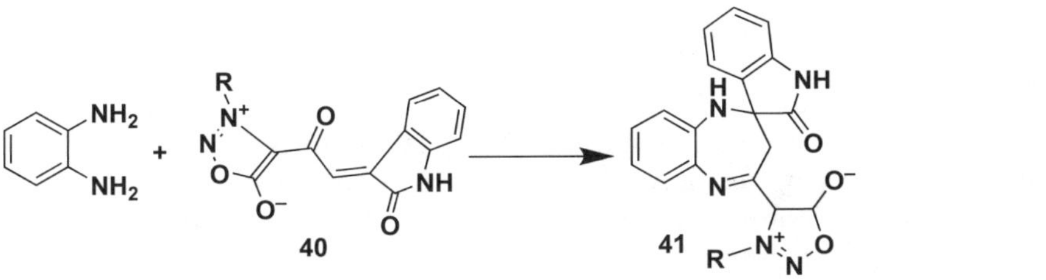

R = Ph, 4-$CH_3C_6H_4$, 4-$CH_3OC_6H_4$, 4-$CH_3OOCC_6H_4$, 4-$ClC_6H_4$, 2-$CH_3O$-4-Cl-$C_6H_4$

**Scheme 4.12**

During the past few years, heterocyclic 1,2-diamines such as derivatives of pyrazole [50, 51], imidazole [52], tetrazole [53], pyridine [54, 55] and pyrimidine [56, 57, 58, 59] were intensively studied as components of the condensation with α,β-unsaturated ketones. In particular, reliable data were presented in [50, 51, 60] on the formation of dihydrodiazepine systems **45** and **46** in the reaction of 5-methyl-2-phenyl-2*H*-pyrazol-3,4-diamine **42** with chalcones **43** and diarylideneacetones **44** (Scheme 4.13).

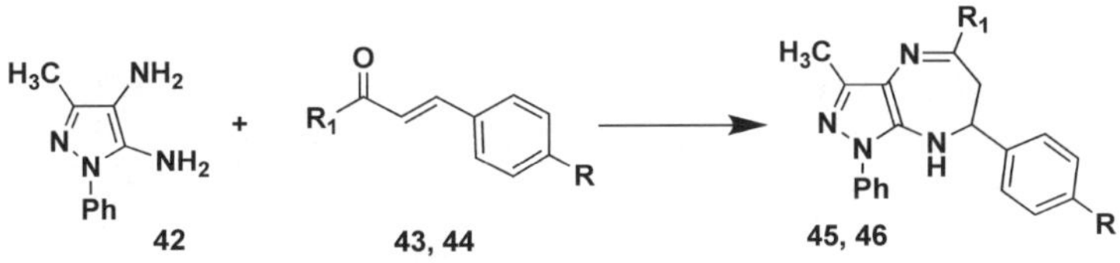

43, 44, 45, 46: R = H, Cl, Br, $CH_3$, $CH_3O$, Ph
43, 45: $R_1$ = H, Ph, 4-$ClC_6H_4$, 4-$BrC_6H_4$, 4-$CH_3C_6H_4$, 4-$CH_3OC_6H_4$, 4-$PhC_6H_4$,
44, 46: $R_1$ = 4-$BrC_6H_4CH$=CH, 4-$ClC_6H_4CH$=CH, 4-$CH_3C_6H_4CH$=CH, 4-$CH_3OC_6H_4CH$=CH

**Scheme 4.13**

The treatment of chalcones **48** with some pyrimidine 1,2-diamines **47**, containing at position 2 either a methoxy or a methylthio group, may also lead to the formation of diazepines **49** [58, 59] (Scheme 4.14).

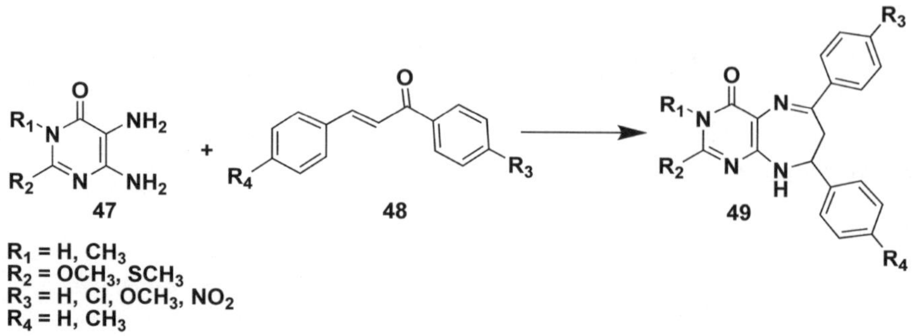


**Scheme 4.14**

On the basis of the literature data [53, 56, 57] of the synthesis of dihydrodiazepine and dihydrotriazepine derivatives by chalcone reaction with diamines such as 1,3-dimethyl-5,6-diaminouracil and diaminotetrazole, the suggested structure of the products of these interactions was later proved to be erroneous [61, 62, 63]. Analysis of the reaction paths actually observed under such conditions is presented in Sect. 4.5.

Synthesis of annelated diazepines based on unsaturated aromatic diazepines may involve the preliminary transformation of ketones into the corresponding 1,3-diaryl-2,3-dibromopropane-3-ones (chalcone dibromides). The interaction between *o*-PDA or some of its substituted analogues with chalcone dibromides leads to the formation of aziridine derivatives [64] (see Chap. 1). However, in the case of 4-nitro-*o*-PDA, either azirenoquinoxalynes **53** or benzodiazepine derivatives **54** may be obtained depending on the reaction conditions [65] (Scheme 4.15). Diazepine derivatives **56** are obtained by the condensation of chalcone dibromides **51** with 5,6-diamino-1,3-dimethyluracil **55** [66], but aziridine derivatives are not isolated in this reaction. It should be noted that compounds **54** and **56** are formed owing to cyclization of the intermediate β-enaminoketones [65, 66, 67] and are easily isolated from the reaction mixture.

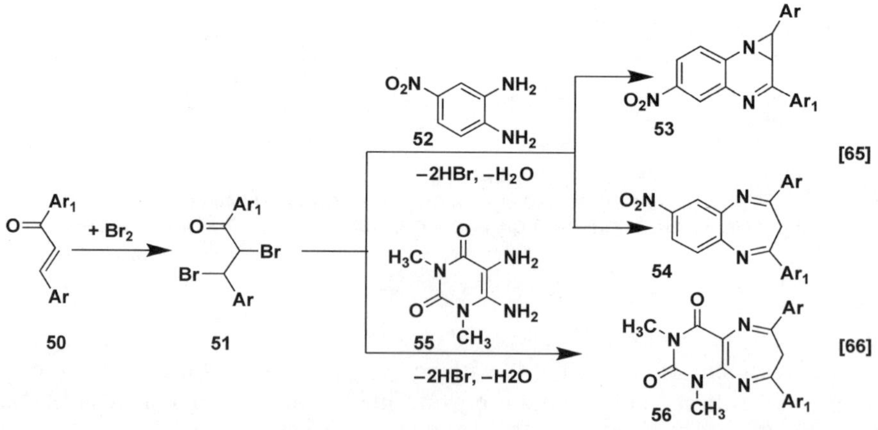


**Scheme 4.15**

Fused diazepines like **54** and **56** can exist in several tautomeric forms (Scheme 4.16)—enamine 1H (**A**), enamine 5H (**C**) and diimine 3H (**B**) (Scheme 4.16).

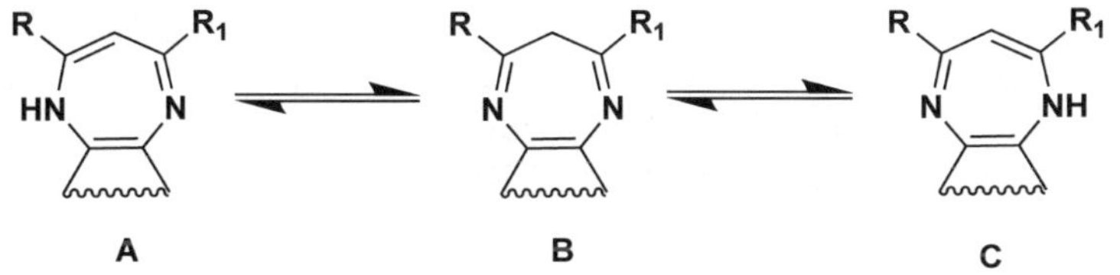


**Scheme 4.16**

The problem of tautomeric equilibriums in annelated 1,5-diazepines was studied in [68] by means of NMR spectroscopy, X-ray analysis and quantum-chemical calculations. It was shown that the electron-withdrawing rings (e.g., pyrimidine moiety) fused with the diazepine cycle increase the stability of the antiaromatic enamine tautomeric forms **A** and **C**, while in the case of benzodiazepine, a diimine tautomer **B** was found to be the most stable. *Ab initio* quantum-chemical calculations and NMR spectroscopic data showed that solvation of seven-membered heterocycles with polar solvents contributes considerably to the stabilization of the enamine forms **A** and **C**. This assumption was also proven by X-ray analysis, which showed that in the solid state these diazepines exist in the diimine form **B**.

## 4.2 Three-Component Condensation of 1,2-Diamines with Ketones

Known from the literature are the heterocycle formation reactions from the interaction of 1,2-diamines with synthetic precursors of unsaturated ketones, i.e., the ones which contain an activated methyl or methylene group (Scheme 4.17). Since such cyclocondensations are obviously related to

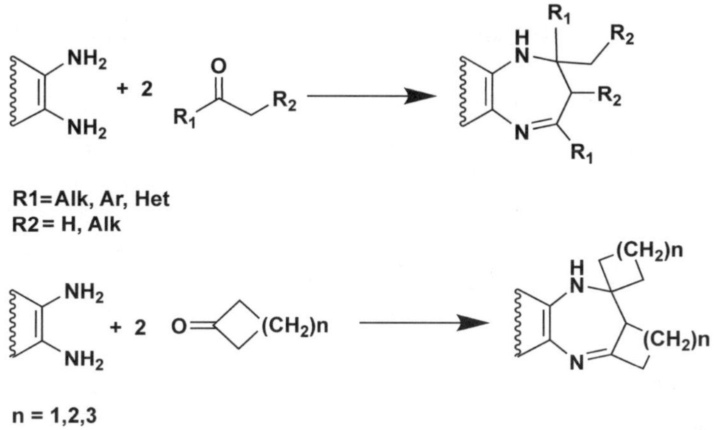


**Scheme 4.17**

heterocyclizations based on unsaturated ketones, it seems appropriate to consider them in the present book.

The multicomponent reaction of a 1,2-diamine with two moles of ketone resulting in the formation of the dihydrobenzodiazepine system was mentioned for the first time in 1951 [69]. This paper was devoted to the reaction of *o*-PDA with different aliphatic ketones, in particular, pentane-2-one **57** in boiling benzene forming diazepine **58** (Scheme 4.18).

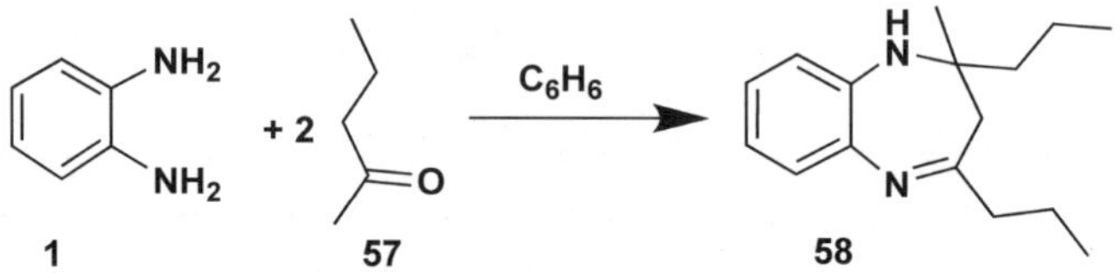


**Scheme 4.18**

Later, reactions of this type were investigated in a large volume of studies. As a rule, cyclocondensation proceeds under the conditions of acid catalysis and has a rather general character. This method was used to obtain various benzo- and heteroannelated dihydroazepines. There was also a synthesis reported based on acyclic diaminomaleonitrile **59** [70] (Scheme 4.19). This reaction is applicable to both aliphatic (acyclic and cyclic) ketones [15, 16, 17, 18, 20, 21, 38, 69, 71, 72, 73, 74, 75, 76, 77, 78, 79, 80, 81, 82, 83, 84, 85, 86, 87, 88, 89, 90, 91, 92, 93, 94] and aromatic ones [16, 17, 18, 20, 21, 25, 76, 77, 78, 79, 80, 81, 82, 83, 84, 89, 90, 92, 93, 94, 95, 96, 97].

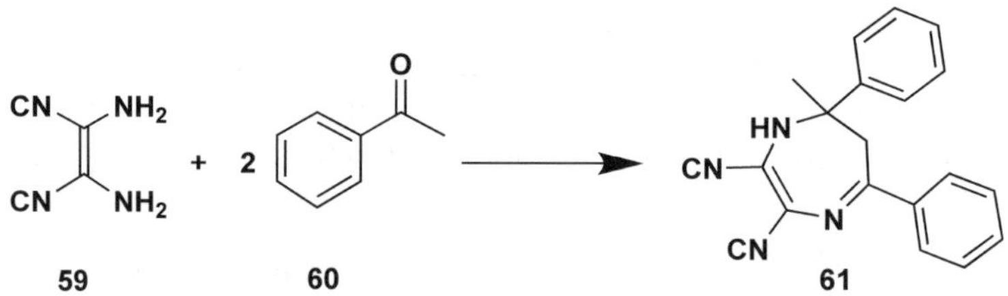


**Scheme 4.19**

As a catalyst both mineral and organic acids can be used. In particular, the reactions between *o*-PDA **1** and acyclic ketones **62** occurred in the presence of polyphosphoric [16] and trifluoroacetic [74] acids (Scheme 4.20). Polyphosphoric

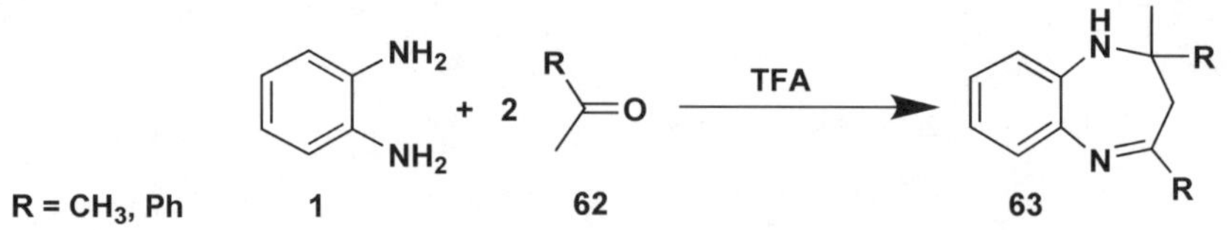


**Scheme 4.20**

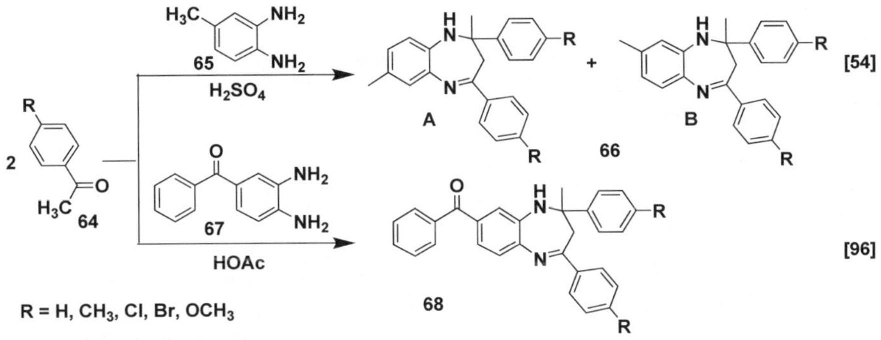


Scheme 4.21

acid can also be used for the reaction of diamines **65** and **67** with aromatic ketones **64** [16]; however, in this case better yield values are obtained using sulfuric [54, 95, 97] or acetic [89, 96] acids as a catalyst (Scheme 4.21). Acetic acid was also used in the reactions of methyl ethyl ketone **70** with a fused aromatic diamine **69** to synthesize dihydrodiazepine **71** [98] (Scheme 4.22).

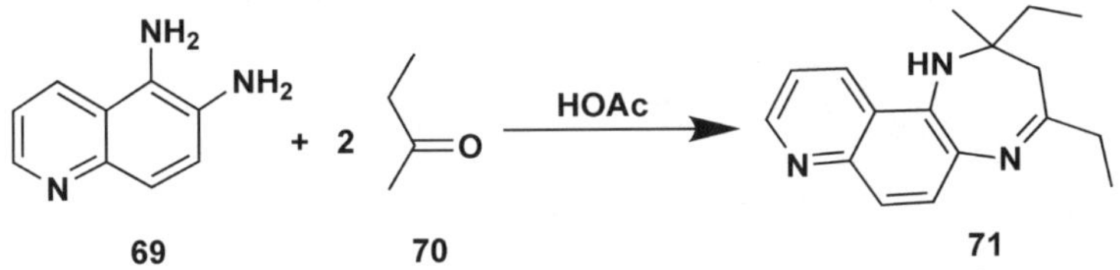


Scheme 4.22

*para*-Toluenesulfonic acid may catalyze the interaction of aliphatic ketones like **57** with diaminofurazan **72** [73] and with *o*-PDA [75] (Scheme 4.23).

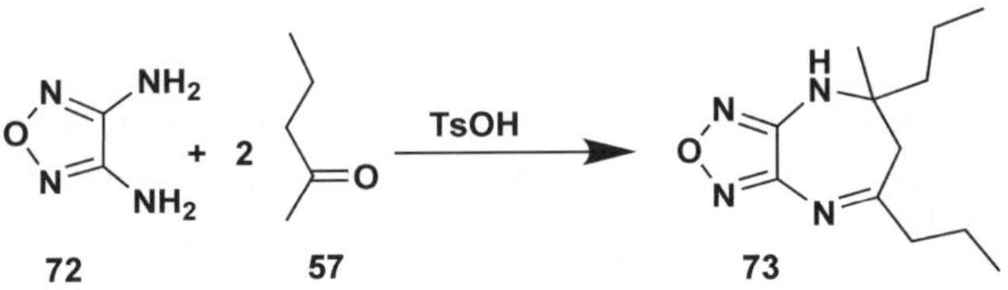


Scheme 4.23

*ortho*-Diamine hydrochloride salts react with ketones without catalysts. Such a method is usually applied in the reactions with aliphatic ketones [72]. However, more prolonged heating of the reaction mixture is required because the yield is reduced in comparison with that for the reactions with acids. At the same time, the said method produces tetrahydrobenzodiazepines **76** without isolation of the corresponding dihydrodiazepines by introduction of sodium tetrahydroborate directly in the reaction mixture containing salt **74** and acetone [15] (Scheme 4.24).

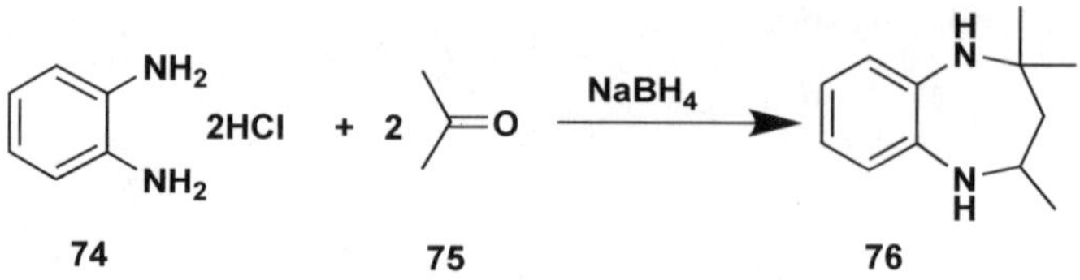


Scheme 4.24

Rather high (80–90%) yields of dihydrodiazepines are observed when the reaction is carried out on a MgO surface in the presence of phosphorus chloroxide [17]. In this manner some dihydrobezodiazepine derivatives **78** and **80** with alkyl substituents are obtained (Scheme 4.25). It should be noted that high product yields are also observed in the reactions of *o*-PDA with cyclic ketones (e.g., cyclopentanone, cyclohexanone and cycloheptanone), which is not typical of routine acid catalysis [14, 99].

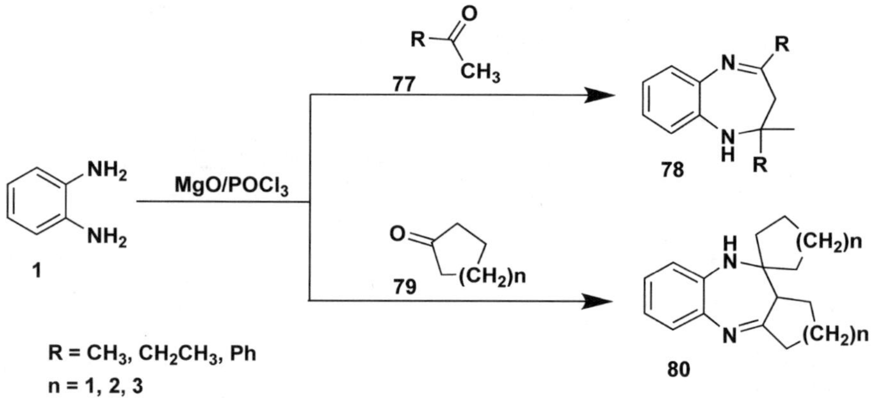


Scheme 4.25

The use of $MoO_2Cl_2$ as a catalyst in the reaction of *o*-PDA **1** with aliphatic ketones **81** produces rather high quantities of dihydrodiazepine hydrochloride salts **82** [71] (Scheme 4.26).

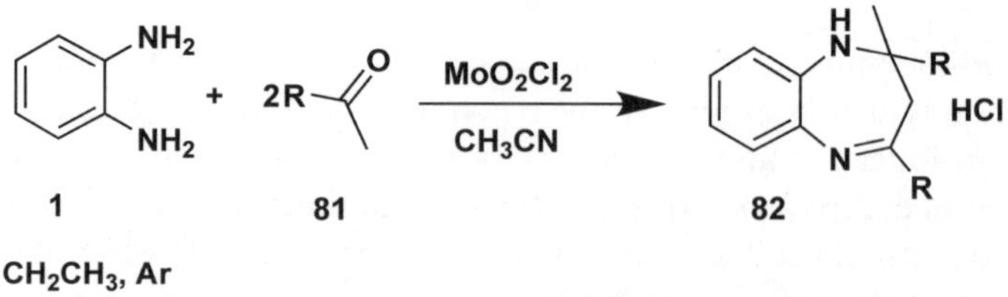


Scheme 4.26

In some recent papers the use of other types of catalysts is described; in particular, catalysts such as ceric ammonium nitrate [76], zinc chloride [77], magnesium perchlorate [78], ytterbium chloride [79], $HClO_4/SiO_2$ [80], iodine [81], indium bromide [82], scandium trifluromethanesulfonate [83], ceric chloride with silica gel [25], $Ag_3PW_{12}O_{40}$ [84], ytterbium trifluromethanesulfonate [18], $(NH_4)H_2PW_{12}O_{40}$ [88], $ZnOCl_2$ $x8H_2O$ [90], trimesic acid [91] or 1,3-*n*-dibutilimidazolium bromide [38].

Microwave-assisted organic synthesis may also be used for carrying out the multicomponent reactions of ketones and 1,2-diamines [20, 21, 92, 100]. For example, the three-component reaction of *o*-PDA **1** with acetoacetic acid ethyl ester **83** and a series of aromatic and heteroaromatic aldehydes **84** proceeds under microwave irradiation with very high yields of diazepines **85** (up to 95%) [100]. Reaction of 2 equiv of cyclohexanone **86** with *o*-PDA **1** was also realized in a microwave field on a basic alumina surface in 4 min [92] (Scheme 4.27).

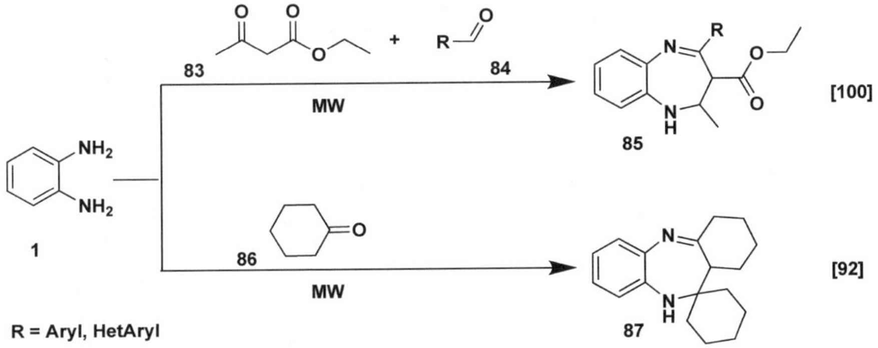


**Scheme 4.27**

As shown in Scheme 4.21, in the reaction between methylene-active ketones and nonsymmetrical *ortho*-diamines, regioisomeric dihydrodiazepines can form. The interaction of nonsymmetrical substituted *o*-PDAs **88** with methyl aryl ketones **89**, which may lead to the formation of the isomeric products **90A** and **90B**, was studied in [54, 65, 95, 97] (Scheme 4.28). It was shown that the

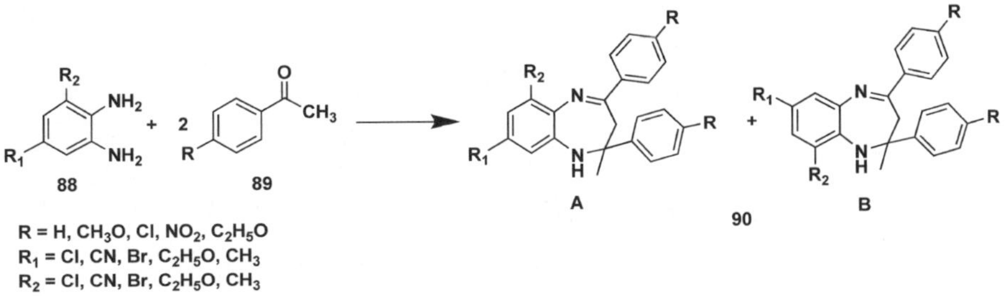


**Scheme 4.28**

isomeric composition of the reaction products depends on the electronic character of R, $R_1$ and $R_2$ substituents. Enhancement of the electron-donor character of the R substituent promotes the formation of dihydrobenzodiazepines, but also increases the probability of the formation of two different isomers. The influence of $R_1$ and $R_2$ substituents is also essential since the introduction of strong electron-donor or electron-acceptor groups in *o*-PDA increases the selectivity of the process. In this case the formation of the azomethine bond in the final dihydrobenzodiazepine predominates owing to the higher nucleophilicity of the nitrogen atom in the diamine group.

The same rule was revealed in [101] while studying the interaction of acetophenones **92** with heterocyclic amines **91** with an acid catalyst. Since the amino groups of the starting diamine are nonequivalent, there exist two possible directions of the reaction but only one of them, i.e., direction A, is observed, with the formation of diazepines **93** but not their regioisomers **94** (Scheme 4.29).

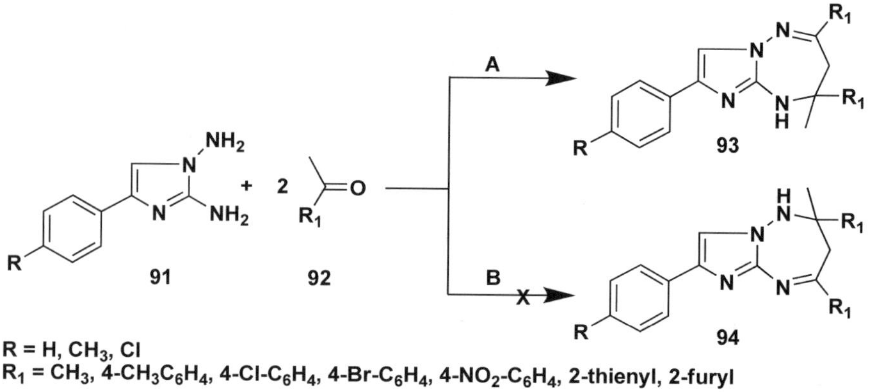


Scheme 4.29

The problems associated with the mechanism of condensation were considered in [14, 70, 99, 102, 103]. From a theoretical viewpoint, there are three possible ways for the diazepine cycle to form and they differ in the sequence of the condensation stages (Scheme 4.30, reactions a–c). No unambiguous

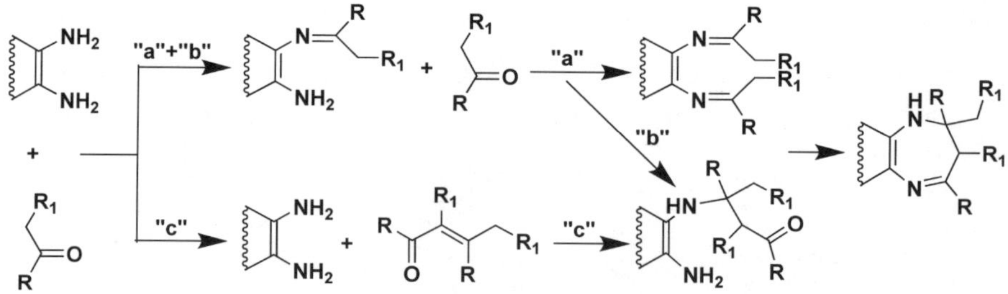


Scheme 4.30

evidence for the said reaction paths is available in the literature. Nevertheless, the experimental facts presented in [14, 70, 99, 102, 103] indicate that the reactions of binucleophiles with both methyl-active ketones and unsaturated ketones are independent of each other. In other words, heterocyclization is not reduced to self-condensation of two ketone molecules leading to formation of an unsaturated carbonyl compound and followed by normal cyclocondensation (i.e., path c is not observed).

In conclusion it should be noted that the reactions of 2 equiv of ketone which result in dihydrodiazepine formation are characteristic not only of 1,2-diamines. For example, described in [104] is the interaction of *ortho*-nitrophenylazide **95** with aliphatic and aromatic ketones (acetone, for example) in the presence of $SmI_2$ which produces dihydrobenzodiazepines at high yields (Scheme 4.31).

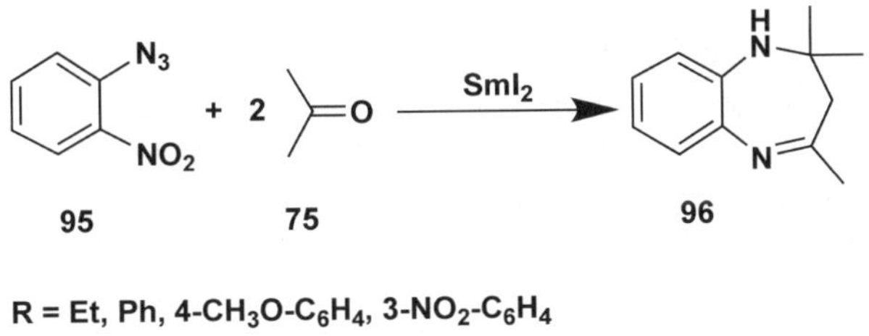


Scheme 4.31

## 4.3 Reactions of Dihydrobenzodiazepines

There is not much literature on the chemical properties of fused dihydrodiazepine systems which are the products of cyclocondensation of chalcones with *ortho*-diamines. In [105, 106] the properties of 2,4-diplhenyl and 2-methyl-2,4-diphenyl substituents of 2,3-dihydro-1*H*-1,5-benzodiazepines, are adequately analyzed but not in a comprehensive manner.

Oxidation of 2,3-dihydrobenzodiazepines by various oxidants ($H_2O_2$, $SeO_2$, peracids, etc.) normally gives resinous reaction products which cannot be easily identified. However, described in [107] is the oxidation of *N*-bezoyl derivatives of benzodiazepines **97** by *meta*-chloroperbenzoic acid in the presence of sodium bicarbonate and triethylbenzyl ammonium chloride (used as an interphase catalyst) which yields compounds **98** (Scheme 4.32).

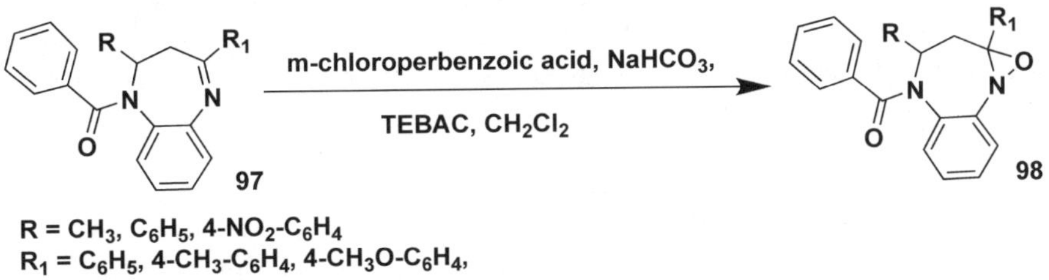


Scheme 4.32

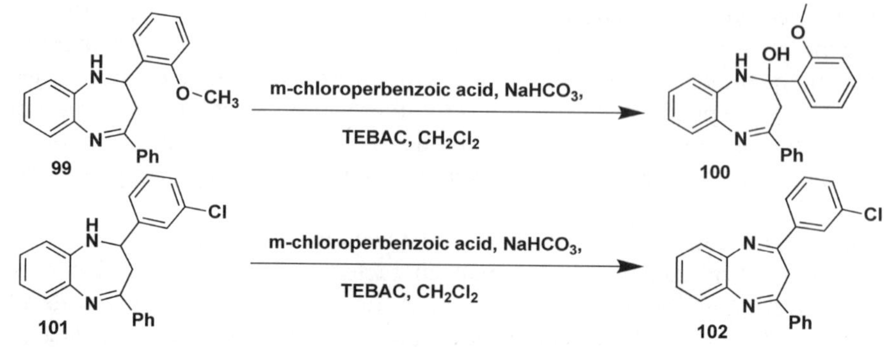


Scheme 4.33

The action of the same reagents on benzodiazepines with unsubstituted amino groups (**99** and **101**) generates hydroxy derivatives **100**, or is accompanied by dehydration (benzodiazepine **102**, Scheme 4.33). 3*H*-1,5-Benzodiazepine **104** can be obtained with satisfactory yields using a milder oxidant, e.g., $K_2S_2O_5$ [105].

Reduction of 2,4-diphenyl-2,3-dihydro-1*H*-1,5-benzodiazepine **103** (R is H, Ar is Ph) by sodium borohydride leads to an equimolar mixture of the diastereomeric tetrahydro derivatives **105** (Scheme 4.34), whereas under analogous

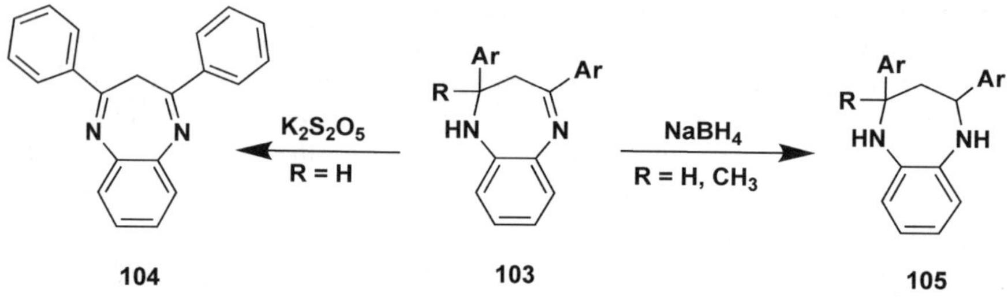


Ar = $C_6H_5$, 4-$CH_3$-$C_6H_4$, 4-$CH_3O$-$C_6H_4$, 4-Cl-$C_6H_4$, 4-Br-$C_6H_4$

Scheme 4.34

conditions reduction of the 2-methyl-2,4-diaryl derivative of this bicycle results in only one of the isomers **105** (R is $CH_3$) being obtained, the configuration of which has not been established thus far [106]. The reduction of 1,5-benzodiazepines was also studied in other publications. As reducing agents, sodium borohydride [90, 108] and Raney nickel [2, 109] were used.

From the reactions characterizing the substituted dihydrobenzodiazepines **106** as cyclic amino derivatives, their acylation and nitrosation were studied. It should be noted that these reactions have different directions for 2,4-disubstituted and 2,2,4-trisubstituted benzodiazepines: in the first case the reactive group is the NH group [105] (compounds **107** and **108**) and in the second case it is the azomethine nitrogen (derivatives **109** and **110**) [14, 106] (Scheme 4.35).

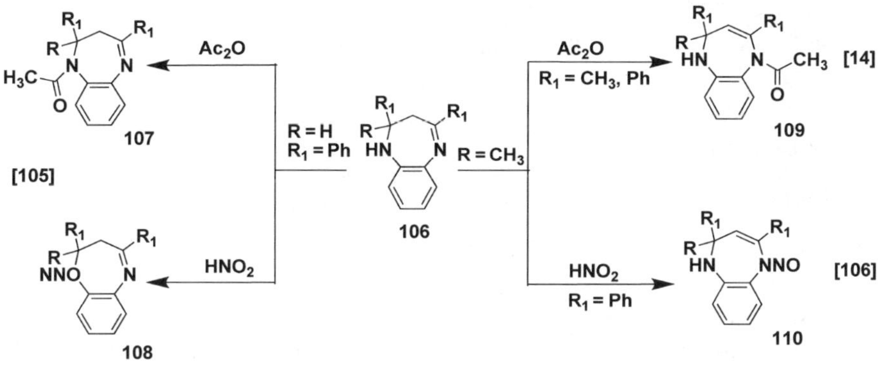

Scheme 4.35

The reaction of succinic acid anhydride **111** with 2-methyl-2,4-diphenyl-2,3-dihydro1*H*-benzo[*b*][1,4]diazepine **112** in toluene in the presence of potassium carbonate leads to 4-(2-methyl-2,4-diphenyl-2,5-dihydro-1*H*-1,5-benzodiazepin-5-yl)-4-oxobutanoic acid **113** in 70% yield [110], while its treatment with 2,4-diphenyl-2,3-dihydro-1*H*-1,5-benzodiazepine **104** follows with chalcone elimination and yields 3-(benzimidazol-2-yl)propionic acid **114** [115] (Scheme 4.36).

The reaction of N-aroylation of 2,3-dihydrobenzodiazepines **115** by aroyl chlorides **116** in the presence of $K_2CO_3$ was studied in detail in [111] (Scheme 4.37).

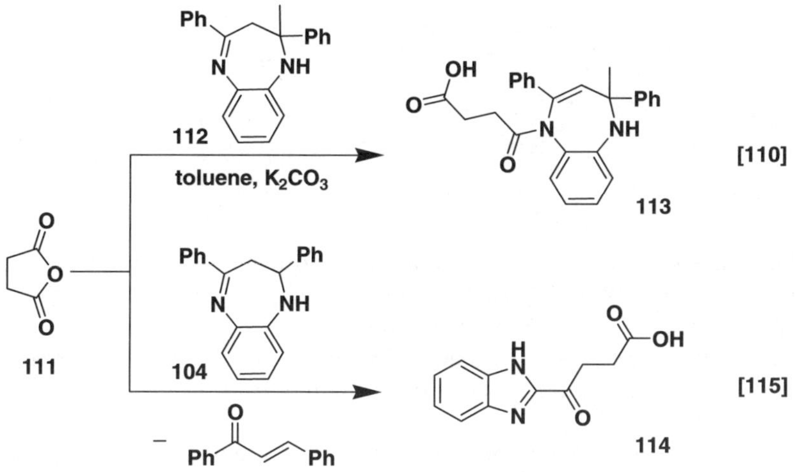

Scheme 4.36

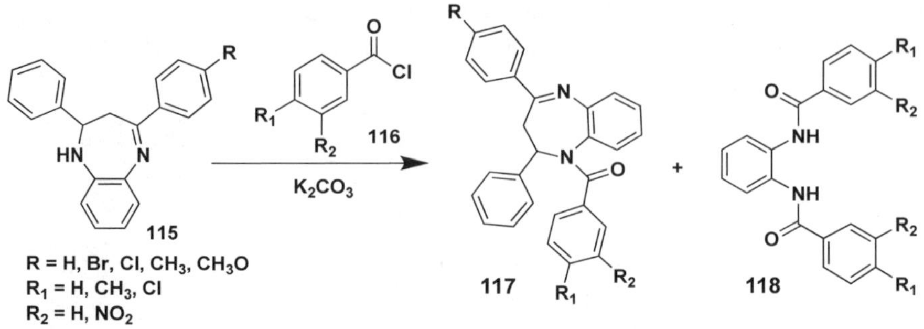


Scheme 4.37

As shown in [111], introduction of any substituent with both a donor and an acceptor nature in the phenyl radical at position 2 of heterocycle **115** impedes the reaction. Besides the desired aroylation products **117**, *N,N′*-diaroyl-1,2-phenylenediamines **118** are isolated.

Some papers are devoted to the modification of 2,3-dihydrodiazepines with respect to the C = N bond. The oxidation reaction involving the azomethine bond and leading to the formation of epoxide [107] was referred to earlier. The interaction of 2,2,4-trimethyl-2,3-dihydrobenzodiazepine **76** with 2-(oxyphenylimino)-1-phenylethanone **119** is described in [112] (Scheme 4.38). The reaction is carried out in benzene at room temperature and apart from the C = N bond involves the 4-methyl group (compound **120**).

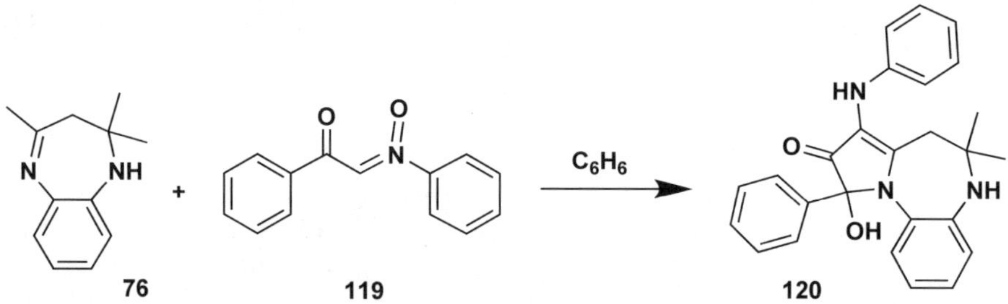


Scheme 4.38

There is also a known 1,3-dipolar cycloaddition of some nitrileamines **122** characterized by high yields at elevated temperatures in tetrahydofuran in the presence of triethylamine [113] (Scheme 4.39). Similar 1,3-dipolar cycloadditions of benzonitrile oxide **125** were studied in [114]. The reaction proceeds under mild conditions at room temperature in methylene chloride (Scheme 4.40).

Interesting results are obtained from the interaction of 2,3-dihydrodiazepines **104** with 2-diazo-1,3-diphenylpropane-1,3-dione **127**. The reaction involves both the secondary amino group and the azomethine bond, thus leading to the formation of compounds **128** and **129** [115] (Scheme 4.41).

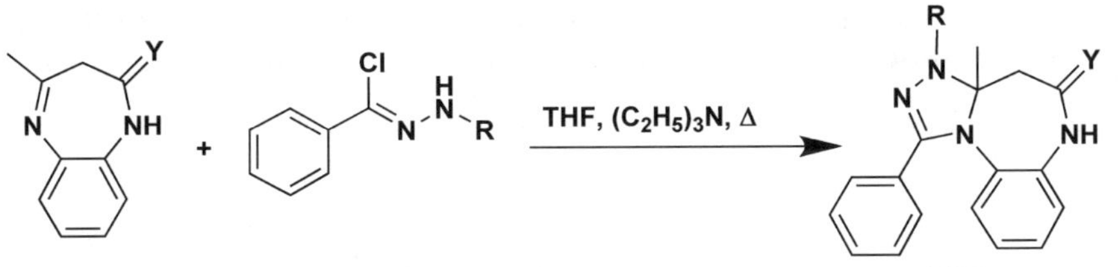


121 122 123

R = $C_6H_5$, $COOC_2H_5$
Y = $C(CH_3)_2$, O

**Scheme 4.39**

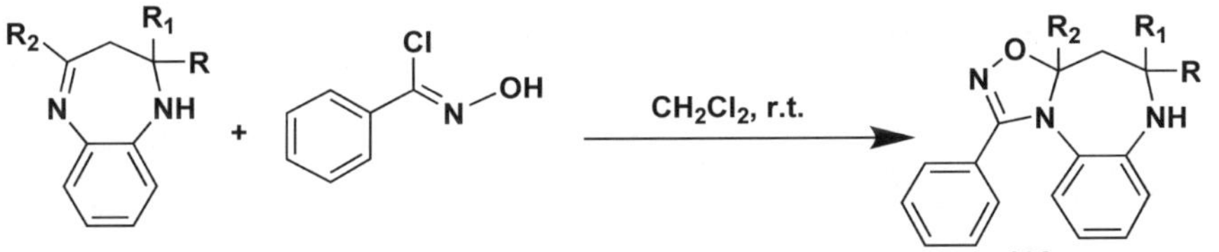


124 125 126

R = H, $CH_3$, $C_6H_5$
$R_1$ = H, $CH_3$, $C_6H_5$
$R_2$ = $CH_3$, $C_6H_5$

**Scheme 4.40**

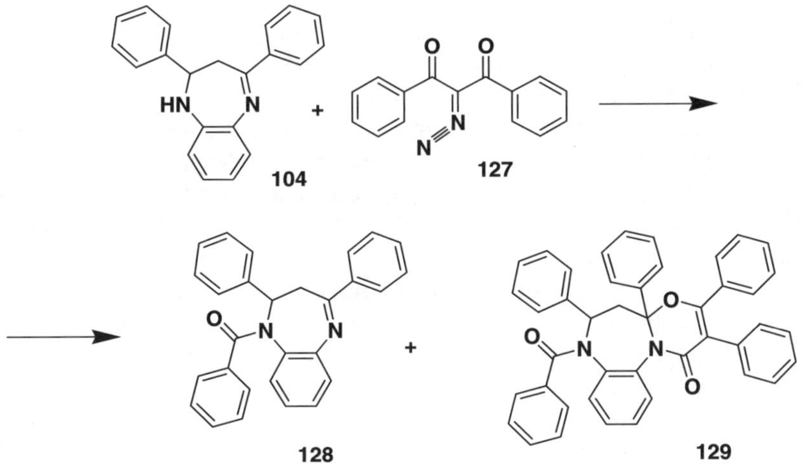

104 127 128 129

**Scheme 4.41**

Similar reactions described in [116] involve the interaction of *o*-PDA **1** with Mannich bases **130** (Scheme 4.42). The authors showed that for a 1:1 ratio of the

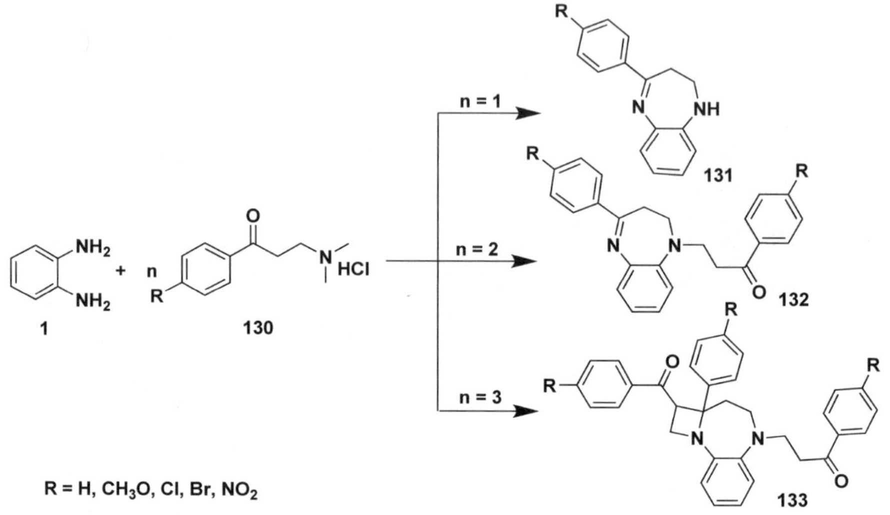


**Scheme 4.42**

initial components 2,3-dihydrobenzodiazepines **131** are formed. Introduction of an additional mole of a Mannich base allows the isolation of N-substituted derivatives of benzodiazepines **132**. At a 1:3 diamine to ketone molar ratio, the reaction also involves the azomethine bond **133**.

There are some known unsuccessful attempts to carry out alkylation (MeI, $Me_2SO_4$), halogenation (*tert*-butyl hypochloride) and nitration of aromatic dihydrobenzodiazepines [7, 105]. Such attempts only resulted in the destruction of the seven-membered heterocycle. As a rule, these destructive processes are typical of dihydrodiazepine systems and often manifest themselves during the synthesis and study of these compounds. Therefore, the results of the destruction of a seven-membered heterocycle are most widespread and include its decomposition into *ortho*-diamine and carbonyl compounds (Scheme 4.43, reactions A and B) [105, 106] and benzimidazole rearrangement accompanied by splitting out of a methyl aryl ketone molecule (Scheme 4.43, reaction C) [117].

Similar to the abovementioned acid-catalyzed transformations into benzimidazole **137**, thermal rearrangement occurs for dihydrobenzodiazepines and the processes of their fragmentation under the action of electron impact (via the corresponding radical and ion-radical intermediates) [117]. The ability to rearrange into imidazoles is essentially influenced by both the substituents in the dihydrocycle and the character of the annelated nucleus. Therefore, some heteroannelated diazepine and triazepine systems often turn out to be stable enough to undergo synthesis either in the presence of acids or at high

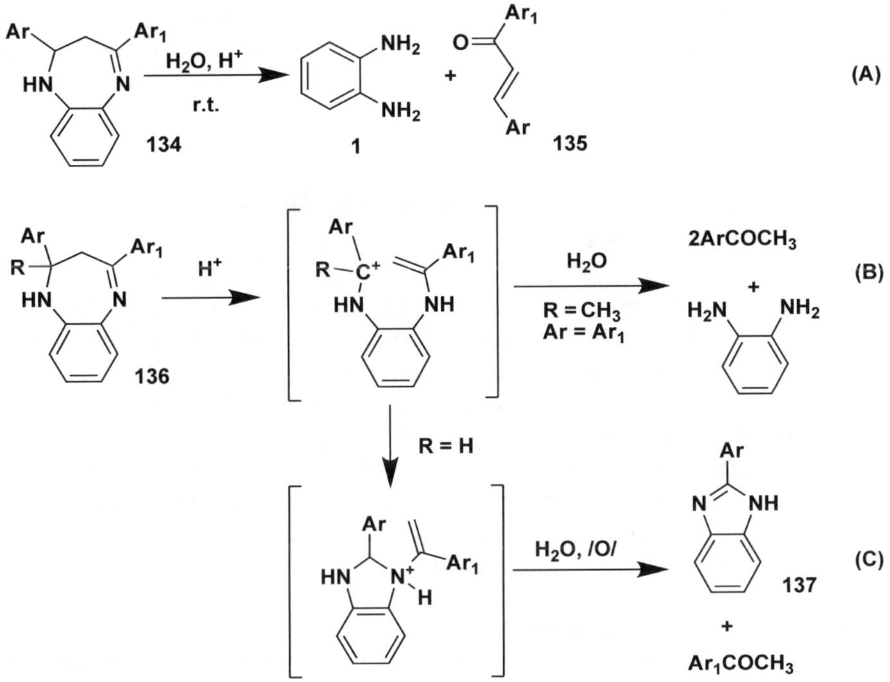

Scheme 4.43

temperatures. CH-acid properties of the methylene unit of dihydrobenzodiazepines illustrate these data [118, 119] concerning the reactions of chalcones with 4-nitro-1,2-phenylenediamine **138** and 4,5-diamino-1-methyl-2-methoxy-1,6-dihydropyrimidin-6-one **139**. In such cases, the initially formed dihydrodiazepines react with the second ketone molecule and finally are transformed into adducts like **140** (Scheme 4.44).

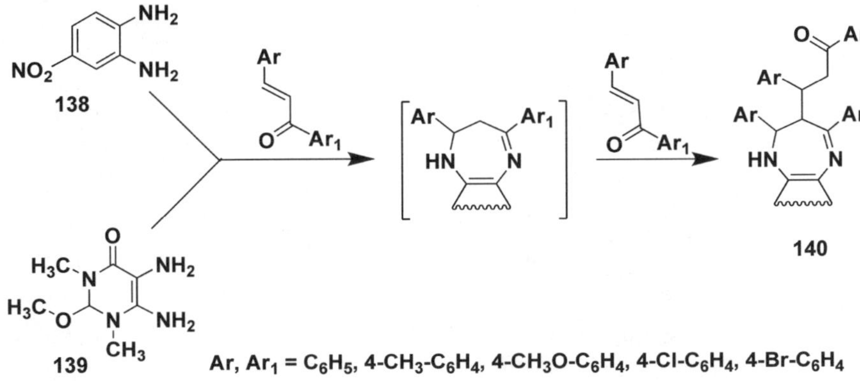

Scheme 4.44

There are no other reactions of the methylene group reported in the literature typical in the case of 1,5-benzodiazepines [120] for annelated dihydro-1,5-diazepines.

## 4.4 Five- and Six-Membered Heterocycles Based on *ortho*-Diamines

As noted already, the reactions of α,β-unsaturated ketones with *ortho*-diamines are often accompanied by various rearrangements and side processes. For instance, the *o*-PDA condensation with chalcone in the presence of an acid catalyst or during overheating of the reaction mixture produces a final product of 2-phenylbenzimidazole [2, 7]; a similar phenomenon is also observed for 2,3-diaminopyridine [54]. Owing to the existence of much more suitable approaches to the synthesis of 2-aryl derivatives of benzimidazole and their heteroannelated analogues (e.g., proceeding from diamines and acids or aldehydes [121]), such a trend in the interaction of chalcones with diamines is unlikely to have any application for this purpose.

In our opinion, the reactions of chalcones **141** with some *ortho*-diamines containing a "hydrazine" amino group (1,2-diaminobenzimidazole **142**, 3,4-diaminotriazoles **143** and 1,2,4-triaminotriazoles **144**) [122, 123, 124] are more interesting from both a practical and a theoretical viewpoint. The possibility to obtain the derivatives of azolopyrimidine **145**—**147** but not the triazepine systems in such reactions was demonstrated [124]. In this case the cyclocondensation is accompanied by the elimination of not only a molecule of water but also a molecule of ammonia. The proposed mechanism of the cyclocondensation [122] implies formation of a dihydroazolopyrimidine system with its subsequent heteroaromatization by amino-group elimination (Scheme 4.45).

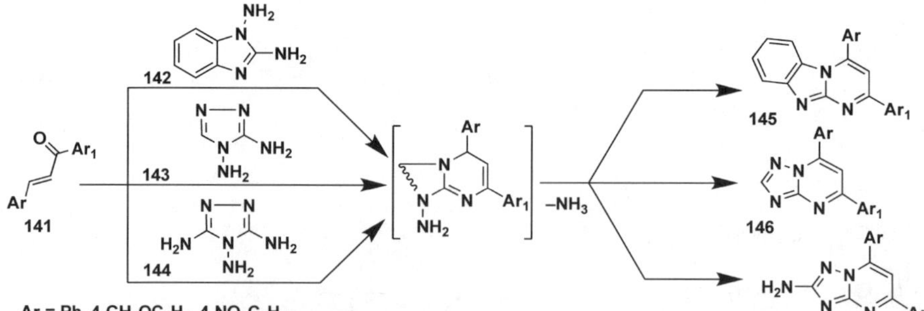


Ar = Ph, 4-CH$_3$OC$_6$H$_4$, 4-NO$_2$C$_6$H$_4$
Ar$_1$ = Ph, 4-CH$_3$C$_6$H$_4$, 4-CH$_3$OC$_6$H$_4$, 4-ClC$_6$H$_4$, 4-NO$_2$C$_6$H$_4$

**Scheme 4.45**

It should be noted that a similar reaction indirectly verifying the said mechanism is also found for 1,2-diamino-4-phenylimidazole **148** [125]. The only difference between the two reactions is the fact that in the latter case it is not the nitrogen atom but is the $\pi$-excessive carbon reaction center that participates in cyclocondensation (Scheme 4.46). Also, subsequent heteroaromatization of the dihydro derivative **150** into imidazopyridine **151** does not require amino-group elimination.

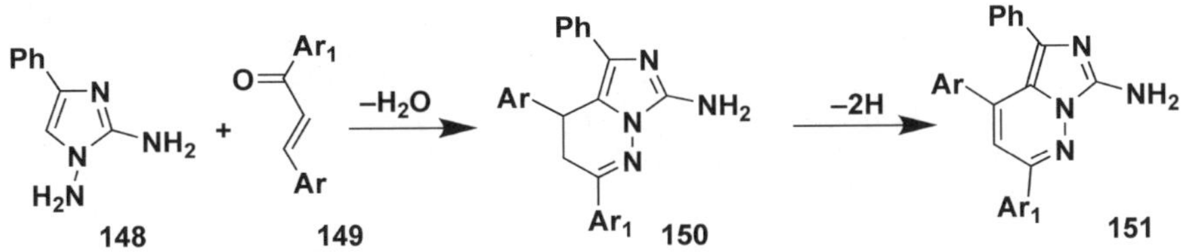

Ar = Ph, 4-$CH_3OC_6H_4$, 4-$BrC_6H_4$, 4-$CH_3C_6H_4$, 3,4-$CH_3OC_6H_3$, 4-$FC_6H_4$,
$Ar_1$ = Ph, 4-$CH_3C_6H_4$, 4-$NO_2C_6H_4$, 4-Br-$C_6H_4$, 2-$OHC_6H_4$

**Scheme 4.46**

The reaction of 4,5-diphenylimidazole-1,2-diamine **152** with substituted chalcones **153** in dimethylformamide for 1 h proceeds with the elimination of the hydrazine amino group, oxidation and formation of the appropriate 2,3,5,7-tetraaryl-5,6-dihydroimidazo[1,2-*a*]pyrimidin-6-ol **154** in low or moderate yields [126] (Scheme 4.47). The treatment of the same diamine with dibromo derivatives **155** produced imidazo[1,2-*a*]pyrimidines **156** [126].

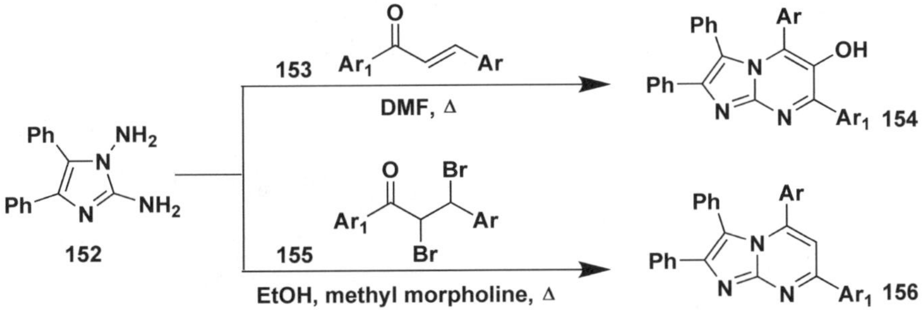

**Scheme 4.47**

In the cyclocondensation, in the second diamine group of 1,2-diamine both the *ortho* and *ipso* heterocycle atoms may react. Such a nontrivial result was obtained while studying the interaction of chalcones **157** and cyclic unsaturated ketones **161** with 5,6-diamino-1,3-dimethyluracil **55** [61, 62]. This refuted the data [57] on the oxazepine **158** structure of the products of this reaction: the formation of the spiro systems **159** and **162** was unambiguously proven

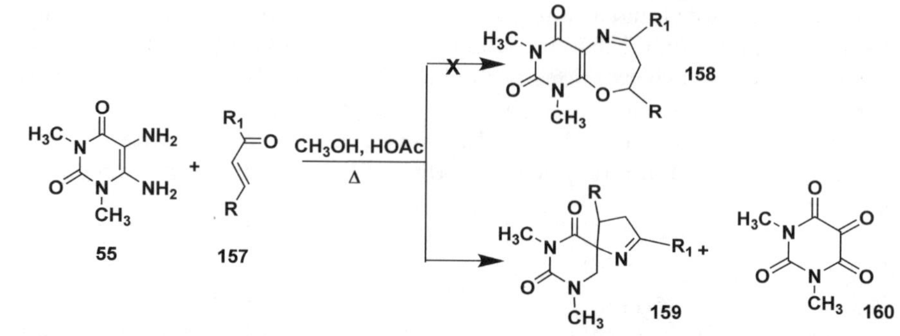


R = Ph, 4-CH$_3$O-C$_6$H$_4$, 4-CH$_3$-C$_6$H$_4$, 4-Br-C$_6$H$_4$, 4-NO$_2$-C$_6$H$_4$, 2-NO$_2$-C$_6$H$_4$
R$_1$ = CH$_3$, Ph, 4-CH$_3$O-C$_6$H$_4$, 4-CH$_3$-C$_6$H$_4$, 4-Br-C$_6$H$_4$, 4-NO$_2$-C$_6$H$_4$,
C$_6$H$_5$CH=CH, 4-CH$_3$O-C$_6$H$_4$CH=CH,

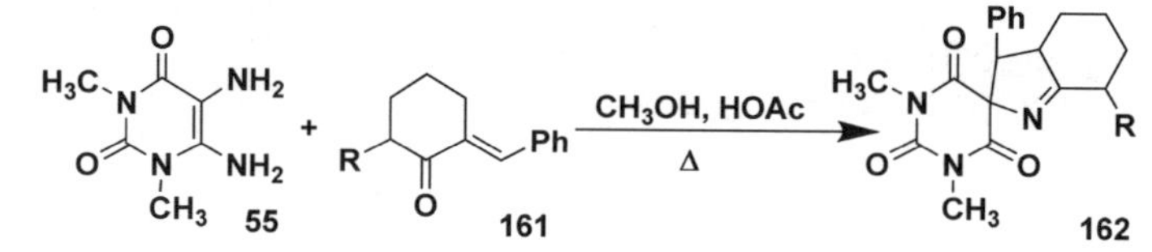


R = H, C$_6$H$_5$CH=

**Scheme 4.48**

by means of X-ray and spectral methods (Scheme 4.48). However, the interaction of cinnamon aldehyde **163** and dibenzoylcyclopentanone **165** with 1,3-dimethyl-5,6-diaminouracil does not lead to spiro compounds but stops at the stage of azomethines **164** and **166** formation [61, 62] (Scheme 4.49).

The direction of the reaction of mesityl oxide **167** with diamine **55** in aprotic solvents (benzene, toluene) also differs from that observed in the case of chalcones:

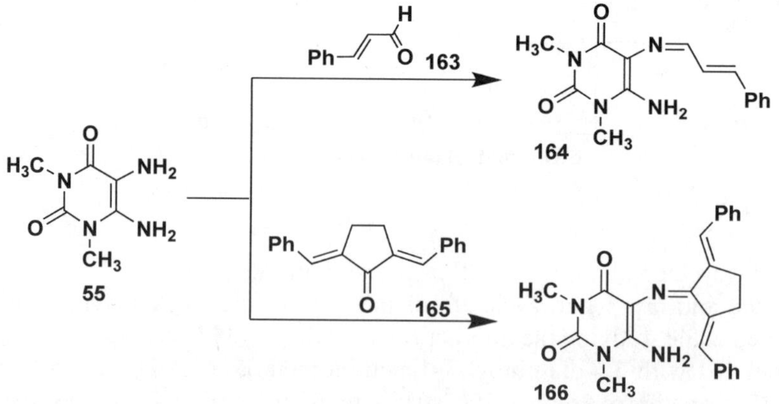


**Scheme 4.49**

the product is dihydrodiazepine. According to the data in [127], the structure of diazepine corresponds to that of **169**, whereas the X-ray data obtained by Chebanov et al. [63] unambiguously prove that the structure is **171** (Scheme 4.50).

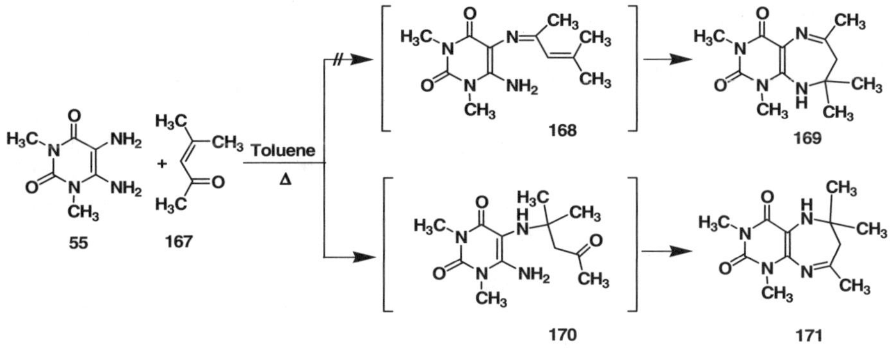

**Scheme 4.50**

The authors believe that the formation of the β-adduct **170** (instead of azomethine **168**) at the first stage of the reaction is caused by higher polarization of the C = C bond of mesityl oxide in comparison with chalcones.

The formation of pyrazoline derivatives **175** (but not dihydrotetrazolotriazepines **173**, as assumed earlier [53]) is brought about by the condensation of chalcones with diaminotetrazole **172** [128, 129]. The structure of the compounds **175** is convincingly verified using X-rays. The mechanism of their formation implies a Dimrothe rearrangement for either the initial diamine or for one of the cyclocondensation intermediates (Scheme 4.51).

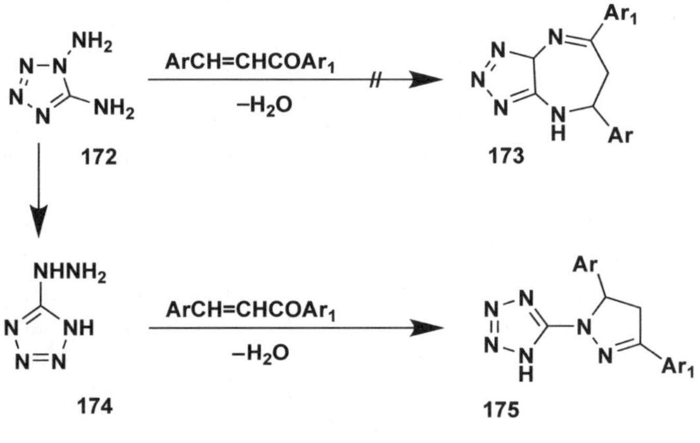

Ar = Ph, 4-$CH_3OC_6H_4$, 4-$BrC_6H_4$, 4-$CH_3C_6H_4$, 3,4-$CH_3OC_6H_3$
$Ar_1$ = Ph, 4-$CH_3C_6H_4$, 4-$NO_2C_6H_4$, 4-Br-$C_6H_4$

**Scheme 4.51**

Reaction of aromatic α,β-unsaturated ketones **176** with *o*-PDA in the presence of $Bi(NO_3)_3/Al_2O_3$ without solvent at 130°C goes in an unusual direction [35]. A mixture of dihydropyridine **177** (major product, 56–76%), dihydrodiazepine **178** (up to 10%) and the appropriate aldehyde **179** was isolated and characterized (Scheme 4.52).

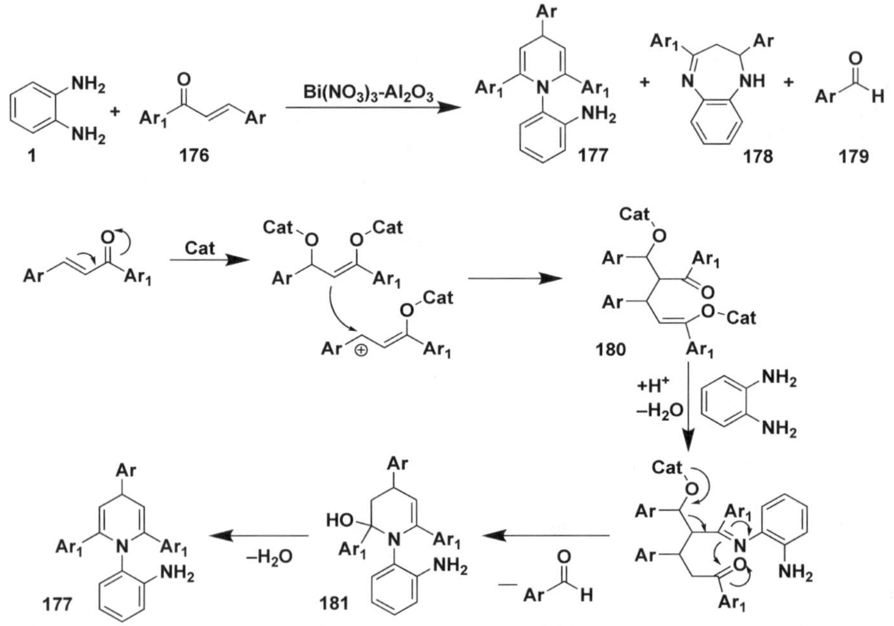


**Scheme 4.52**

According to Verma et al. [35] the mechanism of the reaction may be rationalized as involving β-oxygenation of the bismuth(III) nitrate activated chalcone enolate, which may then undergo a Michael addition to a second α,β-unsaturated ketone (Scheme 4.52) to form a 1,5-diketone enolate adduct **180**. Subsequent heteroannulation with *o*-PDA via condensation and retro-aldol disproportionation may form 2-hydroxy-1,2,4,6-tetraaryl-1,2,3,4-tetrahydropyridine derivatives **181**, which may undergo dehydration to yield 1,2,4,6-tetraaryl-1,4-dihydropyridines **177**.

Investigation of the reactions between *ortho*-diamines and dibenzoylethylene **182** causes many problems for researchers. Owing to the polyelectrophilicity of this ketone molecule as well as the possibility of a redox process in the reaction mixtures, there is a potential in such reactions for the formation of various heterocyclic systems. This leads to difficulties in establishing the structure of the reaction products and gives rise to a number of errors. In particular, during the years 1970–1975, five mutually inconsistent reports devoted to the reaction of dibezoylethylene with *o*-PDA (Scheme 4.53) were published [130, 131, 132, 133, 134]. In these papers five possible structures (**183**, **184**, and **186**–**188**) were

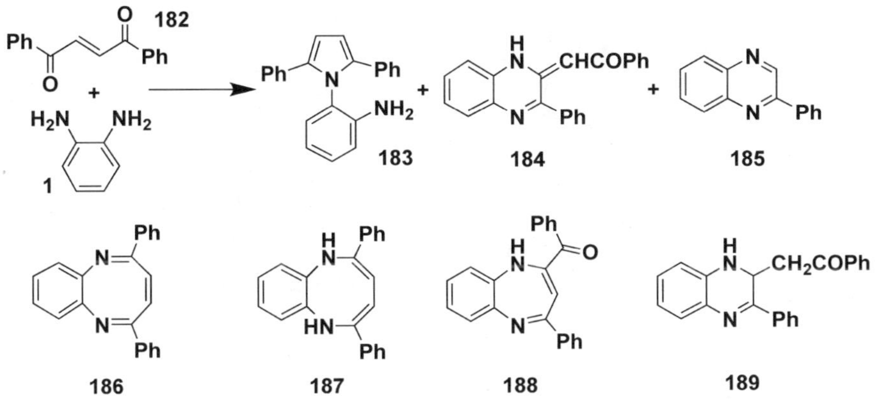


Scheme 4.53

proposed for two heterocyclic reaction products and only for the third condensation product (the earlier-described 2-phenylquinoxaline **185**) was the structure definitively established. Finally, cyclocondensation was shown to result in the formation of pyrrole **183** and quinoxaline derivatives **184** and **185** and the earlier conclusions concerning the diazepine and diazocine structure of the corresponding products were refuted. Moreover, the intermediate of the reactions, i.e., dihydroquinoxaline **189**, was also isolated [134].

The predominant direction of the reaction between *o*-PDA and dibenzoylethylene exclusively depends on the reaction conditions. For instance, boiling of the initial substances in acetic acid leads to the formation of phenacylidene quinoxaline **184** (normally the dominating reaction product), triarylpyrrole **183** and 2-phenylquinoxaline **185** [117]. When the interaction is carried out in boiling methanol in the presence of an acid catalyst, 2-phenylquinoxaline **185** turns out to be the only heterocyclic reaction product. Selective synthesis of dihydroquinoxaline **189** by short-term boiling in methanol was described in [134, 135]. The said method has a general character and is applicable to the synthesis of dihydroquinoxalines based on the derivatives of *o*-PDA and diarylethylene [135].

As in the case of *o*-PDA, ambiguity in the structure of the products is also observed in the reaction of diaroylethylenes with heterocyclic *ortho*-diamines. For example, one of the products of 1,3-dimethyl-5,6-diaminouracil **55** condensation with 4,4-dimethyldibenzoylethylene **190** at first was considered to have the structure of diazepine **196** [136] (Scheme 4.54).

However, the comparison of the spectral characteristics of this compound with those of a model benzoannelated analogue showed the structure of pyrimidopyrazine **195** to be the most probable [134]. Similarly to the case of *o*-PDA, the formation reactions of compound **195** are characterized by a number of condensation processes proceeding in parallel (compounds **191**–**194**).

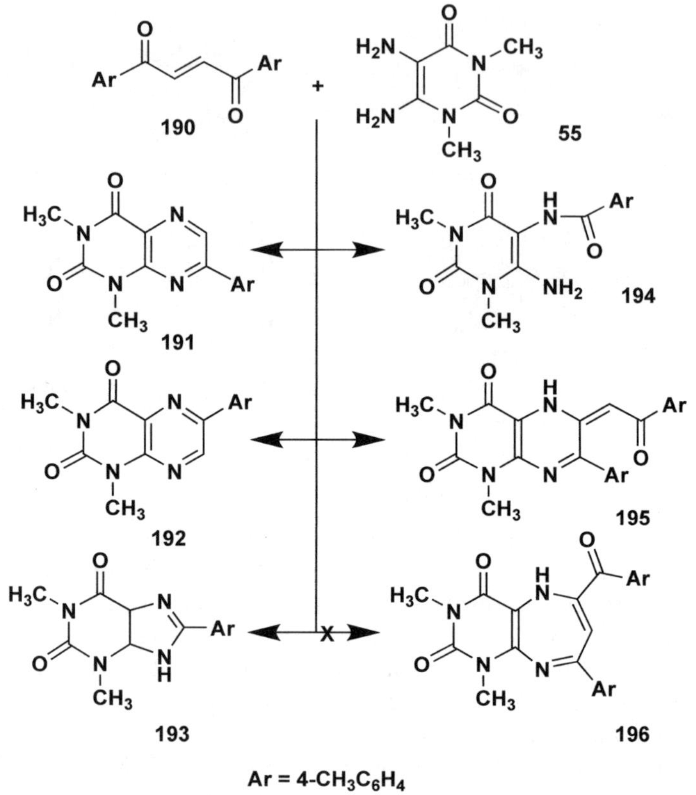


**Scheme 4.54**

Thus, the interaction of *ortho*-diamines with unsaturated 1,4-diketones results in the formation of six- or five-membered heterocycles. This fact shows that the formation of these compounds is more advantageous from the viewpoint of thermodynamics in comparison with the formation of seven-membered dihydrodiazepine systems.

## References


1. Ekeley, J.; Wells, P. *Chem. Ber.*, 1905, *38*, 2259.
2. Ried, W.; Stahlhofen, P. *Chem. Ber.*, 1957, *90*, 815.
3. Mushkalo, L. K. *Nauk. Zap. Kiiv. Derzhav. Univ.*, 1957, *16*, 133.
4. Zhelyazkov, L.; Bizhev A. *God. Vissh. Khim. Tekhnol. Inst.*, 1974, *20*, 251.
5. Nawojski, A.; Nawrocka, W. *Rocz. Chem.*, 1977, *51*, 2117.
6. Yaremenko, F. G.; Orlov, V. D.; Kolos, N. N.; Lavrushin, V. F. *Khim. Geterotsikl. Soedin.*, 1979, 848.
7. Kolos, N. N. *PhD thesis,* Kharkov State University, Kharkov, 1982.

8. Orlov, V. D.; Kolos, N. N.; Yaremenko, F. G.; Lavrushin, V. F. *Khim. Geterotsikl. Soedin.*, 1980, 697.
9. Svetlik, J.; Hanus, V.; Bella, J. *Liebigs Ann. Chem.*, 1989, 91.
10. Nagaraja, G. K.; Vaidya, V. P.; Rai, K. S.; Mahadevan, K. M. *Phosphorus Sulfur Silicon Relat. Elem.*, 2006, *181*, 2797.
11. Prakash, O.; Kumar, A.; Sadana, A.; Singh, S. P. *Synth. Commun.*, 2002, *32*, 2663.
12. Ait-Baziz, N.; Rachedi, Y.; Hamdi, M.; Silva, A. M. S.; Balegroune, F.; Thierry, R.; Sellier, N. *J. Heterocycl. Chem.*, 2004, *41*, 587.
13. Toth, G.; Levai, A.; Szollosy, A.; *Lieb. Ann. Chem.*, 1992, 803.
14. Herbert, J.; Suschitky, H. *J. Chem. Soc. Perkin Trans. I.*, 1974, *23*, 2657.
15. Morals, H. R.; Bulbarela, A.; Contreras, R. *Heterocycles*, 1986, *24*, 135.
16. Jung, D. I.; Choi, T. W.; Kim, Y. Y.; Kim, I. S.; Park, Y. M.; Lee, Y. G.; Jug, D. H. *Synth. Commun.*, 1999, *29*, 1941.
17. Balakrishna, M. S.; Kaboudin B. *Tetrahedron Lett.*, 2001, *42*, 1127.
18. Curini, M.; Epifano, F.; Marcotullio, M. C.; Rosati, O. *Tetrahedron Lett.*, 2001, *42*, 3193.
19. Yadav, J. S.; Reddy, B. V. S.; Eshwaraian, B.; Anuradha, K. *Green Chem.*, 2002, *4*, 592.
20. Kaboudin, B.; Navaee, K. *Heterocycles*, 2001, *55*, 1443.
21. Pozarentzi, M.; Stephanidou, S. J.; Tsoleridis, C. A. *Tetrahedron Lett.*, 2002, *43*, 1755.
22. Kidwai, M.; Ruby; Venkataramanan, R. *Khim. Geterotsikl. Soedin.*, 2004, 745.
23. Ma, Y.; Zhang, Y. *Synth. Commun.*, 2002, *32*, 165.
24. Kodomari, M.; Noguchi, T.; Aoyama, T. *Synth. Commun.*, 2004, *34*, 1783.
25. Sabitha, G.; Reddy, G. S. K. K.; Reddy, K. B.; Reddy, N. M.; Yadav, J. S. *Adv. Synth. Catal.*, 2004, *346*, 921.
26. Kappe, C. O.; Stadler, A. *Microwaves in Organic and Medicinal Chemistry*, Wiley-VCH: Weinheim, 2005.
27. Hayes, B. L. *Aldrichim. Acta*, 2004, *37*, 66.
28. Kappe, C. O.; Dallinger, D. *Nat. Rev. Drug Discov.*, 2006, *5*, 51.
29. Shipe, W. D.; Wolkenberg, S. E.; Lindsley, C. W. *Drug Discov. Today Technol.*, 2005, *2*, 155.
30. Loupy, A. (Ed.). *Microwaves in Organic Synthesis*, vols. 1–2; Wiley-VCH: Weinheim, 2006.
31. Rogers, R. D.; Seddon, K. R. *Science*, 2003, *302*, 792.
32. Sheldon, R. *Chem. Commun.*, 2001, *1*, 2399.
33. Wasserscheid, P.; Keim, W. *Angew. Chem., Int. Ed.*, 2000, *39*, 3772.
34. Welton, T. *Chem. Rev.*, 1999, *99*, 2071.
35. Verma, A. K.; Koul, S.; Razdan, T. K.; Kapoor, K. K. *Can. J. Chem.*, 2006, *84*, 1064.
36. Sucheta, K.; Rao, B. V. *Indian J. Chem. Sect. B*, 2005, *44*, 2152.
37. Kidwai, M.; Mothsra, P. *Synth. Commun.*, 2006, *36*, 817.
38. Jarikote, D. V.; Siddique, S. A.; Rajagopal, R.; Daniel, T.; Lahoti, R. J.; Srinivasan, K. V. *Tetrahedron Lett.*, 2003, *44*, 1835.
39. Du, Y.; Tian, F.; Zhao, W. *Synth. Commun.*, 2006, *36*, 1661.
40. Dawidar, A. M.; Metwally, M. A; Abdel-Galil, F. M.; Berghot, M. A. *Indian J. Chem. B*, 1985, *24*, 1124.
41. Khvat, A. V. *PhD thesis,* Kharkov State University, Kharkov, 1996.
42. Augustin M., Jeschke P. *Synthesis*, 1987, *10*, 937.
43. Abdel-Latif, N. A. *Sci. Pharm.*, 2005, *73*, 195.
44. Hakiki, A.; Mossadak, M.; Mokhles, M.; Rouessac, F.; Duddeck, H.; Mikhova, B. *Tetrahedron*, 1995, *51*, 2293.
45. Curtis, N.F. *Aust. J. Chem.*, 1986, *39*, 239.
46. Douglas, L.; Scheibelin, W.; Hideg, K. *J. Chem. Res. Miniprint*, 1981, *3*, 0838.
47. Hankovszky, O. H.; Hideg, K.; Lloyd, D. *J. Chem. Soc. Perkin Trans. 1*, 1974, 1619.
48. Ried, W.; Gruell, F. *Chem. Ber.*, 1963, *96*, 130.
49. Kavali, J. R.; Badami, B. V. *J. Chem Res. Synop.*, 2000, *12*, 546.
50. Orlov, V. D.; Quiroga, J.; Kolos, N. N. *Khim. Geterotsikl. Soedin.*, 1987, 363.

51. Orlov, V. D.; Quiroga, J.; Marrugo, A.; Kolos, N.N.; Iksanova, S.V. *Khim. Geterotsikl. Soedin.*, 1987, 1563.
52. Orlov, V. D.; Papiashvili, I. Z.; Povstyanoj, M. V.; Idzikovskii, V. A.; Tsyguleva, O. M. *Khim. Geterotsikl. Soedin.*, 1983, 1638.
53. Desenko, S. M.; Orlov, V. D.; Gaponik, P. N.; Karavaj, V. P. *Khim. Geterotsikl. Soedin.*, 1990, 1533.
54. Desenko, S. M. *PhD thesis,* Kharkov State University, Kharkov, 1986.
55. Insuasti, B.; Perez, A.; Valencia, J.; Quiroga, J. *J. Heterocycl. Chem.*, 1997, *34*, 1555.
56. Orlov, V. D.; Papiashvili, I. Z. *Khim. Geterotsikl. Soedin.*, 1985, 241.
57. Orlov, V. D.; Papiashvili, I. Z.; Grigorov, P. A. *Khim. Geterotsikl. Soedin.*, 1983, 671.
58. Insuasty, B.; Ramos, M.; Quiroga, J. *J. Heterocycl. Chem.*, 1994, *31*, 61.
59. Insuasty, B.; Insuasty, H.; Quiroga J. *J. Heterocycl. Chem.*, 1999, *36*, 933.
60. Insuasty, B.; Quiroga, J.; Abonia, R.; Insuasty, H.; Mosquera, M.; Cruz, S.; Nogueras, M.; Sortino, M.; Zacchino, S. *Heterocycl. Commun.*, 2004, *10*, 103.
61. Kolos, N. N.; Chebanov, V. A.; Orlov, V.D. *Khim. Geterotsikl. Soedin.*, 1999, 1230.
62. Kolos, N. N.; Orlov, V. D.; Chebanov, V. A.; Shishkin, O. V.; Kuznetsov, V. P.; Kulikov, A. Y. *Khim. Geterotsikl. Soedin.*, 1996, 978.
63. Chebanov, V. A.; Kolos, N. N.; Shishkin, O. V.; Shishkina, S. V.; Orlov, V. D. *Funct. Mater.*, 2003, *10*, 55.
64. Orlov, V. D.; Yaremenko, F. G.; Kolos, N. N.; Lavrushin, V. F. *Dokl. Akad. Nauk USSR, Ser. B*, 1978, 245.
65. Kolos, N. N.; Orlov, V. D.; Yaremenko, F. G.; Vorob'eva, N. P.; Shishkin, O. V.; Struchkov, Y. T.; Ivkov, S. M. *Khim. Geterotsikl. Soedin.*, 1995, 950.
66. Orlov, V. D.; Kolos, N. N.; Tueni, M.; Yur'eva, E. Y.; Ivkov, S. M. *Khim. Geterotsikl. Soedin.*, 1992, 947.
67. Kolos, N. N.; Shishkin, O. V.; Orlov, V.D.; Struchkov, Y.T. *Khim. Geterotsikl. Soedin.*, 1995, 551.
68. Chebanov, V. A.; Desenko, S. M.; Shishkin, O.V.; Kolos, N.N.; Komykhov, S. A.; Orlov, V. D.; Meier, H. *J. Heterocycl. Chem.*, 2003, *40*, 25.
69. Elderfild, McC. *J. Am. Chem.* Soc., 1951, *73*, 975.
70. Begland, R. W.; Harter, D. R.; Jones, F. N.; Sam, D. J.; Sheppard, W. A.; Webster, O. W.; Weigert, F. J. *J. Org. Chem.*, 1974, *39*, 2341.
71. Abramenko, V. L.; Sergienko, V. S.; Garnovsky, A. D.; Poraj-Koshits, M. A.; *Zh. Obsch. Khim.*, 1990, *60*, 1593.
72. Kaupp, G.; Pogodda, U.; Schmeyers, J. *Chem. Ber.*, 1994, *127*, 2249.
73. Fischer, J. W.; Atkins, R. L.; Nissan, R. A.; Lowe-Ma, C. K.; Decker, D. L. *J. Heterocycl. Chem.*, 1986, *23*, 1519.
74. Zelenin, K. N.; Ukraintsev, I. V.; Alekseev, V. V. *Khim. Geterotsikl. Soedin.*, 1998, 363.
75. Guzen, K. P.; Cella, R.; Stefani, H. A. *Tetrahedron Lett.*, 2006, *47*, 8133.
76. Varala, R.; Enugala, R.; Nuvula, S.; Adapa, S. R.; *Syn. Lett.*, 2006, 1009.
77. Pasha, M. A.; Jayashankara, V. P. *Heterocycles*, 2006, *68*, 1017.
78. Zhang, Z.-H.; Yang, S.-T.; Lin, J. *Synth. Commun.*, 2006, *36*, 1645.
79. Wu, J.; Xu, F.; Zhou, Z.; Shen, Q. *Synth. Commun.*, 2006, *36*, 457.
80. Das, B.; Reddy, M. R.; Ramu, R.; Reddy, K. R.; Geethangili, M. *J. Chem. Res. Synop.*, 2005, 598.
81. Chen, W.-Y.; Lu, J. *Syn. Lett.*, 2005, 1337.
82. Yadav, J. S.; Reddy, B. V. S.; Praveenkumar, S.; Nagaiah, K. *Synthesis*, 2005, 480.
83. De, S. K.; Gibbs, R. A. *Tetrahedron Lett.*, 2005, *46*, 1811.
84. Yadav, J. S.; Reddy, B. V. S.; Praveenkumar, S.; Nagaiah, K.; Lingaiah, N.; Saiprasad, P. S. *Synthesis*, 2004, 901.
85. Shukla, S.; Misra, U.; Bhalla, M.; Mukerjee, D.; Saxena, A. K.; Sinha, J. N.; Shanker, K. *Boll. Chim. Farm.*, 2001, *140*, 53.

86. Dhasmana, A.; Mehrotra, S.; Gupta, T. K.; Bhargava, K. P.; Parmar, S. S.; Barthwal, J. P. *Arzneim. Forsch.*, 1984, *34*, 943.
87. Rao, P. S.; Veeranagaiah, V. *Indian J. Chem. Sect. B*, 1990, *29*, 529.
88. Giri, B. Y.; Devi, B. L. A. P.; Lakshmi, K. V.; Prasad, R. B. N.; Lingaiah, N.; Prasad, P. S. S. *Synth. Commun.*, 2006, *36*, 3797.
89. Ponnuswamy, S.; Murugadoss, R.; Jeyaraman, R.; Thiruvalluvar, A.; Parthasarathy, V. *Indian J. Chem. Sect. B*, 2006, *45*, 2059.
90. Nagawade, R. R.; Shinde, D. B. *Mendeleev Commun.*, 2006, 113.
91. Thakuria, H.; Pramanik, A.; Borah, B. M.; Das, G. *Tetrahedron Lett.*, 2006, *47*, 3135.
92. Sivamurugan, V.; Deepa, K.; Palanichamy, M.; Murugesan, V. *Synth. Commun.*, 2004, *34*, 3833.
93. Bandgar, B. P.; Bettigeri, Sampada V.; Joshi, Neeta S. *Synth. Commun.*, 2004, *34*, 1447.
94. Reddy, B.; Sreekanth, P. M. *Tetrahedron Lett.*, 2003, *44*, 4447.
95. Quiroga, J.; Insuasty, B.; Abonia, R.; Baumann, W. *J. Heterocycl. Chem.*, 1994, *31*, 813.
96. Shanthan, R. P.; Venkataratnam, R. V. *Indian J. Chem. Sect. B*, 1992, *31*, 733.
97. Insuasty, B.; Abonia, R.; Quiroga, J.; M-eier, H. *J Heterocycl. Chem.*, 1993, *30*, 229.
98. Reddy, A. P. R.; Veeranagaiah, V. *Indian J. Chem. Sect. B*, 1986, *25*, 527.
99. Reddi, P. S.; Patham, C. V.; Rao, S. N. V. *Indian J. Chem.*, 1972, *10*, 982.
100. Kidwai, M.; Mothsra, P. *Indian J. Chem. Sect. B*, 2006, *45*, 2330.
101. Papiashvili, I. Z. *PhD thesis*, Kharkov State University, Kharkov, 1983.
102. Orlov, V. D.; Desenko, S. M.; Kolos, N. N. *Khim. Geterotsikl. Soedin.*, 1984, 147.
103. Orlov, V. D.; Desenko, S. M. *Khim. Geterotsikl. Soedin.*, 1985, 1673.
104. Zhong, W.; Zhang, Y.; Chen, X. *J. Chem. Res. Synop.*, 2000, 532.
105. Orlov, V. D.; Kolos, N. N.; Desenko, S. M.; Lavrushin, V. F. *Khim. Geterotsikl. Soedin.*, 1982, 830.
106. Orlov, V. D.; Desenko, S.M.; Pivnenko, N.S. *Khim. Geterotsikl. Soedin.*, 1987, 1122.
107. Xu, J.; Chen, L. *Heteroat. Chem.*, 2000, *11*, 158.
108. Nawrocka, W.; Sztuba, B.; Opolski, A.; Wietrzyk, J.; Kowalska, M. W.; Glowiak, T. *Arch. Pharm. (Weinheim)*, 2001, *334*, 3.
109. Kalyanam, N.; Manjunatha, S. G. *Heterocycles*, 1991, *32*, 1131.
110. Chebanov, V. A.; Muravyeva, E. A.; Kobzar, K. M. *Khim. Geterotsikl. Soedin.*, 2003, 441.
111. Kolos, N. N.; Chebanov, V. A.; Orlov, V. D. *Fiziol. Akt. Rechov.*, 2000, *29*, 14.
112. Aversa, M. C.; Gianetto, P.; Ferlazzo, A.; Bruno, G. *J. Chem. Soc. Perkin Trans. 2*, 1986, 1533.
113. Aversa, M. C.; Ferlazzo, A.; Gianetto, P.; Kohneke F.H. *Synthesis*, 1986, 230.
114. Chimirri, A.; Grasso, S.; Ottana, R.; Romeo, G.; Zappala, M. *J. Heterocycl. Chem.*, 1990, *27*, 371.
115. Xu, J.; Jin, S.; Xing, Q. *Phosphorus Sulfur Silicon Relat. Elem.*, 1998, *141*, 57.
116. Insuasty, B.; Abonia, R.; Quiroga, J.; Salcedo, A.; Kolshorn, H.; Meier, H. *Eur. J. Org. Chem*, 2000, 1973.
117. Orlov, V. D.; Kolos, N. N.; Zolotarev, B. M.; *Khim. Geterotsikl. Soedin.*, 1983, 390.
118. Insuasty, B.; Ramos, M.; Moreno, R.; Quiroga, J.; Sanchez, A.; Nogueras, M.; Hanold, N.; Meier, H. *J. Heterocycl. Chem.*, 1995, *32*, 1229.
119. Kolos, N. N.; Orlov, V. D.; Arisa, D.; Shishkin, O. V.; Struchkov, Y. T.; Vorob'eva, N. P. *Khim. Geterotsikl. Soedin.*, 1996, 87.
120. Barltrop, J. A.; Richards, C. G.; Russel, D. M.; Ryback, G. *J. Chem. Soc.*, 1959, 132.
121. Barton D., Ollis W. D. (Eds.). *Comprehensive Organic Chemistry: The Synthesis and Reactions of Organic Compounds*, vol. 4. *Heterocyclic Compounds*, Pergamon: Oxford, 1979.
122. Desenko, S. M.; Orlov, V. D. *Khim. Geterotsikl. Soedin.*, 1989, 1071.
123. Desenko, S. M.; Orlov, V. D.; Lipson, V.V.; Estrada, J. *Khim. Geterotsikl. Soedin.*, 1991, 1215.
124. Orlov, V. D.; Desenko, S. M.; Kruglenko, V. P.; Gnidec, V. P.; Kluev, N. A.; Povstyanoj, M. V. *Khim. Geterotsikl. Soedin.*, 1986, 1136.
125. Kolos, N. N.; Orlov, V. D.; Paponov, B. V.; Baumer, V. N. *Khim. Geterotsikl. Soedin.*, 1999, 1388.

126. Kolos, N. N.; Paponov, B. V.; Orlov, V. D. *Khim. Geterotsikl. Soedinin.*, 2003, 899.
127. Quan, D. Q.; Gaujolle, R.; Dang, T. B. T. *Compt. Rend. C*, 1972, *274*, 885.
128. Kolos, N. N.; Paponov, B. V.; Orlov, V. D. *Khim. Geterotsikl. Soedin.*, 2003, 310.
129. Kolos, N. N.; Paponov, B. V.; Orlov, V. D.; Lvovskaya, M. I.; Doroshenko, A. O.; Shishkin, O. V. *J. Mol. Struct.*, 2006, *785*, 114.
130. Shevchuk, M. I.; Tolochko, A. F.; Dombrovsky, A. V. *Zh. Org. Khim.*, 1970. *6*, 1108.
131. Bindra, A.; Le Goff, E. *Tetrahedron Lett.*, 1974, *16*, 1523.
132. Trattner, R. B.; Perlmutter, H. D.; *J. Heterocycl. Chem.*, 1974, *11*, 89.
133. Perlmutter, H. D.; Trattner, R. B., *J. Heterocycl. Chem.*, 1974, *11*, 847.
134. Bass, R. G.; Crichton, D. D.; Mets, H. K.; Johnson, A. F. *Tetrahedron Lett.*, 1974, *15*, 2073.
135. Orlov, V. D.; Desenko, S. M.; Insuasty, B.; Potekhin, Y. T.; Struchkov, Y. T. *Khim. Geterotsikl. Soedin.*, 1987, 523.
136. Wawzonek, S. *J. Org. Chem.*, 1976, *46*, 310.

# Addendum: Selected Synthetic Procedures

## A.1 Synthesis of Three-Membered Heterocycles

### *A.1.1 2-Benzoyl-3-phenylaziridine*

#### A.1.1.1 Method A

(Orlov, V. D.; Yaremenko, V. F.; Kolos, N. N.; Pivnenko, N. S.; Lavrushin, V. F. *Khim. Geterotsikl. Soedin.*, 1980, 544)

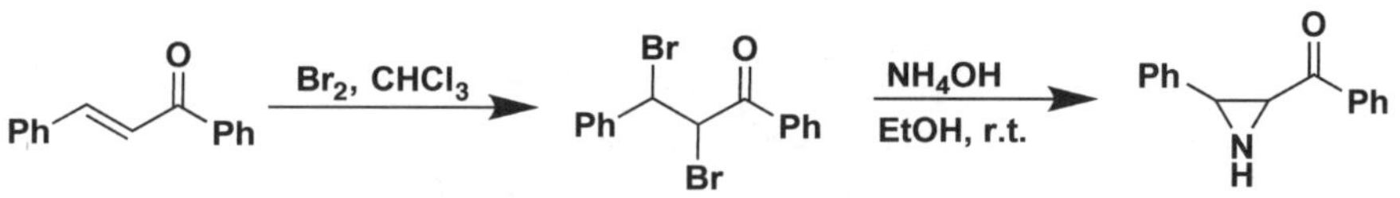


1 2 3

Scheme A.1

**2,3-Dibromo-1,3-diphenylpropan-1-one**

Chalcone **1** (2.08 g, 0.01 mol; Scheme A.1) was dissolved in 10 ml of chloroform and a solution of 1.6 g (0.01 mol) bromine in 10 ml of chloroform was added dropwise with vigorous stirring (each new portion after decoloration of the reaction mixture). The mixture was allowed to stand overnight and the precipitate of 2,3-dibromo-1,3-diphenylpropan-1-one **2** formed was filtered. Yield 80%. Melting point 160–162°C.

**2-Benzoyl-3-phenylaziridine**

The suspension of 2,3-dibromo-1,3-diphenylpropan-1-one **2** (3.6 g, 0.01 mol) in 30 ml of 95% ethanol and 12–15 ml of concentrated ammonium hydroxide was stirred for 7 h and filtered. 2-Benzoyl-3-phenylaziridine **3** was recrystallized from ethanol. Yield 77%. Melting point 100–101°C.

### A.1.1.2 Method B

(Jin, X. L.; Sugihara, H.; Daikai, K.; Tateishi, H.; Jin, Y. Z.; Furuno, H.; Inanaga, J. *Tetrahedron*, 2002, *58*, 8321)

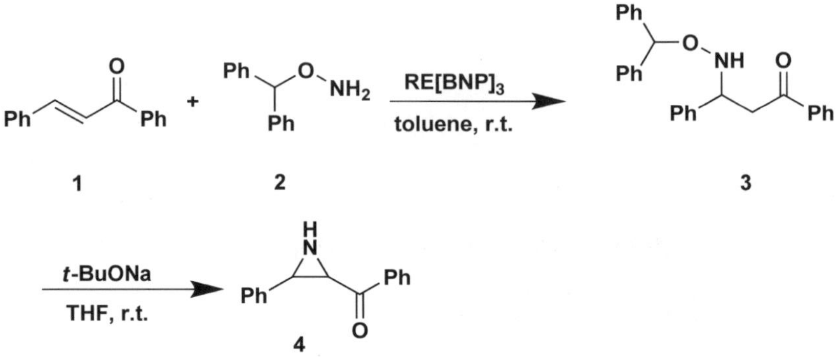


**Scheme A.2**

(RE is rare earth; BNP is 1,1′-binaphthyl-2,2′-diyl phosphate)

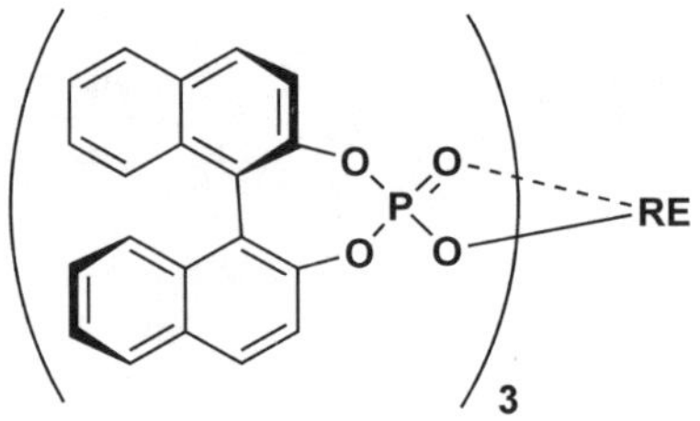


#### (3*S*)-1,3-Diphenyl-3-diphenylmethoxyaminopropan-1-one

To a suspension of scandium [(*R*)-1,1′-binaphthyl-2,2′-diyl phosphate]$_3 \cdot 3H_2O$ (10.9 mg, 0.01 mmol) in dry toluene (1 ml) was successively added **1** (20.8 mg, 0.1 mmol) and **2** (31.9 mg, 0.16 mmol) under an argon atmosphere, and the mixture was stirred for 24 h at room temperature (Scheme A.2). The reaction mixture was passed through a short column of silica gel. The eluent was evaporated and the residue was purified by column chromatography on silica gel (hexane–AcOEt 4:1) to give (3*S*)-1,3-diphenyl-3-diphenylmethoxyaminopropan-1-one **3** (40.4 mg, 99%) as colorless needles. Melting point 79.5–79.8°C.

#### (2*S*,3*R*)-2-Benzoyl-3-phenylaziridine

To a solution of **3** (20.35 mg, 0.05 mmol) in 0.5ml of dry tetrahydrofuran (THF) was added NaO-*t*-Bu (0.1 M in THF, 50 ml, 0.005 mmol) under an argon atmosphere and the mixture was stirred for 1.5 h at room temperature. The reaction mixture was passed through a short column of silica gel, the eluent was evaporated and the residue was purified by chromatography on silica gel (AcOEt–hexane 1:8) to give (2*S*,3*R*)-2-benzoyl-3-phenylaziridine **4** (11.2 mg, 100%) as a colorless solid.

## *A.1.2 (1R,6S)-4,4-Dimethyl-2-(4-nitrophenyl)-6-phenyl-3,5-diazabicyclo[3.1.0]hex-2-ene*

(Heine, H. W.; Weese, R. H.; Cooper, R. A. *J. Org. Chem.*, 1967, *32*, 2708)

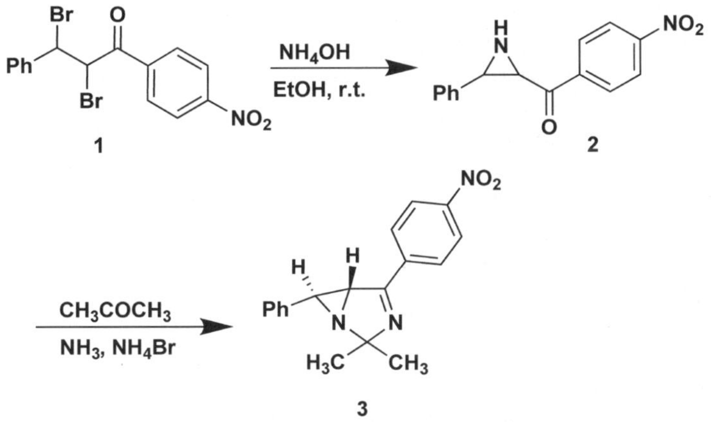


**Scheme A.3**

### A.1.2.1 *trans*-2-Phenyl-3-(4-nitrobenzoyl)aziridine

A suspension of 2,3-dibromo-1-(4-nitrophenyl)-3-phenylpropan-1-one **1** (5 g, 12.2 mmol) in 50 ml of 95% ethanol and 12–15 ml of concentrated ammonium hydroxide was stirred for 6.5 h and filtered (Scheme A.3). *trans*-2-Phenyl-3-(4-nitrobenzoyl)aziridine **2** was recrystallized from ethanol. Melting point 122–122.5°C.

### A.1.2.2 4,4-Dimethyl-2-(4-nitrophenyl)-6-phenyl-3,5-diazabicyclo[3.1.0]hex-2-ene

*trans*-2-Phenyl-3-(4-nitrobenzoyl)aziridine **2** (0.26 g, 1 mmol) was dissolved or suspended in 20 ml of commercial absolute ethanol. A large excess of acetone (about 3 ml) was added together with 0.1 g of ammonium bromide and the reaction mixture was saturated with ammonia. The reaction flask was closed and kept at room temperature for a minimum of 3 days. The 4,4-dimethyl-2-(4-nitrophenyl)-6-phenyl-3,5-diazabicyclo[3.1.0]hex-2-ene **3** which had gradually precipitated during this time was filtered off and dried. Melting point 151–152°C.

## *A.1.3 7-(4-Nitrophenyl)-2-phenyl-3,6-diazabicyclo[4.1.0]hept-2-ene*

(Heine, H. W.; Henzel, R. P. *J. Org. Chem.*, 1969, *34*, 171)

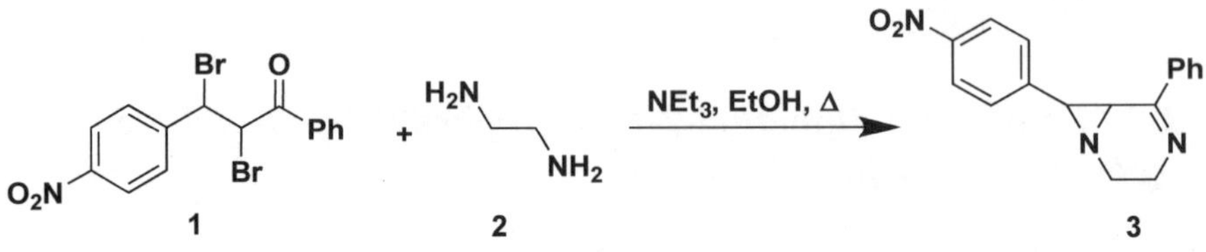


**Scheme A.4**

To a suspension of l-phenyl-2,3-dibromo-3-(4-nitrophenyl)-l-propanone **1** (12.4 g, 30.2 mmol) in 1,000 ml of 95% ethanol was added 1.80 g (30 mmol) of ethylenediamine and 6.06 g (60 mmol) of triethylamine (Scheme A.4). The mixture was stirred and heated to boiling. After the dibromide had dissolved, the reaction mixture was allowed to stand overnight at room temperature. Evaporation of most of the solvent precipitated crude 7-(4-nitrophenyl)-2-phenyl-3,6-diazabicyclo[4.1.0]hept-2-ene **3**, which was filtered and washed with 95% ethanol and with water. A yield of 7.05 g (80%) was obtained. Melting point 137–140°C with decomposition.

## *A.1.4 1,1a-Dihydro-1-aryl-2-phenylazirino[1,2-*a*]quinoxaline*

(Heine, H. W.; Henzel, R. P. *J. Org. Chem.*, 1969, *34*, 171)

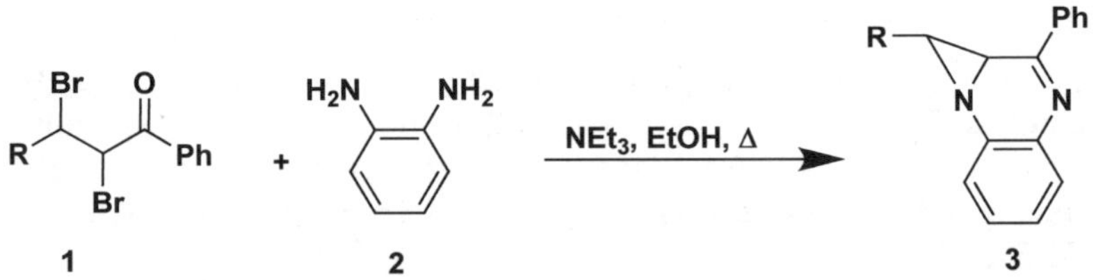


**Scheme A.5**

### A.1.4.1 1,1a-Dihydro-1-(4-nitrophenyl)-2-phenylazirino[1,2-*a*]quinoxaline

A mixture of 2-bromo-1-phenyl-3-(4-nitrophenyl)-2-propen-1-one **1** (0.50 g, 1.2 mmol), *ortho*-phenylenediamine **2** (0.163 g, 1.5 mmol) and triethylamine (0.152 g, 1.5 mmol) in 30 ml of 95% ethanol was heated to boiling (Scheme A.5). The mixture was allowed to stand overnight and then was filtered to remove a small amount of unreacted starting reagents. Most of the solvent was evaporated, and the mixture was cooled and filtered to give 0.21 g (48%) of crude 1,1a-dihydro-1-(4-nitrophenyl)-2-phenylazirino[1,2-*a*]quinoxaline **3** (R is 4-$NO_2C_6H_4$). Melting point 135–137°C.

### A.1.4.2 1,1a-Dihydro-1,2-diphenylazirino[1,2-*a*]quinoxaline

1,1a-Dihydro-1,2-diphenylazirino[1,2-*a*]quinoxaline **3** (R is $C_6H_4$) was prepared in the same manner as in the previous method by reacting 9.72 g (90 mmol) of *ortho*-phenylenediamine **2**, 11.04 g (30 mmol) of 1,3-diphenyl-2,3-dibromo-1-propanol **1** and 6.06 g (60 mmol) of triethylamine. The reaction mixture was allowed to stand for 1 week at room temperature and compound **3** was isolated with yield 4.3 g (48%). Melting point 123–125°C.

## *A.1.5 3,3-Dimethyl-1-(4-nitrophenyl)-5-(4-phenylphenyl)-3,5a-dihydro-1*H*-azireno[1,2-*c*]imidazole*

(Zbruyev, A. I.; Vashchenko, V. V.; Andryushchenko, A. Y; Desenko, S. M.; Musatov, V. I.; Knyazeva, I. V.; Chebanov, V. A. *Tetrahedron*, 2007, *63*, 4297)

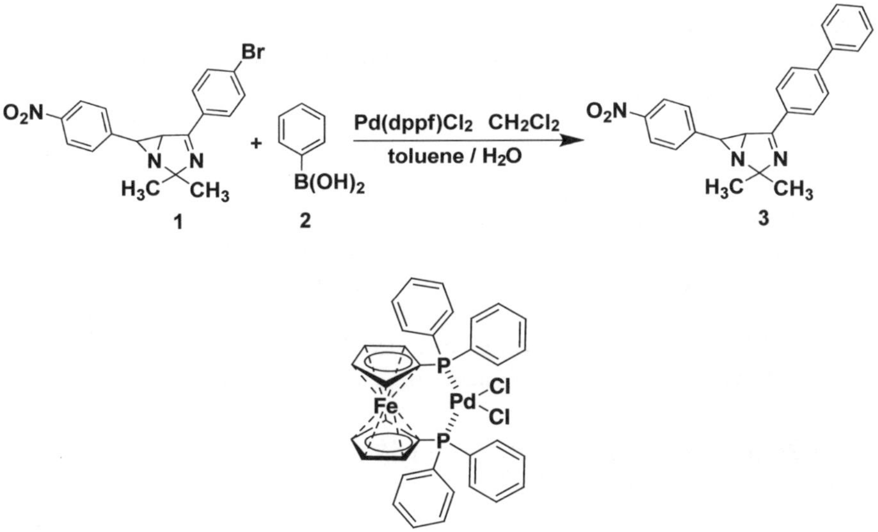


Pd[dppf]$Cl_2$:

**Scheme A.6**

### A.1.5.1 4-Phenylboronic acid

Grignard reagent was obtained from magnesium (0.169 g, 7.0 mmol) and 4-bromobiphenyl (1.5 g, 6.4 mmol) in absolute THF (25 ml), refluxed for 4 h, and cooled to 0°C. Trimethylborate (2.2 ml, 19.2 mmol) was added dropwise to the solution obtained. The reaction mixture was stirred for 1 h, allowed to warm up to 20°C and to stand overnight. Then it was decomposed with 16% HCl solution (130 ml) and mixed for 2 h. The precipitate of 4-phenylboronic acid **2** obtained was filtered off, washed with water and heptane and dried in a vacuum. Yield 0.847 g (67%) of white crystals.

#### A.1.5.2 3,3-Dimethyl-1-(4-nitrophenyl)-5-(4-phenylphenyl)-3,5a-dihydro-1*H*-azireno[1,2-*c*]imidazole (Nonmicellar Procedure)

A mixture of aziridine **1** (1.00 g, 2.59 mmol), 4-phenylboronic acid **2** (0.379 g, 3.1 mmol), toluene (30 ml), water (15 ml) and 1-butanol (3 ml) in a three-necked flask was degassed in a vacuum and flashed with argon under vigorous stirring, then Pd(dppf)$Cl_2 \cdot CH_2Cl_2$ (56 mg, 0.069 mmol) was added (Scheme A.6). The degassing was repeated, and the mixture was heated to reflux under an argon atmosphere and a solution of $NaHCO_3$ (870 mg, 10.36 mmol) in water (15 ml) was directly added. Refluxing continued and samples of the reaction mixture were taken for high-performance liquid chromatography analysis at 3, 5, 10, 30, 70 and 140 min after addition of the base.

#### A.1.5.3 3,3-Dimethyl-1-(4-nitrophenyl)-5-(4-phenylphenyl)-3,5a-dihydro-1*H*-azireno[1,2-*c*]imidazole (Micelle-Catalyzed Procedure)

This procedure with micellar catalysis is similar to the one described above, except for the addition of a surfactant to the mixture of reactants. Bromo-substituted aziridines **1** (2.59 mmol), 4-phenylboronic acid **2** (0.379 g, 3.1 mmol), sodium dodecyl sulfate (0.300 g), toluene (30 ml), water (15 ml), 1-butanol (3 ml), Pd(dppf)$Cl_2 \cdot CH_2Cl_2$ (56 mg, 0.069 mmol), and a solution of $NaHCO_3$ (870 mg, 10.36 mmol) in water (15 ml) were used. After addition of the base, the reaction mixture was refluxed for 5 min and cooled to room temperature. The organic layer was separated, the water layer was extracted with $CH_2Cl_2$ (2.15 ml) and the combined organic extract was washed with water (50 ml), dried over $Na_2SO_4$, filtered through a short plug of silica gel (20 ml), evaporated to dryness and stored in vacuum desiccator to produce 3,3-dimethyl-1-(4-nitrophenyl)-5-(4-phenylphenyl)-3,5a-dihydro-1*H*-azireno[1,2-*c*]imidazole **3**.

### *A.1.6 2,4,6-Triphenyl-3,5-diazabicyclo[3.1.0]hex-2-ene (Multicomponent Procedure)*

(Risitano, F.; Grassi, G.; Foti, F.; Moraci, S. *Synlett*, 2005, *10*, 1633)

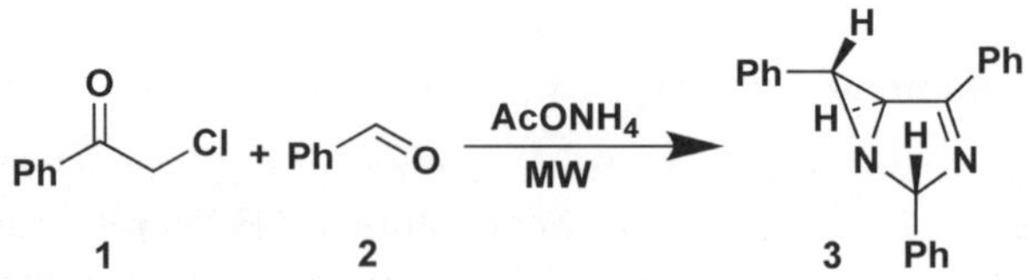


**Scheme A.7**

A mixture of phenacyl chloride **1** (0.46 g, 3 mmol), aldehyde **2** (0.64 g, 6 mmol), ammonium acetate (1.55 g, 20 mmol) and glacial acetic acid (10 ml) in *n*-PrOH (20 ml) in the presence of molecular sieves (4 Å) was irradiated for 5–10 min in a self-tunable CEM microwave synthesizer at 90 °C (Scheme A.7). After the reaction had been cooled to room temperature, the solvent was removed under vacuum and the residue was crystallized from ethanol to give 2,4,6-triphenyl-3,5-diazabicyclo[3.1.0]hex-2-ene **3** as colorless crystals. Melting point 155–156°C.

## A.2 Synthesis of Five-Membered Heterocycles

### *A.2.1 3-Benzoyl-4-phenyl-$\Delta^1$-pyrazoline*

(Smith, L. I.; Pings, W. B. *J. Org. Chem.*, 1937, *2*, 23)

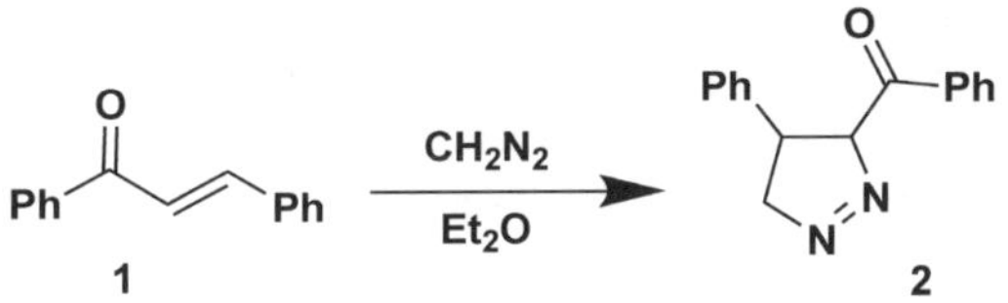


**Scheme A.8**

To the cold ethereal diazomethane solution prepared from 10 g of nitrosomethylurea (about 2.8 g $CH_2N_2$, 0.067 mol; volume 100 ml) was added a solution of benzalacetophenone **1** (14.5 g, 0.07 mol) in 150 ml of ether (Scheme A.8). Within 5 min a precipitate formed. After 1 h at −14°C filtration yielded 13.5 g of the colorless 3-benzoyl-4-phenyl-$\Delta^1$-pyrazoline **2**. Partial evaporation of the ether gave an additional 3 g of the material. The total yield, based on 13.5 g of benzalacetophenone (0.065 mol), was 100%. The material may be crystallized from methyl or ethyl alcohol, a mixture of ethyl acetate and petroleum ether, a mixture of chloroform and petroleum ether, or carbon tetrachloride. The best results were obtained using methyl alcohol. Melting point 92–93°C.

Upon standing for about 4 months, this material decomposed to oil, from which no crystalline material could be isolated.

### *A.2.2 (4′-Aryl-4′,5′-dihydro-1′H-pyrazol-3′-yl)-(4-aryl-1H-pyrrol-3-yl)methanone*

(Padmavathi, V.; Reddy, B. J. M.; Subbaiah, D. R. C. V. *New J. Chem.*, 2004, *28*, 1479)

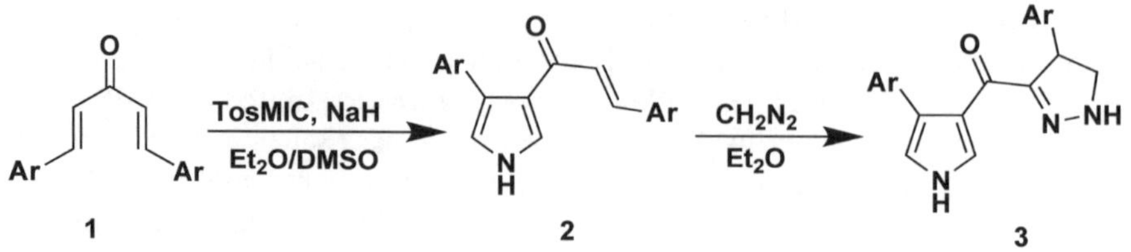


TosMIC = Tosyl methyl isocyanide

Scheme A.9

#### A.2.2.1 3′-Aryl-1′-(4-aryl-1*H*-pyrrol-3-yl)-prop-2′-enone: General Procedure

A mixture of tosyl methyl isocyanide (5 mmol) and bischalcone **1** (5 mmol) in $Et_2O$–dimethyl sulfoxide (2:1) was added dropwise under stirring to a suspension of NaH (50 mg) in $Et_2O$ (10 ml) at room temperature (Scheme A.9). Stirring was continued for about 5 h. Then, water was added and the product was extracted with $Et_2O$. The ethereal fraction was dried over anhydrous $Na_2SO_4$. The solvent was removed in vacuo. The resulting solid, 3′-aryl-1′-(4-aryl-1*H*-pyrrol-3-yl)-prop-2′-enone **2**, was purified by recrystallization from methanol.

#### A.2.2.2 (4′-Aryl-4′,5′-dihydro-1′*H*-pyrazol-3′-yl)-(4-aryl-1*H*-pyrrol-3-yl)methanone: General Procedure

To a cooled solution of **2** (5 mmol) in dichloromethane (20 ml), an ethereal solution of diazomethane (40 ml, 0.4 M) and triethylamine (0.12 g) was added (Scheme A.9). The reaction mixture was kept at −20 to −15 °C for 40–48 h. The solvent was removed under reduced pressure. The resulting solid, (4′-aryl-4′,5′-dihydro-1′*H*-pyrazol-3′-yl)-(4-aryl-1*H*-pyrrol-3-yl)methanone **3**, was purified by recrystallization from methanol.

### *A.2.3 4′-Aryl-1,2,3,4,4′,5′-hexahydro-3′*H*-naphthalene-2-spiro-3′-pyrazol-1-ones : General Procedure*

(Molchanov, A. P.; Korotkov, V. S.; Kostikov, R. R. *Zh. Org. Khim.*, 2004, *40*, 501)

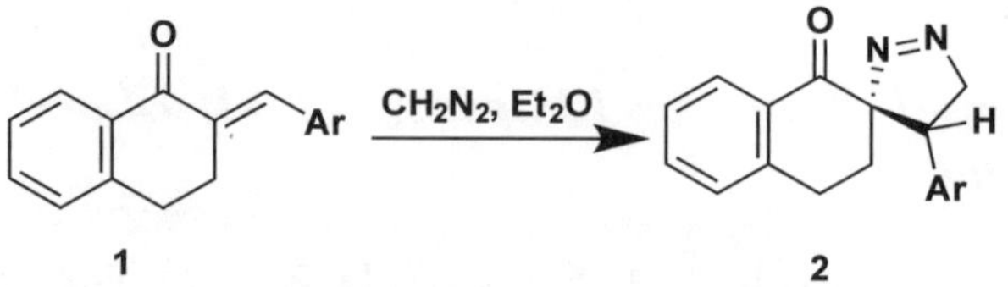


Scheme A.10

A solution of diazomethane in diethyl ether, prepared from 5 g (49 mmol) of *N*-nitrosomethylurea, was added to a cold solution of 4 mmol of the corresponding 2-arylmethylene-1,2,3,4-tetrahydronaphthalen-1-one **1** in 10 ml of benzene, and the mixture was left overnight (Scheme A.10). The solvent was evaporated and the residue was recrystallized from ethanol.

## *A.2.4 5-(4′-Methoxyphenyl)-l,3-diphenyl-4-pyrazoline*

(Kidwai, M.; Misra, P. *Synth. Commun.*, 1999, *29*, 3237)

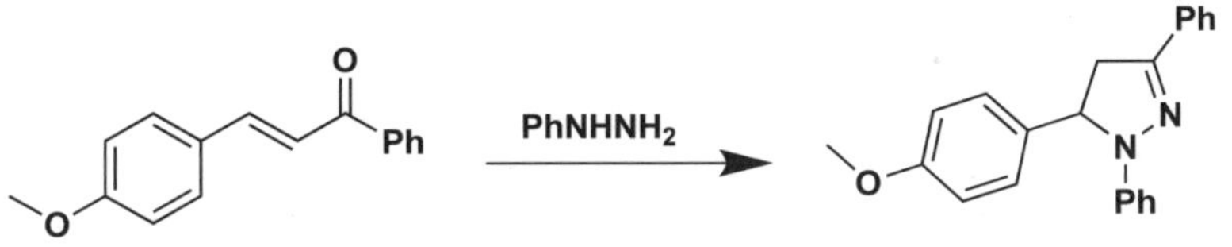


**Scheme A.11**

### A.2.4.1 Method A

Phenylhydrazine (1.4 g, 0.013 mol) was added to 4-methoxychalcone **1** (2.38 g, 0.01 mol) in acetic acid (15 ml) and the mixture was refluxed under constant stirring for 3 h (Scheme A.11). The reaction mixture was diluted with ice-cold water. The solid was collected, washed with water and recrystallized from ethanol as pale-yellow plates. The yield of **2** was 68%. Melting point 124–126°C.

### A.2.4.2 Method B

Phenylhydrazine (1.4 g, 0.013 mol) was added to compound **1** (2.38 g, 0.01 mol) in acetic acid (15 ml) and the contents were subjected to microwave irradiation for an appropriate time and worked up as described in method A. The yield of **2** was 78%.

### A.2.4.3 Method C

Basic alumina (18 g) was added to the solution of phenylhydrazine (1.4 g, 0.013 mol) and chalcone **1** (2.38 g, 0.01 mol) dissolved in dichloromethane (5 ml) at room temperature. The reaction mixture was thoroughly mixed and the adsorbed material was dried in air (beaker) and placed in an alumina bath inside a microwave oven. Upon completion of the microwave reaction (1–2 min), based on thin-layer chromatography (TLC) examination, the mixture was cooled to room temperature and then the product was extracted into dichloromethane (4 × 15 ml). Removal of the solvent under reduced pressure gave the product, which was purified by crystallization from a mixture of methanol and dichloromethane. The yield of **2** was 82%.

### *A.2.5 1-Thiocarbamoyl-3,5-diaryl-2-pyrazoline derivatives*

(Turan-Zitouni, G.; Chevallet, P.; Kilic, F. S.; Erol, K. *Eur. J. Med. Chem. Chim. Ther.*, 2000, *35*, 635)

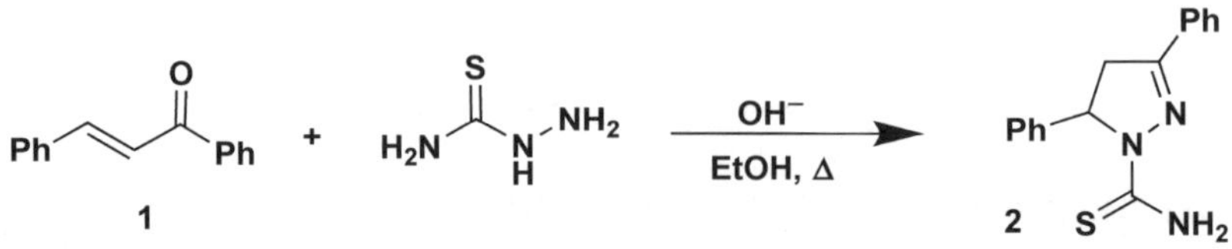

**Scheme A.12**

A mixture of chalcone (2.08 g, 0.01 mol), thiosemicarbazide (0.91 g, 0.01 mol) and NaOH (1 g, 0.025 mol) was refluxed in ethanol (25 ml) for 8 h (Scheme A.12). The solution was poured into ice-water. The precipitate was filtered off and crystallized from methanol. Melting point 204–206°C.

### *A.2.6 One-Pot Synthesis of 1-Methyl-5-(4-nitrophenyl)-3-phenyl-4,5-dihydro-1H-pyrazole*

(Müller, T. J. J.; Ansorge, M.; Aktah, D. *Angew. Chem. Int. Ed.*, 2000, *39*, 1253)

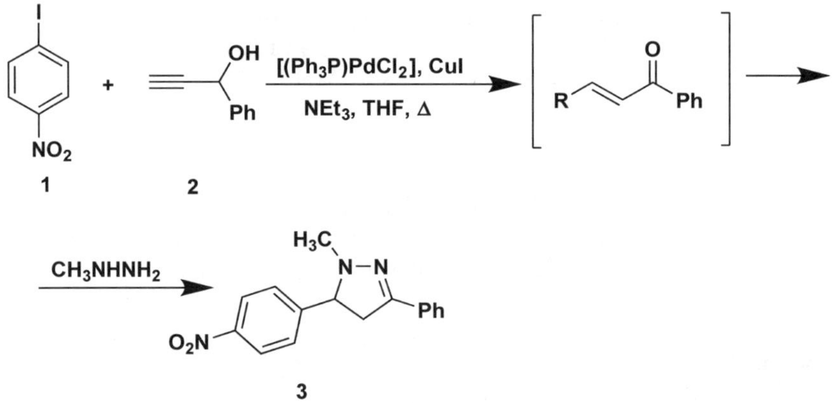

**Scheme A.13**

To a solution of 1-iodo-4-nitrobenzene **1** (0.25 g, 1.00 mmol), [Pd($PPh_3$)$_2$$Cl_2$] (22 mg, 0.02 mmol) and CuI (2 mg, 0.01 mmol) in anhydrous THF (10 ml) and triethylamine (5 ml) under nitrogen, a solution of 1-phenylprop-2-yn-1-ol **2** (0.14 g, 1.05 mmol) in THF (10 ml) was added dropwise over a period of 30 min (Scheme A.13). Then the reaction mixture was heated to reflux for 10 h. After the mixture had been cooled, *N*-methylhydrazine (0.15 ml, 3.76 mmol) was added and heating was continued for 5 h. After removal of the solvents in vacuo, the crude product was flash-chromatographed on silica gel (length 10 cm, diameter 1 cm) with diethyl ether–pentane to produce **3** (0.26 g; (92%). Melting point 118–119°C.

### *A.2.7 2,3-Diphenyl-7-benzylidene-3,3a,4,5,6,7-hexahydro-2*H*-indazole and 1,3-diphenyl-7-benzylidene-3,3a,4,5,6,7-tetrahydroindazole*

(Gella, I. M.; Yaya, A. R.; Orlov, V. D. *Kharkov Univ. Bull.*, 2001, *30*, 103)

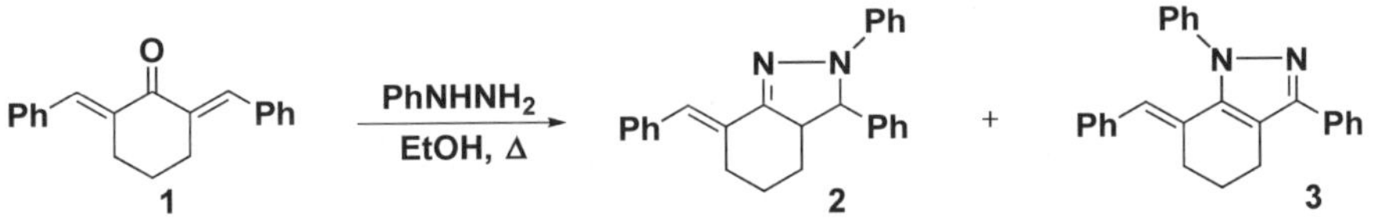


**Scheme A.14**

To a suspension of 1.7 g (6.2 mmol) 2,6-dibenzylidenecyclohexanone **1** in 70 ml of ethanol 0.7 ml (6.2 mol) of phenylhydrazine was added and the mixture was refluxed for 2–3 h until the initial ketone had completely disappeared, then the solvent was evaporated and the residue was diluted with benzene and chromatographed on an $Al_2O_3$ column with benzene (Scheme A.14). The first fraction eluted exhibited luminescence and contained 2,3-diphenyl-7-benzylidene-3,3a,4,5,6,7-hexahydro-2*H*-indazole **2**. After evaporation of the solvent and crystallization from a mixture of ethanol and benzene, 1.35 g (60%) of **2** with a melting point of 157–159°C was obtained. The next fraction was evaporated and 0.56 g (25%) of 1,3-diphenyl-7-benzylidene-3,3a,4,5,6,7-tetrahydroindazole **3** with a melting point of 175–177°C was obtained.

## A.3 Synthesis of Six-Membered Heterocycles

### *A.3.1 7,7-Dimethyl-2,4-diphenyl-7,8-dihydroquinolin-5(6*H*)-one*

(Powers, D. G.; Casebier, D. S.; Fokas, D.; Ryan, W. J.; Troth, J. R.; Coffen, D. L. *Tetrahedron*, 1998, *54*, 4085)

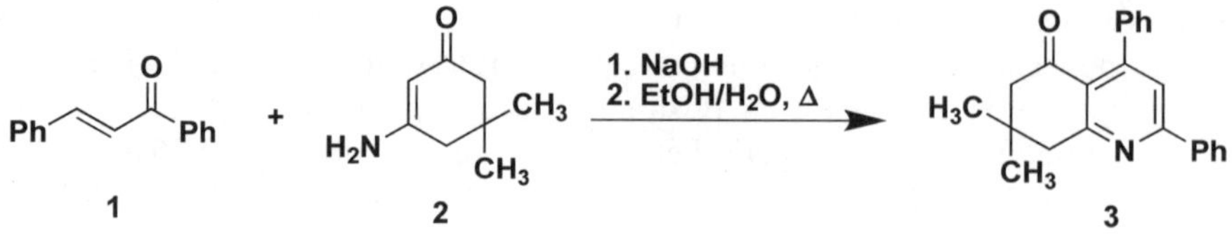

Scheme A.15

To a reaction vial containing chalcone **1** (0.1 g, 0.5 mmol), enamine **2** (0.02 g, 0.5 mmol) as a solid and 4 ml of absolute ethanol, 2 ml of a 0.25 M solution of NaOH in absolute ethanol was added (Scheme A.15). The reaction mixture was capped, shaken to ensure mixing and then heated at 80°C for 12 h. Upon completion, the reaction mixture was cooled to room temperature and quenched with 1 ml of 0.5 M HCl in ultrapure water. The reaction mixture was shaken and then concentrated to dryness in a vacuum to yield the product as a solid. Yield 65%. Melting point 125–127°C.

## *A.3.2 8-Benzoyl-5-hydroxy-5-methyl-1,2,3,5,6,7-hexahydroimidazo[1,2-*a*]pyridine*

(Zhang, J.; Wang, M.; Huang, Z. *J. Chem. Soc. Perkin Trans. 1*, 1999, *15*, 2087)

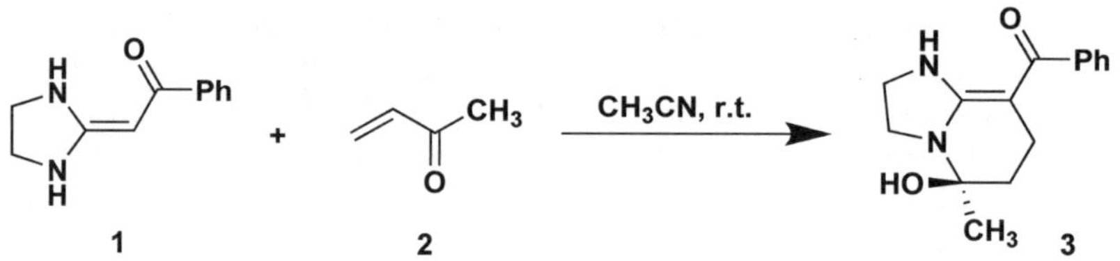

Scheme A.16

A mixture of cyclic enamine **1** (0.38 g, 2 mmol) and unsaturated ketone **2** (0.14 g, 2 mmol) in acetonitrile or ethanol (15 ml) was stirred at room temperature for 15 h (Scheme A.16). After removal of the solvent under vacuum, the residue was recrystallized from a suitable solvent to give pure **3** (75%). Melting point 95–96°C.

## *A.3.3 4,6-Diphenyl-4,5-dihydropyrimidine-2-thione*

### A.3.3.1 Method A

(Simon, D.; Lafont, O.; Farnoux, C. C.; Miocque, M. *J. Heterocycl. Chem.*, 1985, *22*, 1551)

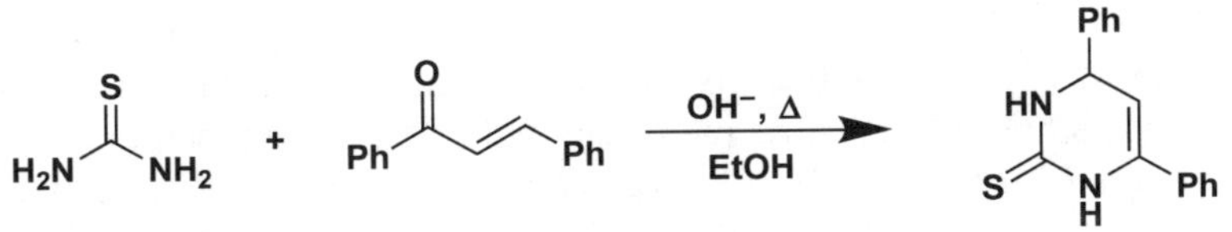

1 2 3

Scheme A.17

To a solution of chalcone **2** (2.08 g, 0.01 mol) and thiourea **1** (1.52 g, 0.02 mol) in 100 ml of absolute ethanol under reflux, a solution of potassium hydroxide (0.8 g, 0.02 mol) in 100 ml of ethanol was added dropwise with agitation (Scheme A.17). The mixture was refluxed with agitation for 4 h, the solvent was evaporated and the residue was purified by recrystallization. Yield 68%. Melting point 184°C.

### A.3.3.2 Method B

(Zigeuner, G.; Brunetti, H.; Ziegler, H.; Bayer, M. *Monatsh. Chem.*, 1970, *101*, 1767)

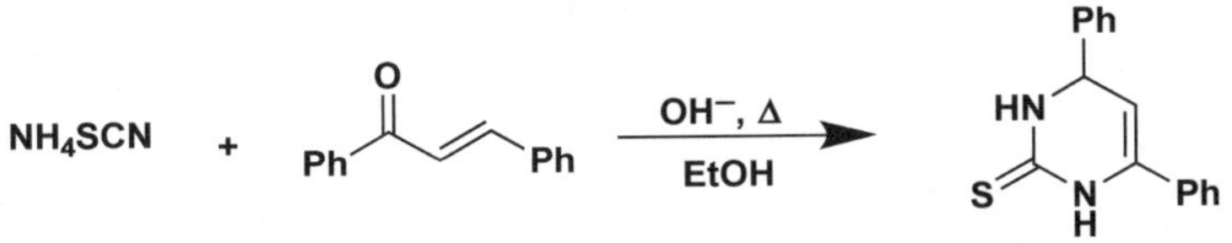

4 2 3

Scheme A.18

1-Phenyl-1-buten-3-one **2** (12 g, 0.06 mol) was refluxed with $NH_4SCN$ (6 g, 0.07 mol) in 35 ml of benzene and 2.5 g cyclohexanol for 5–6 h with water separation (Scheme A.18). The residue was filtered off, washed with ethanol and ether and crystallized from ethanol. Yield 9 g. Melting point 198–199°C.

## *A.3.4 2-Methyl-2,4-diphenyl-1,4-dihydropyrimidine*

### A.3.4.1 Method A

(Weis, A. L.; Mamaev, V. P. *Izv. Sib. Otd. Akad. Nauk SSSR Ser. Khim.*, 1975, *6*, 148)

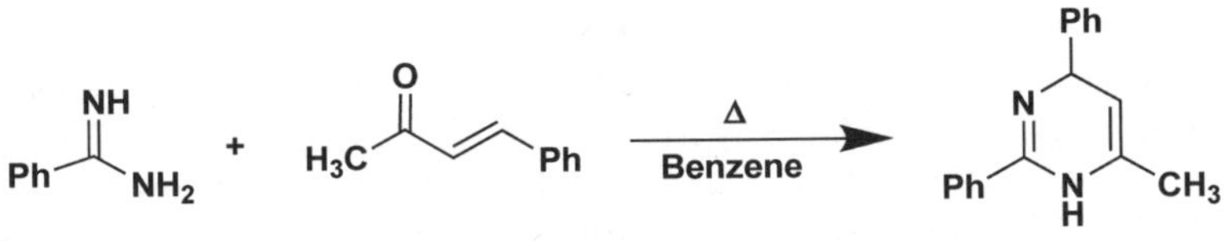

1 2 3

Scheme A.19

A solution of 6.3 g (0.0525 mol) of benzamidine **1**, prepared from 8.3 g (0.0525 mol) of benzamidine hydrochloride and an equimolar amount of sodium in methanol with subsequent removal of NaCl and evaporation of the solvent, and 7.3 g (0.05 mol) of benzalacetone **2** in 100 ml of dry benzene was heated under reflux (Scheme A.19) with a Dean–Stark trap until the calculated volume of water had separated (after about 4 h). After allowing it stand overnight at 5°C, 8.6 g (78%) of 2-methyl-2,4-diphenyl-1,4-dihydropyrimidine **3** was filtered off; melting point 145–146°C. An additional portion of dihydropyrimidine could be isolated from the filtrate.

### A.3.4.2 Method B

(Weis, A. L. *Synthesis*, 1985, 528)

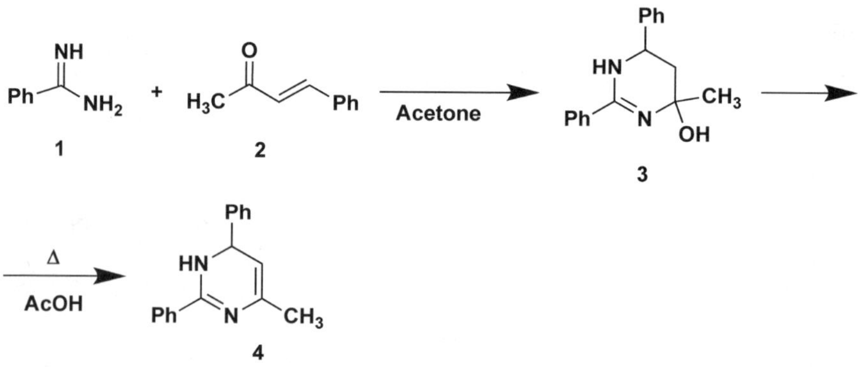


**Scheme A.20**

**6-Hydroxy-6-methyl-2,4-diphenyl-1,4,5,6-tetrahydropyrimidine**

A solution of benzalacetone **2** (2.9 g, 0.02 mol) in dry acetone (50 ml) was added dropwise (30 min) at 0–10°C to a solution of amidine **1** (2.4 g, 0,02 mol) in dry acetone (50 ml) with constant magnetic stirring in a dry inert atmosphere (Scheme A.20). Following the addition, the mixture was stirred for another 30 min, after which the cooling bath was removed and the mixture was stirred at room temperature. The reaction was monitored by TLC. A white solid started to precipitate after the first hour. When the starting materials had disappeared almost completely (TLC), the white copious precipitate was filtered and washed with dry diethyl ether. The mother liquor was evaporated to dryness, the crude residue was triturated with a small amount of dry diethyl ether, and an additional portion of undissolved tetrahydropyrimidine was filtered off. The white solids were combined, dried and recrystallized. The yield of 6-hydroxy-6-methyl-2,4-diphenyl-1,4,5,6-tetrahydropyrimidine **3** was 82%. Melting point 80–81°C.

**2-Methyl-2,4-diphenyl-1,4-dihydropyrimidine**

A solution of 6-hydroxy-6-methyl-2,4-diphenyl-1,4,5,6-tetrahydropyrimidine **3** (5.3 g, 0.02 mol) in glacial acetic acid (15–20 ml) was heated at 80°C, for approximately 48 h. After cooling, ice (20 g) was added followed by dichloromethane (40 ml). A 25% aqueous solution of ammonia (20–25 ml) was added slowly with constant stirring until the mixture became basic. The dichloromethane layer was separated and the upper layer was extracted with dichloromethane (3 × 50 ml). The organic layers were combined, dried with magnesium sulfate and the solvent was evaporated in vacuo. The residue of 2-methyl-2,4-diphenyl-1,4-dihydropyrimidine **4** was recrystallized from ethyl acetate. Yield 98%. Melting point 145°C.

## *A.3.5 2,4-Di(2-hydroxyphenyl)-9-isopropyl-1,5,9-triazaspiro[5.5]undec-1-ene*

(Muravyova, E. A.; Desenko, S. M.; Musatov, V. I.; Knyazeva, I. V.; Shishkina, S. V.; Shishkin, O. V.; Chebanov, V. A. *J. Comb. Chem.*, 2007, *9*, 798)

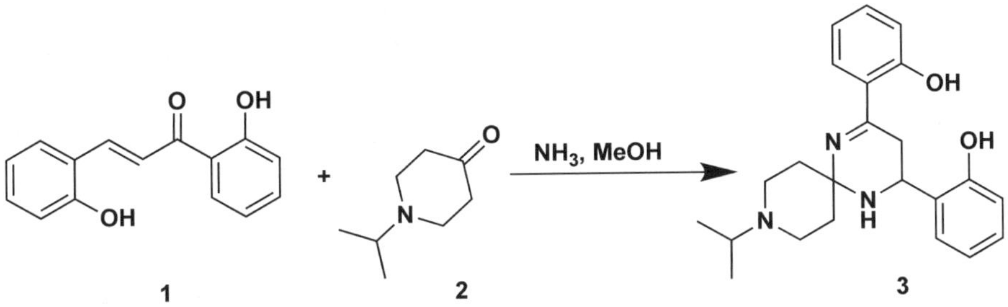


**Scheme A.21**

### A.3.5.1 Method A

A mixture of the α,β-unsaturated ketone **1** (0.24 g, 0.001 mol) and 1-isopropyl-piperidin-4-one **2** (0.14 g, 0.001 mol) in methanol saturated with ammonia (2 ml) was sonicated at room temperature for 90 min in a round-bottomed flask equipped with a condenser (Scheme A.21). The reaction mixture was allowed to stand for 2 h at room temperature and was then filtered to give the solid tetrahydropyrimidine products **3** and these were washed with methanol and dried in air. The reaction products were obtained in high purity (more than 98% by $^1$H NMR) and did not require further purification by recrystallization. Melting point 239–240°C.

### A.3.5.2 Method B

A mixture of the α,β-unsaturated ketone **2** (0.24 g, 0.001 mol) and 1-isopropyl-piperidin-4-one **2** (0.14 g, 0.001 mol) in methanol saturated with ammonia (2 ml) was shaken with an orbital shaker (or stirred on a magnetic stirrer) at room temperature for 48 h in a sealed conical flask. The reaction mixture was filtered to give the solid tetrahydropyrimidine products **3** and these were washed with methanol. The reaction products were crystallized from propan-2-ol and dried in air. Melting point 238–239°C.

## *A.3.6 2-Amino-6-(2-bromophenyl)-4-(1H-indol-3-yl) nicotinonitrile*

(Manna, F.; Chimenti, F.; Bolasco, A.; Bizzarri, B.; Filippelli, W. *Eur. J. Med. Chem. Chim. Ther.*, 1999, *34*, 245)

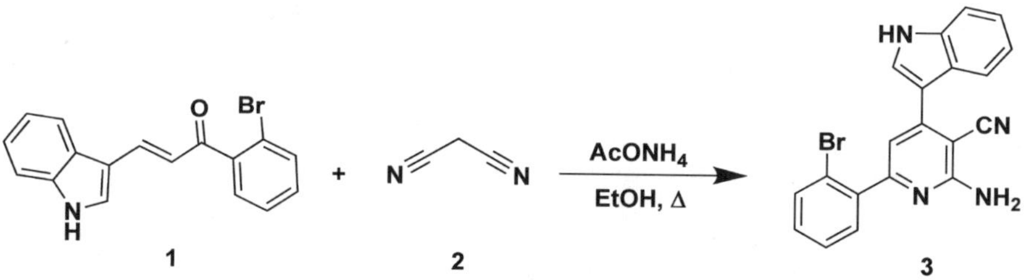


**Scheme A.22**

A solution of indolic chalcone **1** (18.2 g, 0.056 mol), malononitrile **2** (3.7 g, 0.056 mol) and ammonium acetate (39.3 g, 0.38 mol) in 80 ml of anhydrous ethanol was refluxed at 110 °C with stirring for 24 h (Scheme A.22). The resulting solid was collected by filtration, water-washed and purified with warm ethanol. The yield of **3** was 74%. Melting point 223–225°C.

## *A.3.7 6-Phenyl-4-(pyridin-3-yl)-2-thioxo-1,2-dihydropyridine-3-carbonitrile*

(Krauze, A.; Germane, S.; Eberlins, O.; Sturms, I.; Klusa, V.; Duburs, G. *Eur. J. Med. Chem. Chim. Ther.*, 1999, *34*, 301)

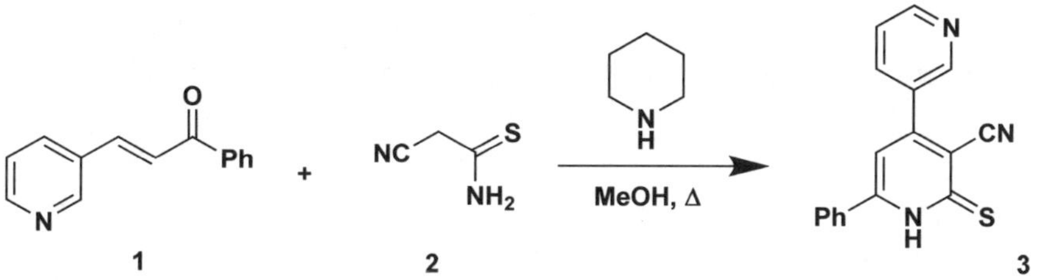

**Scheme A.23**

A mixture of 1-phenyl-3-(3′-pyridyl)-2-propen-1-one **1** (4.19 g, 0.02 mol), cyanothioacetamide (2.0 g, 0.02 mol) and piperidine (0.85 g, 0.01 mol) in 25 ml of ethanol was refluxed for 30 min (Scheme A.23). Then 5 ml of acetic acid was added and the reaction mixture was stirred for 2 h at room temperature and cooled to 0°C. The precipitated dark-yellow crystals were filtered off and washed with 20 ml of cold ethanol and 20 ml of water to yield 4.63 g (80%) of **3**. Melting point 232–234°C.

### *A.3.8 5,7-Diphenyl-4,7-dihydro-1,2,4-triazolo[1,5-*a*]pyrimidine*

(Orlov, V. D.; Desenko, S. M.; Potekhin, K. A.; Struchkov, Y. T. *Khim. Geterotsikl. Soedin.*, 1988, 229)

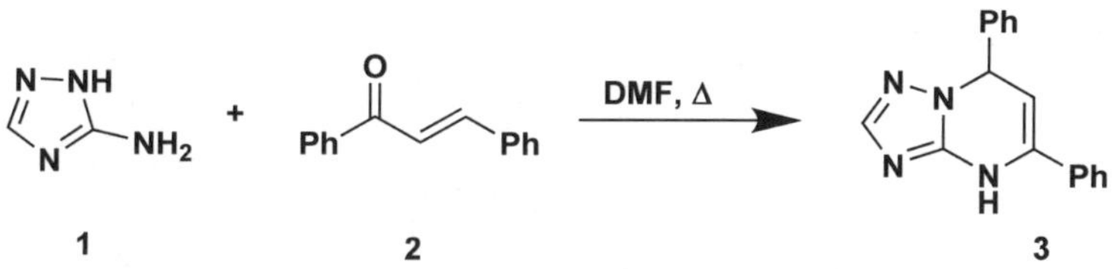

**Scheme A.24**

A solution of 0.83 g (4 mmol) of chalcone **2** and 0.4 g (4.7 mmol) of 3-amino-1,2,4-triazole **1** in 1 ml of dimethylformamide (DMF) was refluxed for 1 h (Scheme A.24). The mixture after cooling was diluted with benzene (30 ml) and the precipitate was filtered off. The 0.75 g (68%) of **3** obtained had a melting point of 227°C (from a benzene–DMF mixture).

### *A.3.9 5-Phenyl-4,7-dihydro-1,2,4-triazolo[1,5-*a*]pyrimidine*

(Desenko, S. M.; Orlov, V. D.; Lipson V. V.; Shishkin, O. V.; Potekhin, K. A., Struchkov, Y. T. *Khim. Geterotsikl. Soedin.*, 1999, 109)

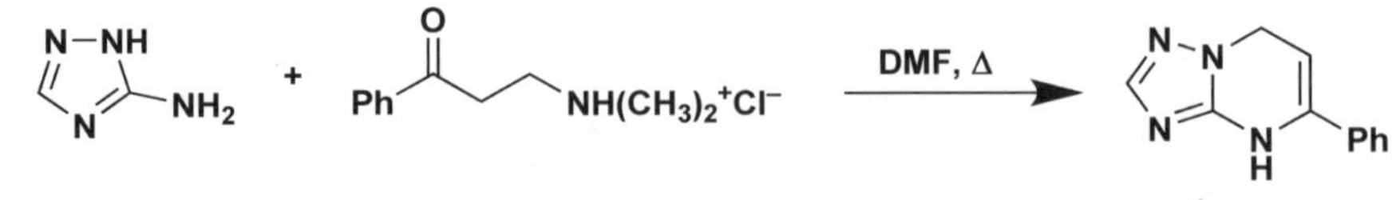


**Scheme A.25**

A solution of 1.1 g (5 mmol) of $\beta$-dimethylaminopropiophenone hydrochloride **2** and 0.42 g (5 mmol) of 3-amino-1,2,4-triazole **1** in 1 ml of DMF was refluxed for 30 min (Scheme A.25). The mixture was cooled and the precipitate formed was filtered off. The 0.7 g (70%) of **3** obtained had a melting point of 201–203°C (from 2-propanol).

## *A.3.10 7-(4-Methoxyphenyl)-3-(phenylcarbamoyl)-4,7-dihydropyrazolo[1,5-*a*]pyrimidine-5-carboxylic acid*

(Chebanov, V. A.; Sakhno, Y. I.; Desenko, S. M.; Chernenko, V. N.; Musatov, V. I.; Shishkina, S. V.; Shishkin, O. V.; Kappe, C. O. *Tetrahedron*, 2007, *63*, 1229)

### A.3.10.1 Method A

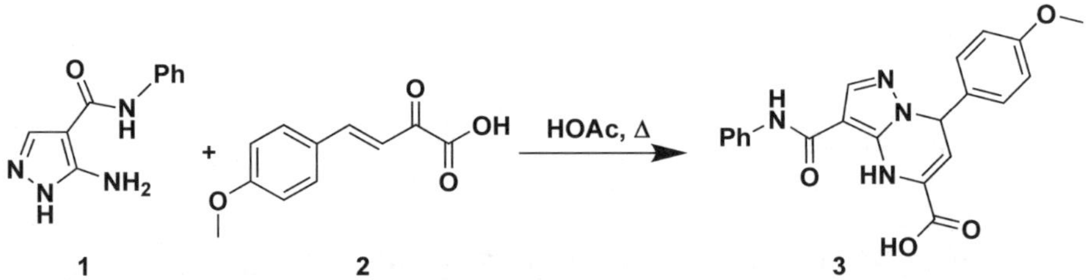


**Scheme A.26**

A mixture of 5-amino-*N*-phenyl-1*H*-pyrazole-4-carboxamide **1** (0.2 g, 1 mmol) and arylidenepyruvic acid **2** (0.2 g, 1 mmol) was refluxed in 5 ml of acetic acid for 10–20 min until a solid started to precipitate (Scheme A.26). After cooling, the crystals formed were removed by filtration, washed with ethanol and dried in air. If required, the products were crystallized from ethanol. Yield 68%. Melting point 254–256°C.

### A.3.10.2 Method B

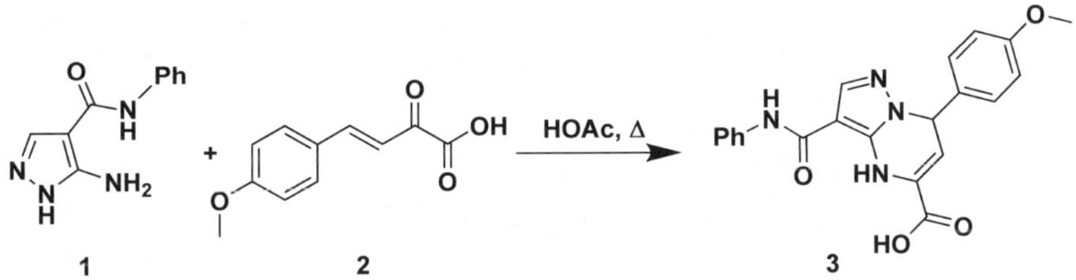

1 2 3

Scheme A.27

A mixture of 5-amino-*N*-phenyl-1*H*-pyrazole-4-carboxamide **1** (0.2 g, 1 mmol), pyruvic acid **4** (0.08 g, 1 mmol) and its aldehyde **5** (0.13 g, 1 mmol) was refluxed in 3 ml of acetic acid for 2 h (Scheme A.27). After cooling, the crystals precipitated were removed by filtration, washed with ethanol and air-dried. If required, the products were crystallized from ethanol. Yield 46%.

## A.4 Reactions of α,β-Unsaturated Carbonyls with 1,2-Diamines

### *A.4.1 2,3-Dihydro-2,4-diphenyl-1*H*-1,5-benzodiazpine*

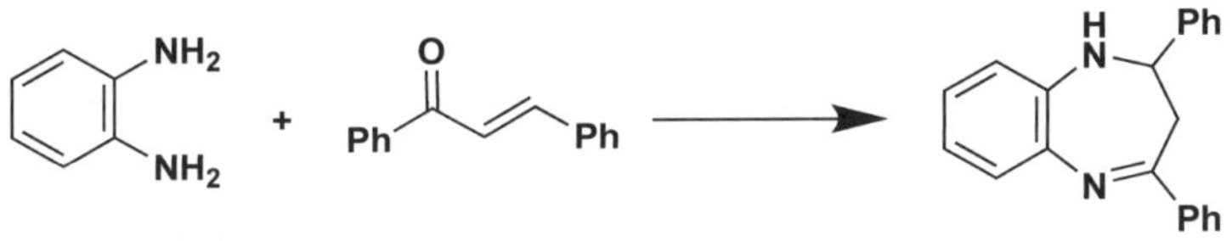

1 2 3

Scheme A.28

### A.4.1.1 Method A

(Orlov, V. D.; Kolos, N. N.; Yaremenko, F. G.; Lavrushin, V. F. *Khim. Geterotsikl. Soedin.*, 1980, 697)

A mixture of chalcone **2** (2.8 g, 0.01 mol) and *ortho*-phenylenediamine **1** (1.62 g, 0.015 mol) in 40 ml of methanol was refluxed for 10 h (Scheme A.28). The reaction mixture was concentrated and after cooling, the target diazepin **3** was filtered off and recrystallized from methanol. Yield 72%. Melting point 129–130°C.

#### A.4.1.2 Method B (Solvent-Free)

(Kodomari, M.; Noguchi, T.; Aoyama, T. *Synth. Commun.*, 2004, *34*, 1783)

Chalcone **2** (0.28 g, 1 mmol) and *ortho*-phenylenediamine **1** (0.16 g, 1.5 mmol) were dissolved in diethyl ether (20 ml). Alumina (neutral, 2 g) was then added to the mixture and the mixture was stirred for a while, followed by removal of the solvent under reduced pressure. The mixture was stirred at 80°C for 3 h under a nitrogen atmosphere. Alumina was separated by filtration after eluting the product with ethyl acetate (20 ml). After the solvent had been evaporated under reduced pressure, the residue was purified by column chromatography on silica gel (hexane–ethyl acetate 10:1) to yield **3** (81%) as a yellow solid.

### *A.4.2 10-Methyl-11-oxa-2,9-diazatetracyclo-[8.7.1.0$^{3,8}$.0$^{12,17}$]octadeca-3,5,7,12,14,16-hexaene*

(Svetlik, J.; Hanus, V.; Bella, J. *Liebigs Ann. Chem.*, 1989, 91)

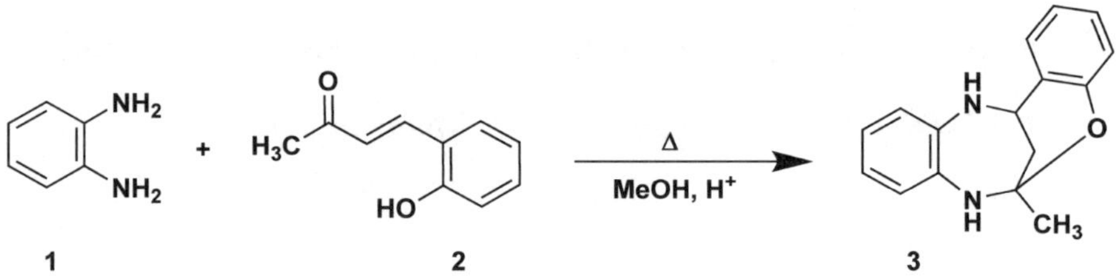


**Scheme A.29**

A solution of *ortho*-hydroxybenzalacetone **2** (1.62 g, 0.01 mol) and *ortho*-phenylenediamine **1** (1.1 g, 0.011 mol) in methanol (30 ml) containing two drops of concentrated hydrochloric acid was refluxed for 1 h (TLC control; Scheme A.29). After evaporation of the solvent, the oil formed was chromatographed on silica gel ($CHCl_3$) and then compound **3** was crystallized from diethyl ether–light petroleum ether as colorless prisms (0.35 g, 14%). Melting point 126–127°C (dec.).

### *A.4.3 Tetracyclic 1,5-Diazepines*

(Tóth, G.; Lévai, A.; Szőllősy, Á. *Liebigs Ann. Chem.*, 1992, 803)

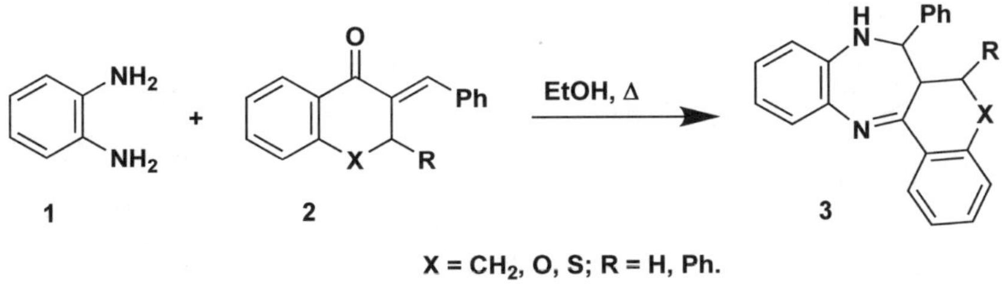


Scheme A.30

A mixture of the appropriate cyclic unsaturated ketone **2** (10 mmol) and *ortho*-phenylenediamine **1** (20 mmol) in 50 ml of ethanol was refluxed for 48 h, the solvent was evaporated under reduced pressure and the residue was crystallized from ethanol to obtain compound **3** (Scheme A.30).

## *A.4.4 (10*R*),(14*R*)-10,14-Dimethyl-2,9-diazatricyclo[8.5.1.0^{3,8}]hexadeca-1,3(8),4,6-tetraene*

(Hakiki, A.; Mossadak, M.; Mokhles, M.; Rouessac, F.; Duddeck, H.; Mikhova, B. *Tetrahedron*, 1995, *51*, 2293)

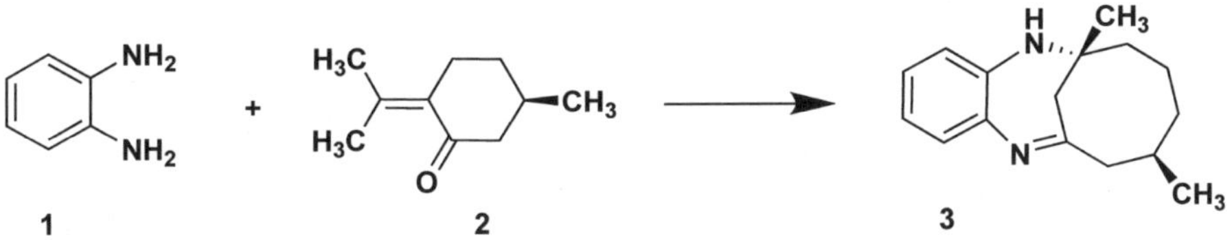


Scheme A.31

(*R*)-(+)-Pulegone **2** (2.0 g, 13.14 mmol) and *ortho*-phenylenediamine **1** (1.42 g, 13.14 mmol) were dissolved in 30 ml of dry toluene (Scheme A.31). Subsequently, the mixture was refluxed, the solvent evaporated and the residue subjected to column chromatography over silica gel (cyclohexane–ethyl acetate 7:3), producing 2.16 g (68%) of compound **3**. Melting point 105°C.

## *A.4.5 1,3,6,6,8-Pentamethyl-6,7-dihydro-1*H*-pyrimido[4,5-*b*][1,4]diazepine-2,4-dione*

(Chebanov, V. A.; Kolos, N. N.; Shishkin, O. V.; Shishkina, S. V.; Orlov, V. D. *Funct. Mater.*, 2003, *10*, 55)

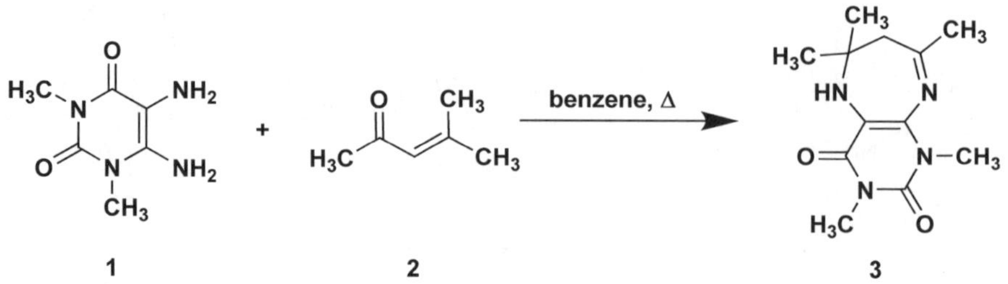


Scheme A.32

A mixture of 1,3-dimethyl-5,6-diaminouracil **1** (0.4 g, 2.3 mmol) and mesityl oxide **2** (0.22 g, 2.3 mmol) in 40 ml of dry benzene was refluxed for 18 h (Scheme A.32). The solvent was evaporated completely and the crude residue was crystallized from a benzene–hexane mixture (1:1). The yield of diazepine **3** was 61%. Melting point 170°C.

### *A.4.6 7,9-Dimethyl-2,4-diphenyl-1,7,9-triazaspiro[4,5] dec-1-ene-6, 8,10-triols*

(Kolos, N. N.; Orlov, V. D.; Chebanov, V. A.; Shishkin, O. V.; Kuznetsov, V. P.; Kulikov, A. Y. *Khim. Geterotsikl. Soedin.*, 1996, 978)

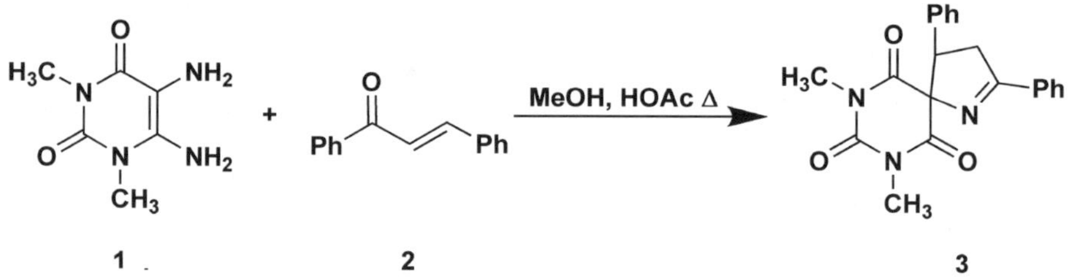


Scheme A.33

A mixture of 5,6-diamino-1,3-dimethyl-uracil **1** (0.2 g, 1.2 mmol), chalcone **2** (0.25 g, 1,2 mmol) and 0.4 ml of glacial acetic acid in 20 ml of methanol was refluxed for 3 h (Scheme A.33). After cooling, the colorless crystals of **3** precipitated and were filtered. Yield 82%. Melting point 229–231°C.

## References

Chebanov, V. A.; Kolos, N. N.; Shishkin, O. V.; Shishkina, S. V.; Orlov, V. D. *Funct. Mater.*, 2003, *10*, 55.

Chebanov, V. A.; Sakhno, Y. I.; Desenko, S. M.; Chernenko, V. N.; Musatov, V. I.; Shishkina, S. V.; Shishkin, O. V.; Kappe, C. O. *Tetrahedron*, 2007, *63*, 1229.

Desenko, S. M.; Orlov, V. D.; Lipson V. V.; Shishkin, O. V.; Potekhin, K. A., Struchkov, Y. T. *Khim. Geterotsikl. Soedin.*, 1999, 109.
Gella, I. M.; Yaya, A. R.; Orlov, V. D. *Kharkov Univ. Bull.*, 2001, *30*, 103.
Hakiki, A.; Mossadak, M.; Mokhles, M.; Rouessac, F.; Duddeck, H.; Mikhova, B. *Tetrahedron*, 1995, *51*, 2293.
Heine, H. W.; Henzel, R. P. *J. Org. Chem.*, 1969, *34*, 171.
Heine, H. W.; Weese, R. H.; Cooper, R. A. *J. Org. Chem.*, 1967, *32*, 2708.
Jin, X. L.; Sugihara, H.; Daikai, K.; Tateishi, H.; Jin, Y. Z.; Furuno, H.; Inanaga, J. *Tetrahedron*, 2002, *58*, 8321.
Kidwai, M.; Misra, P. *Synth. Commun.*, 1999, *29*, 3237.
Kodomari, M.; Noguchi, T.; Aoyama, T. *Synth. Commun.*, 2004, *34*, 1783.
Kolos, N. N.; Orlov, V. D.; Chebanov, V. A.; Shishkin, O. V.; Kuznetsov, V. P.; Kulikov, A. Y. *Khim. Geterotsikl. Soedin.*, 1996, 978.
Krauze, A.; Germane, S.; Eberlins, O.; Sturms, I.; Klusa, V.; Duburs, G. *Eur. J. Med. Chem. Chim. Ther.*, 1999, *34*, 301.
Manna, F.; Chimenti, F.; Bolasco, A.; Bizzarri, B.; Filippelli, W. *Eur. J. Med. Chem. Chim. Ther.*, 1999, *34*, 245.
Molchanov, A. P.; Korotkov, V. S.; Kostikov, R. R. *Zh. Org. Khim.*, 2004, *40*, 501.
Müller, T. J. J.; Ansorge, M.; Aktah, D. *Angew. Chem. Int. Ed.*, 2000, *39*, 1253.
Muravyova, E. A.; Desenko, S. M.; Musatov, V. I.; Knyazeva, I. V.; Shishkina, S. V.; Shishkin, O. V.; Chebanov, V. A. *J. Comb. Chem.*, 2007, *9*, 798.
Orlov, V. D.; Yaremenko, V. F.; Kolos, N. N.; Pivnenko, N. S.; Lavrushin, V. F. *Khim. Geterotsikl. Soedin.*, 1980, 544.
Orlov, V. D.; Kolos, N. N.; Yaremenko, F. G.; Lavrushin, V. F. *Khim. Geterotsikl. Soedin.*, 1980, 697.
Orlov, V. D.; Desenko, S. M.; Potekhin, K. A.; Struchkov, Y. T. *Khim. Geterotsikl. Soedin.*, 1988, 229.
Padmavathi, V.; Reddy, B. J. M.; Subbaiah, D. R. C. V. *New J. Chem.*, 2004, *28*, 1479.
Powers, D. G.; Casebier, D. S.; Fokas, D.; Ryan, W. J.; Troth, J. R.; Coffen, D. L. *Tetrahedron*, 1998, *54*, 4085.
Risitano, F.; Grassi, G.; Foti, F.; Moraci, S. *Synlett*, 2005, *10*, 1633.
Simon, D.; Lafont, O.; Farnoux, C. C.; Miocque, M. *J. Heterocycl. Chem.*, 1985, *22*, 1551.
Smith, L. I.; Pings, W. B. *J. Org. Chem.*, 1937, *2*, 23.
Svetlik, J.; Hanus, V.; Bella, J. *Liebigs Ann. Chem.*, 1989, 91.
Tóth, G.; Lévai, A.; Szőllősy, Á. *Liebigs Ann. Chem.*, 1992, 803.
Turan-Zitouni, G.; Chevallet, P.; Kilic, F. S.; Erol, K. *Eur. J. Med. Chem. Chim. Ther.*, 2000, *35*, 635.
Weis, A. L. *Synthesis*, 1985, 528.
Weis, A. L.; Mamaev, V. P. *Izv. Sib. Otd. Akad. Nauk SSSR Ser. Khim.*, 1975, *6*, 148.
Zbruyev, A. I.; Vashchenko, V. V.; Andryushchenko, A. Y; Desenko, S. M.; Musatov, V. I.; Knyazeva, I. V.; Chebanov, V. A. *Tetrahedron*, 2007, *63*, 4297.
Zhang, J.; Wang, M.; Huang, Z. *J. Chem. Soc. Perkin Trans. 1*, 1999, *15*, 2087.
Zigeuner, G.; Brunetti, H.; Ziegler, H.; Bayer, M. *Monatsh. Chem.*, 1970, *101*, 1767.

# Index

*Ab initio* calculations, 131, 157
σ-Acceptor, 30
σ-Acceptor group, 105
Acetophenone, 76, 87
Acid anhydrides, 11
1,2-Addition, 1, 5
1,4-Addition, 1, 75
β-Addition, 2
Aldimine, 16
Aliphatic diamines, 150
Alkaloid sesbasine, 129
Alkylphenols, 117
Ambident electrophile, 1
Amidine, 131–133, 194
Amidine type tautomerism, 131, 133
β-Amination, 150
Aminoazines, 100, 102, 106
Aminoazole, 83–100, 102, 109
2-Amino-6-(2-bromophenyl)-4-(1*H*-indol-3-yl)nicotinonitrile, 196
2-Aminobenzimidazole, 83, 90, 94, 95, 98
Aminohydrine, 16
2-Aminoimidazoles, 83
2-Aminoindole, 84
3-Amino-5-methylthio-1,2,4-triazole, 90
5-Aminopyrazoles, 83, 90, 91, 96–98
5-Aminotetrazole, 83, 94–98
3-Amino-1,2,4-triazole, 83, 85, 86, 88, 90, 94
6-Aminouracil, 102
Ammonium acetate, 17, 77, 187, 196
Angular dihydropyrimidines, 91
Anilines, 100–107
Arsenic acid, 101
Arylboronic acids, 11
(4'-Aryl-4',5'-dihydro-1'*H*-pyrazol-3'-yl)-(4-aryl-1*H*-pyrrol-3-yl) methanone, 40, 187–188
4'-Aryl-1,2,3,4,4',5'-hexahydro-3'*H*-naphthalene-2-spiro-3'-pyrazol-1-ones, 188–189
Aryliden-1,3-cyclonones, 90
Aryliden derivatives of
  chromanone, 40
  cyclohexanone, 46
  flavanone, 40
  indanone, 40
  tetralone, 40
  thiochromanone, 40
2-Arylidenindan-1,3-diones, 46
Arylidenpyruvic acids, 94, 96–98, 101, 102
2-Aryliden-1-tetralones, 46
Arylpyruvic acids, 100
Azaheterocycles, 1
Azirenoimidazole formation mechanism, 16
Azirenoimidazoles, 15
  multicomponent synthesis, 16
Azirenoquinoxalyne, 156
Aziridine ring-opening mechanism, 27
Aziridines
  chemical properties, 24
  protonation, 21
Aziridinyl ketone
  acylation reactions, 11
  cis/trans-isomerization, 15
  N-cyclohexyl derivative, 7
  one-pot synthesis, 10
  transformations, 11
Aziridinyl ketones, 5–14, 16
Azirinochalcones, 12
Azomethine, 89, 116
Azomethine bond
  1,3-dipolar cycloaddition, 166
  oxidation, 166

Azomethine nitrogen, 165
Azomethine ylides, 14, 23, 32

Benzalcyclohexane, 151
Benzamidine, 73
1,5-Benzodiazepine, 151
Benzonitrileoxide, 166
Benzopyrane, 129
Benzothiazepine, 152
8-Benzoyl-5-hydroxy-5-methyl-1,2,3,5,6,7-hexahydroimidazo[1,2-*a*] pyridine, 192
2-Benzoyl-3-phenylaziridine, 181
3-benzoyl-4-phenyl-$\Delta^1$-pyrazoline, 187
2-Benzylquinoxaline, 21
1,2-Bielectrophile, 5
1,4-Bielectrophile, 5
1,2-Binucleophiles, 37
Binucleophiles, 5
Bipolar ylide, 14
α-Bromochalcones, 18
Bromo chalcones, 7

C-alkylation, 113
Carbonyl group
  reduction11
Catalytic hydrogenation, 111, 112
Cation-radical, 109
Ceric ammonium nitrate, 161
CH-acid, 1, 61, 76, 80, 82, 169
CH-acidic nucleophilic center, 61
Chalcone, 1, 2, 8, 10, 37–40, 71, 77, 78, 84, 104, 116, 149
Charge control, 113
Chloroanhydrides, 11
Cyanoacetamide, 76, 79
Cyanoacetic acid, 76, 79
3-Cyanopyridines, 77
Cyanothioacetamide, 76, 79
Cyclic emanine, 63
Cyclic enamine, 61
Cyclic thioureas, 70
Cycloaddition reactions, 24
Cycloalkanones, 76
Cyclododec-2-enone, 106
Cyclonon-2-enone, 106
Cyclooct-2-enone, 106
Cycloundec-2-enon, 106

2,4-Diamine-1,3,5-triazine, 73
1,2-Diaminoanthraquinone, 150
1,2-Diaminobenzimidazole, 170
Diaminomaleonitrile, 158
1,2-Diamino-4-phenylimidazole, 171
2,3-Diaminopyridine, 170
Diaminotetrazole, 156
3,5-Diamino-1,2,4-triazole, 90
3,5-Diaminotriazoles, 86
2,6-Diaminouracil, 103
Diaroylacetone, 40
1,3-Diaryl-2,3-dibromopropan-1-one, 6
1,3-Diaryl-2,3-dibrom-propane-3-ons, 156
2,6-Diarylidencyclohexanones, 46
3,5-Diaryliden-4-piperidones, 46
Diarylpyrrolidine-2,3-dione, 102
Diazepine, 149
Diazocine, 175
2-Diazo-1,3-diphenylpropane-1,3-dione, 166
Diazolalkanes, 37
Dibenzoylethylene, 86, 174
2,3-Dibromo-1,3-diphenylpropan-1-one, 181
1,5-Dicarbonyl compounds, 61
Dihalocarbene addition, 125
Dihydroacetic acid, 151
1,1a-Dihydro-l-aryl-2-phenylazirino[1,2-*a*] quinoxaline, 184–185
Dihydroazines
  [2 + 2]-cycloadditions, 121
  acylation, 113
  alkylation, 113
  building-blocks, 130
  cycloaddition reactions, 120
  disproportionation, 109
  heteroaromatization, 108
  hydrolytic cleavage, 117
  olefin cycloaddition, 122
  oxidative addition, 118
  reduction, 111
  stereoselective reduction, 111
  stereoselective vicinal diamination, 118
  substituent elimination, 108
Dihydroazolopyrimidine
  synthesis, 83
Dihydroazolopyrimidines
  oxidation, 86
Dihydrobenzodiazepine, 149, 158
Dihydrobenzodiazepines
  acylation, 164
  alkylation, 168
  N-aroylation, 165
  nitrosation, 164
  oxidation, 163
  reduction, 164
  thermal rearrangement, 168
1,1a-Dihydro-1,2-diarylazireno [1,2-*a*]quinoxaline, 18

Dihydrodiazepines
chemical properties, 163
2,3-Dihydro-2,4-diphenyl-1*H*-1,5-benzodiazpine, 199
1,2-Dihydropyridine, 81
Dihydro-1*H*-azireno[1,2-*c*]imidazole, 186
micellar catalyzed procedure, 186
non-micellar procedure, 186
3,5a-Dihydro-1*H*-azireno[1,2-*c*] imidazoles, 13
synthesis, 15
Dihydropyridines
chemical properties, 108
synthesis, 61
Dihydropyrimiden-2-thione, 68
Dihydropyrimidines
chemical properties, 108
oxidation, 68
synthesis, 65
Dihydrotriazepine, 156
Dihydrotriazolopyrimidines
mechanism of multicomponent formation, 89
2,4-Di(2-hydroxyphenyl)-9-isopropyl-1,5,9-triazaspiro[5]undec-1-ene, 195
Diimine tautomer, 157
Dimedone, 90
Dimerizations of dihydropyridines, 122
1,3-Dimethyl-6-aminouracil, 106, 107
1,3-Dimethyl-5,6-diaminouracil, 156, 172
4-Dimethyldibenzoylethylene, 175
7-Dimethyl-2,4-diphenyl-7,8-dihydroquinolin-5(6*H*)-one, 191–192
7,9-Dimethyl-2,4-diphenyl-1,7,9-triazaspiro[4,5]dec-1-ene-6,8,10-triols, 202
3-Dimethyl-1-(4-nitrophenyl)-5-(4-phenylphenyl)-3,5a-dihydro-1*H*-azireno[1,2-*c*]imidazole, 186
4-Dimethyl-2-(4-nitrophenyl)-6-phenyl-3,5-diaza-bicyclo[3.1.0]hex-2-ene, 183
2,3-Diphenyl-7-benzylidene-3,3a,4,5,6,7-hexahydro-2*H*-indazole (2) and 1,3-diphenyl-7-benzylidene-3,3a,4,5, 6,7-tetrahydroindazole, 191
2,4-Diphenyl-2,3-dihydro-1*H*-1,5-benzodiazepine, 150
5,7-Diphenyl-4,7-dihydro-1,2,4-triazolo [1,5-a]pyrimidine, 197
4,6-Diphenyl-4,5-dihydropyrimidine-2-thione, 192
1,3-Diphenyl-3-diphenylmethoxyaminopropan-1-one, 182
4,5-Diphenyl-imidazole-1,2-diamine, 171
Dipolar addition, 10
1,3-Dipolar cycloaddition, 120
Dipolar cycloaddition, 23, 41
Direct aziridination, 11

α-Effect, 2
Electroluminescent polymers, 37
Elimination of one aryl group, 104
Enamine, 61, 100
Enamine intermediate, 91
Enaminoester, 65
β-Enaminoketones, 156
Enaminonitriles, 64
Enediimine, 27
Enemine intermediate, 89
Ethyl cyanoacetate, 76, 80
Ethylene diamine, 154
Ethylenediamine, 18
Exocyclic unsaturated ketones, 40

First-order reaction, 133
Friedel-Crafts reaction, 115
Furan-2-one, 102
Furyl analogues of chalcones, 38
Fused aziridines, 121

Gabriel reaction, 5, 7, 9
Guanidine, 61, 65, 71

Hantzch synthesis, 61, 62
Heteroannulation, 174
Heteroaromatization of azine
mechanism, 109
Heteroaromatized derivatives of azoloazines
reagioselectivity, 109
Heterocyclic analogues of chalcones, 40
Hexahydro-1,2-oxazepine, 49
HOMO, 2
Hydrazine, 13, 41, 43, 125
Hydroxylamine, 7, 41, 49, 108, 116
6-Hydroxy-6-methyl-2,4-diphenyl-1,4,5,6-tetrahydropyrimidine, 194
Hydroxyurea, 49

Imidazopyridine, 171
Imine-enamine tautomeric equilibrium, 131
Imine tautomeric form, 131
Indium bromide, 161
Indium chloride, 101

Individual tautomers, 134
Internal nucleophilic substitution, 119
Intramolecular addition, 65, 74, 122
Intramolecular cyclization, 120, 126, 128, 150
Intramolecular reaction, 78
Iodine-amine complex, 8
Ionic liquids, 151
Isocyanates, 11
Isonicotinyl hydrazide, 42
Isothiocyanates, 14

Ketimine, 16

Lead tetraacetate, 10
Lewis acids, 122
Light-emitting diodes, 37
Lithium diisopropilamide, 129
LUMO, 2

Magnesium perchlorate, 161
Malonic acid, 76–77
Malonoamide, 76, 79
Malononitrile, 76
Malonothioamide, 79
Mannich bases, 83, 168
*M*-chloroperbenzoic acid, 127
Mechanism of isoxazoline formation, 49
Mesityl oxide, 69, 73, 101, 149, 172
*Meta*-chloroperbenzoic acid, 163
Metal alcoholates, 7
5-(4'-Methoxyphenyl)-l,3-diphenyl-4-pyrazoline, 189
7-(4-Methoxyphenyl)-3-(phenylcarbamoyl)-4,7-dihydropyrazolo[1,5-a] pyrimidine-5-carboxylic acid, 198–199
2-Methyl-2,4-diphenyl-1,4-dihydropyrimidine, 193–194
Methylethylketone, 159
Methylhydrazine, 43, 44
1-Methyl-5-(4-nitrophenyl)-3-phenyl-4,5-dihydro-1*H*-pyrazole, 190
5-Methyl-2-phenyl-2*H*-pyrazol-3,4-diamine, 155
Michael addition, 174
Michael adduct, 93
Michael-type addition, 49, 116
Microwave assisted procedures, 42
Microwave irradiation, 67, 92, 99, 101, 107, 151, 152, 161, 189
Multicomponent diazepine formation mechanism, 163
Multicomponent reaction, 74, 76, 88, 91, 92, 94, 97, 102, 106, 107

NADH analogues, 121
N-bromocyclohexylamine, 7
Nitrile oxides, 120
Nitroethandiamines, 64
7-(4-Nitrophenyl)-2-phenyl-3,6-diaza-bicyclo[4.1.0]hept-2-ene, 184
N-methyl-quinolin-2-one, 152
Nodal nitrogen atom, 45
Non-cyclic ketosulfones, 94
Nonspecific solvation, 136
Nonsymmetrical *ortho*-diamines, 161
N-phenylmaleimide, 24
N-substituted thiourea, 68

*O*-hydroxybenzylideneacetone, 65
Oligochalcones, 43
Orbital control, 113
Orbital symmetry rules, 26
Organic luminophores, 37
*Ortho*-aminothiophenol, 152
*Ortho*-diamines, 149, 163
  hydrocloride salts, 159
*Ortho*-hydroxychalcone, 150
*Ortho*-nitrophenylazide, 163
*Ortho*-oxybenzalacetone, 150
*Ortho*-phenylenediamine, 149
1,2,4-Oxadiazepin-3-one, 49
Oxazepine, 171
Oxazine ring
  reductive opening, 54
Oxazolidines, 128
Oxidative agents, 108
Oxidative cyclization, 44

Pentane-2-one, 158
1,3,6,6,8-Pentamethyl-6,7-dihydro-1*H*-pyrimido[4,5-*b*][1,4]diazepine-2,4-dione, 201–202
Perturbation of the molecular orbitals theory, 2
Phenoxazine, 151
2-Phenylbenzimidazole, 150, 170
5-Phenyl-4,7-dihydro-1,2,4-triazolo[1,5-*a*] pyrimidine, 197
1,2-Phenylenediamine, 18
Phenylhydrazine, 41
4-Phenylphenylboronic acid, 185

6-Phenyl-4-(pyridin-3-yl)-2-thioxo-1,2-dihydropyridine-3-carbonitrile, 196
Photoactivated intermolecular dimerization, 14
Photochromic transformations, 31
Photochromism, 29, 32
Piperidine, 79
Polycyclic lactones, 127
Potassium cyanide, 119
Potassium methoxide, 66
Prototropic rearrangements, 39
Puligone, 154
Pyrazoles, 37
Pyrazoline, 37
  acylation, 51
  azocoupling, 51
  chemical properties, 50
  cross-conjugated fragments, 50
  cyanoethylation, 52
  diarylhydrazine π-system, 50
  diastereomerization, 54
  dimerization, 51
  electronic density, 50
  electrophilic reactions, 51
  formylation, 51
  nitration, 51
  oxidation, 51
  photocyclization, 54
  physicochemical properties, 50
  qualitative analysis, 52
  stereoisomers, 46
  sulfonation, 51
Pyrazoline formation
  kinetics, 45
  mechanism, 45
2-Pyrazolines, 37
Pyrazoloquinazoline, 92
Pyrazoloquinoline, 91
Pyridinecarboxamide, 79
Pyrimidinones, 71
Pyrimido[1,2-*a*]pyrimidine, 73
Pyruvic acid, 86, 94, 96

Quinoline, 62, 101
Quinolizinones, 92

Radical reaction of hydrazine, 44
Radiochromic compounds, 32
Reaction of urea
  mechanism, 75
Reactions of arylidenpyruvic acids
  mechanism, 98
Reactions of cyanoacetic acid derivatives
  mechanism, 82
Reductive cyclization, 14
Regioselectivity, 19
Retro-aldol disproportionation, 174
Room temperature ionic liquids, 153

Scandium triflate161
Scintillating materials, 37
Second-order reaction, 133
Semi-aminales, 53
Semicarbazide, 48
Semiempirical calculations, 131
Single-electron oxidation, 109
Sodium azide, 119
Sodium borohydride, 164
Sodium ethoxide, 65, 70, 78, 129
Sodium methoxide, 65, 80
Sodium tetrahydroborate, 159
Solid-phase reaction, 101
Solid-supported synthesis, 107
Solvent-free reaction, 151
Sonicated reaction, 74
Sonication, 93
Sonogashira coupling, 43
Southwick reaction, 8
Spiropyrazolines, 40
*S,S*-diphenylsulfimide, 9
Stereochemistry, 45
Stereoselective ring opening, 119
Steric tensions, 131
Suzuki–Miyaura reaction, 32
Synthetic precursors of nsaturated carbonyls, 157
Synthetic precursors of unsaturated carbonyls, 76

Tautomeric transition, 133
Tautomerism, 130, 133, 135
  kinetic investigations, 133
*Tert*-butyl lithium, 116
1,2,4,6-Tetraaryl-1,4-dihydropyridines, 174
Tetracyanoethylene, 28
Tetrahydrobenzodiazepines, 159
1,3,4,6a-Tetrahydroazireno[1,2-a]pyrazines, 18
1,2,5,6-Tetrahydropyrimidine, 74
Tetrahydrothiazepine, 154
Thermo- and photoinduced cycloaddition, 23
Thioamides, 11
Thioamidine, 73

2-Thio-6-aminouracile, 104
1-Thiocarbamoyl-3,5-diaryl-2-pyrazoline derivatives, 190
Thiosemicarbazide, 44
Thiourea, 61, 65
Three-component reactions, 74
Titanium tetrachloride, 123
T*rans*-2-phenyl-3-(4-nitrobenzoyl) aziridine, 183
Transfer of hydride ion, 109
1,2,4-Triaminotriazoles, 170
Triarylpyrrole, 175
Triazepine, 149
Triethylamine, 80
Triethylbenzyl ammonium chloride, 163
Trimesic acid, 161
1,3,5-Triphenylbenzene, 76
2,4,6-Triphenyl-3,5-diazabicyclo [3.1.0]hex-2-ene, 187

Ultrasonic irradiation, 43, 74, 94, 100
Unsaturated ketoacids, 94
Urea, 61, 65, 70
Urea-like binucleophiles, 71
UV irradiation, 65

Vilsmeier reaction, 115

Ytterbium chloride, 161
Ytterbium triflate, 161

Zinc chloride, 101, 161